DRAWING THE MAP OF LIFE

DRAWING THE MAP OF LIFE

Inside the Human Genome Project

VICTOR K. McELHENY

A MERLOYD LAWRENCE BOOK

BASIC BOOKS
A Member of the Perseus Books Group

Published as a Merloyd Lawrence Book
By Basic Books
A Member of the Perseus Books Group

Designed by Brent Wilcox
Set in 11.5 point Minion by the Perseus Books Group

Library of Congress Cataloging-in-Publication Data
McElheny, Victor K.
Drawing the map of life : inside the Human Genome Project / Victor K. McElheny.
p. ; cm.
"A Merloyd Lawrence book."
Includes bibliographical references and index.
ISBN 978-0-465-04333-0 (alk. paper)
1. Human Genome Project—History. I. Title.
[DNLM: 1. Human Genome Project. 2. Human Genome Project. 3. Genomics. QZ 50 M4776d 2010]
QH447.M355 2010
611'.0181663—dc22

2010003339

1 2 3 4 5 6 7 8 9 10

For Ruth

CONTENTS

"Two opposing laws seem to me now in contest. The one, a law of blood and death, opening out each day new modes of destruction, forces nations to be always ready for the battle. The other, a law of peace, work, and health, whose only aim is to deliver man from the calamities which beset him . . . [O]f this we may be sure, that science, in obeying the law of humanity, will always labor to enlarge the frontiers of life."

Louis Pasteur,
remarks read at the dedication of
the Institut Pasteur, 14 November 1888;
quoted in *Encyclopaedia Britannica,*
Eleventh Edition (1911) 20, 894

PREFACE

Ten years ago, in the East Room of the White House, U.S. President Bill Clinton celebrated the completion of a "most wondrous map" of the DNA that makes up the human genome. He compared it to the map of the vast American West that Meriwether Lewis and William Clark presented to Thomas Jefferson in the White House nearly two centuries earlier. This book outlines the events leading to that ceremony and chronicles the torrent of efforts since then to fill in the genomic map and make sense of it.

This narrative of the human genome project and its consequences appears just as commercial firms are preparing to sequence the entire DNA of tens of thousands of people—only ten years after scientists managed to spell out the order of the adenines, thymines, cytosines, and guanines of just one human. The history sketched here is recent, vast, and exploding. The pages ahead summarize a continuing revolution that began in fundamental biology and now, more and more, is entering our lives to create a genomic age. It is one of the greatest positive achievements of our scientific civilization.

This book does not focus on what *ought* to be happening or *might* happen in a still-hidden future but instead concentrates on what *has* happened and *is* happening. Its fundamental question is, How did we get here?

In a limited space, only a few major topics can be explored. The story begins with the creation of the techniques and tools that enabled the genome project to begin. As attention to human genetic diseases rapidly grew, the project was organized in the late 1980s amid turmoil among biologists. Remarkably, the project did not originate from the top down, as did both the atomic-bomb and moon programs, but rather out of grassroots consultations over about three years. In the 1990s, after years of often-rocky preparation, decisions were reached about the methods to determine the sequence of human DNA. Then a highly publicized race between nonprofit and for-profit teams achieved rough drafts of the human genome in 2000.

But now, the question was: What did that sequence mean? The 3 billion subunits of human DNA were like a string of letters, with no gaps to show where the words began and ended, filling every page of a stack of telephone books. How could the sequence be interpreted to find both protein blueprints and a largely unknown array of controls? It was clear that biology had raced to the starting gate. To use the sequence to illuminate the actual sources of disease would require enormous additional effort in the years and decades beyond.

The research of the last ten years has brought a shower of scientific surprises that overturned many previous notions of what a gene or a genome is and does. Genomic knowledge is spreading into the fields of agriculture, energy, and environmental protection, all of which have the most profound implications for human health. With the appearance of new tools—growing in capacity at an amazing rate—there is a dramatic speedup in the range of problems that can be tackled.

During the race to the human sequence, the focus was on how alike individuals are. Now the focus is on differences. Researchers across the world have begun spelling out the myriad of changes in DNA that make one person more susceptible to a disease than another. To help probe more and more deeply into these predisposing mutations, thousands upon thousands of human volunteers are donating their DNA for science. Before the end of the first decade of the twenty-first century, medicine was beginning to use genomic techniques to determine which drugs would benefit particular patients.

All of these developments have taken place in the culture of doubt that pervades science. While striving to pick important problems that are amenable to experiment, scientists are enjoined to view skeptically any finding, no matter how attractive, as a possible artifact, not a real discovery. And the scientists struggle with a corollary principle: openness. The rule, never easy and not always observed, has always been, Tell all, tell it right away, and take your lumps. Famous exponents of this ruthless doctrine were two refugee founders of modern biology, Max Delbrück from Germany and Salvador Luria from Italy. Each summer in the 1940s, at Cold Spring Harbor Laboratory near Long Island Sound, they inducted small bands of researchers into a study of life at so elemental a level that the mechanics of heredity could be seen in simple terms. The organisms they worked with—and talked about so often while sitting on a lawn sloping down toward the harbor—were bacteria and the tiny viruses, called bacteriophages, that preyed on them.

Openness was at the core of the phage ethos, and it soon propagated to the genetic research systems of the future, such as yeast, the nematode *C. elegans*, and the fruit fly, *Drosophila melanogaster*. It became a dominant theme of genomics, where data are posted every day on the Internet. And even as genomics has matured and pervaded all biology, the demands of openness continue to trigger debates. One concerns continued universal access to research findings. Another concerns publishing analytical reports in such a way that any scientist can see them, even if the individual researcher or academic institutions cannot afford print or electronic subscriptions to a particular journal.

The march of genomics is taking place alongside a background of ethical and spiritual anxieties. Would the information, particularly the theories of evolution that are the core of all biology and medicine, assault people's religious faith? Would all people benefit regardless of income, in rich countries and poor, whether they suffer from infectious diseases of greatest impact on children or diseases like cancer that fall more heavily on older people? Would individuals or the medical professionals who help them comprehend information that does not constitute fate but a set of probabilities and risks? Would individuals act on information indicating they should change their lifestyles? Would the finding of susceptibility to a disease for which there is no treatment devastate individuals or families? Would the information be misused by employers and insurers—or by people wishing to design their offspring? Questions like these, which actually represent the classical challenges of all measures to protect public or individual health, constantly face humanity. Although some cite them as reasons not to perform genomic research, others have insisted vehemently that failure to learn new ways to alleviate human suffering is unacceptable. To date, this latter view has prevailed.

Some ask whether this churning world of genomics is a race between commercial and academic motives. Surely, this is a simplistic view. Commercial firms and research institutions have, in effect, cooperated by competing—just as they did in the atomic-bomb and lunar-landing programs decades ago. One side has driven the other in the heavily technological, even industrial, genomic enterprise. At many points, there was blindness on one side or the other. Of the immense quantities of sophisticated equipment and chemicals, the lion's share has come from private firms. But smart and anxious customers in academic laboratories have put the manufacturers through the wringer in tackling numerous imperfections. The major nonprofit genome centers have had to learn how to run industrial operations alongside their laboratories. And industry has had to learn how to build and finance biotechnology and genomics companies of suitable size and sustained capacity for innovation.

Competition remains equally hot in the governmental and philanthropic spheres. In the United States, the National Institutes of Health have found themselves under pressure from the Department of Energy, the National Science Foundation, and the Department of Agriculture, each of which has strong motivations to conduct genomic research. Agencies in other countries, like the huge Wellcome Trust in Britain, a private charity, have funded an important share of the total cost and stiffened the spine of other agencies at dicey moments. Volunteer associations focusing on particular illnesses, such as heart disease or cancer, have competed for many years to influence scientists, legislators, and bureaucrats to push harder for their particular concerns. Because many of the underlying biological problems, such as development from egg to adult, are general, the lobbying has helped expand health research in general. For many decades, wealthy individuals have come forward in large numbers to support biomedical research that they hope will be "transformative."

It is worth mentioning that this seemingly arcane process is not an alien implant. Genomic research grows organically from a century and a half of work on human health and nutrition. It is merely the latest phase of a biological and medical revolution that has been gathering strength continually, precipitating at least a fivefold increase in humanity's numbers. The succession of rapid achievements that seem miraculous excites the hope, the expectation—the demand—for more ways to reduce pain and hunger and to save and extend human life.

The work on genomes has grown more heavily computational and more intricately collaborative than ever before in the history of biological and medical research. It has acquired a momentum of its own, pervading biology and medicine and agriculture. As Neil Risch of the University of California, San Francisco, recalled in 2007, "It changes the way people think." We are confronting a force to be reckoned with in our lives. We have to understand the genomic enterprise to understand our times and our future.

CHAPTER ONE

Building the Toolbox

"Progress in science depends on
new techniques, new discoveries and
new ideas, probably in that order."

—Sydney Brenner,
The Times (London), October 29, 1986

Restriction Enzymes: Molecular Scissors

In the spring of 1968, Hamilton O. Smith of Johns Hopkins was intrigued by the possibility of removing a technical obstacle that troubled him and many of his fellow molecular biologists. Perhaps there really was a "chemical scalpel," an enzyme that could cut DNA into manageable pieces so that their sequences of subunits could be "read." Despite a string of triumphs over twenty years, including Robert Holley's painstaking decipherment of transfer RNA, biologists still lacked the tools to "see" into genes. In the late 1960s, the totality of DNA forming the genomes of humans and every other form of life were a jumble of black boxes.

Smith and hundreds of fellow molecular biologists around the world knew the outlines of genetically controlled life. They knew there is a code that spells out the makeup of the scores of thousands of proteins that do the main work of a living cell. They also knew that the code carries the vast swarm of stop-and start-signals that control which proteins exist and do their work at a given time.

Further, they knew that the code consists of an alphabet of four chemicals called bases, or nucleotides, strung along the twin sugar-phosphate backbones of the DNA double helix. Deoxyribonucleic acid (DNA) gets its name from the fact that its ribose sugar lacks one oxygen atom that is found in the related, and possibly more ancient, sister compound ribonucleic acid (RNA). The four nucleotides form links between the backbones, whose alternating sugars and phosphates repeat continually. Jutting out at right angles to the backbones, these information-bearing components are called adenine, thymine, guanine, and cytosine. These are the famous A, T, G, and C of a thousand textbooks. By means of relatively loose hydrogen bonds, the bases form complementary pairs, A always bonding with T, and G with C.

The sequences of these chemical "letters" are read in triplets, specifying another sequence: the twenty different kinds of amino acids that make up proteins. Conceptually, the chains of amino acids are parallel to the chains of DNA. The triplet ACG, for example, stands for the amino acid threonine. A mutation in the triplet CTT causes most cases of cystic fibrosis by changing just one amino acid subunit of one protein.

In the late 1960s, the explosively expanding community of molecular biologists was growing more and more impatient about obstacles to manipulating DNA so that it could be completely dissected and understood. The delay threw up a barrier to progress in research that was transforming biology into a queen of the sciences and attracting the best and the brightest. Smith, thirty-seven years old and working at the Johns Hopkins University Medical School, was at least as impatient as his colleagues. He knew very well that technical capabilities, as much as ideas, drive science. This fundamental point is at least as true in biology as in astronomy, high-energy physics, oceanography, or space exploration.

Smith was born in 1931 in New York City, where his father, an assistant professor at the University of Florida, was completing work on a doctorate in education at Columbia. Although the family shuttled back and forth from Florida, Smith's childhood memories of New York were stronger. He wrote that his parents entertained him and his elder brother, a year older, "with arithmetic problems and a small Gilbert chemistry set." After his father became a professor at the University of Illinois, Smith spent most of his boyhood in Champaign-Urbana, in what he described as "an atmosphere of intense intellectualism." Smith and his brother spent the money from their newspaper-delivery routes on equipment for their basement laboratory, where Smith laid down the habits of a craftsman, even an artist, in science. Such qualities won wide admiration. In 1999, the geneticist David Botstein said, "Ham Smith is one of the premier craftsmen of molecular biology." Smith's colleague Mark Adams said then, "A lot of us are users of the molecular biology toolkit. Ham has been, to a large extent, a creator of that toolkit." Biochemist Mark Ptashne said, "I'm not sure people realize how smart he is. He has a razor-like, laser-like intelligence."[1]

As a college student at Berkeley in the early 1950s, Smith was fascinated when George Wald of Harvard, a later Nobel Prize winner, lectured on the biochemistry of the retina. After medical school at Johns Hopkins and an internship in St. Louis, where he met and married his wife, Liz, he spent two relaxed years with the U.S. Navy in San Diego. There he read of research on aberrations in human chromosomes and plunged into the study of genetics. The interest grew stronger during a medical residency in Detroit, where he first heard of the work of Max Delbrück and Salvador Luria on bacterial viruses in the 1940s and of the discovery by Francis Crick and James Watson of a structure for DNA in 1953.[2]

During five years at the University of Michigan, Smith made discoveries about a process known as lysogeny. He found out how a particular virus called P22 inserted its own DNA into a bacterium and dwelt there silently, instead of taking over its host and converting its machinery into a factory for more virus. It was then that he heard of Werner Arber's research in Switzerland into how bacteria defeat a viral invasion by cutting up the alien DNA (while protecting their own DNA from the same process).

Thus, Smith had been pulled into a maelstrom of new knowledge about the molecular details of genetics, which had grown rapidly since the 1940s and 1950s. A major step was establishing that DNA—long dismissed as too uncomplicated to play a commanding role in the living cell—is the "transforming principle." In 1944, Oswald Avery and colleagues at the Rockefeller Institute proved it with pneumococcus microbes, and in 1952 Alfred Hershey and Martha Chase of Cold Spring

Harbor proved it with bacteria-attacking viruses. DNA was the seat of hereditary information in all cells and even most viruses. But how could DNA control heredity? The answer came from Watson and Crick's 1953 double-helix model, in which the two sugar phosphate strands wound around each other and were cross-linked by the side-group As, Ts, Gs, and Cs—A always with T, and G always with C. The two complementary strands could pull apart and be copied into "daughter" DNA for the next generation of cells.

Also in 1953, and also in Cambridge, other momentous discoveries were made. Frederick Sanger worked out the sequence of the amino acids of a relatively simple protein, insulin. Max Perutz and John Kendrew hit on a method for deciphering the three-dimensional structure of the much larger proteins, myoglobin and hemoglobin. Thus, windows were opening into how the form of proteins dictated their function. Soon after, scientists in the United States, including Arthur Kornberg, Severo Ochoa, and Marianne Grunberg-Manago, discovered specific enzymes that could copy DNA or its sister chemical RNA. The ribose sugar of RNA contains the oxygen atom missing in DNA. RNA can take several forms, such as working copies of genes, escorts of protein subunits to the place of assembly, portions of ribosome structures where proteins are stitched together, and a myriad of tiny RNAs that can interfere with stages of protein synthesis.[3]

How could the information in the DNA be converted into proteins? In a remarkably short time, about a decade, the main mechanisms were discovered. Crick, Sydney Brenner, Marshall Nirenberg, and others deciphered the "genetic code" in which the triplets of A, T, G, and C spelled out one of the twenty amino acids. Meanwhile, Brenner and Francois Jacob, working in Pasadena, California, and rivals in Watson's lab in Cambridge, Massachusetts, including Walter Gilbert, found the short-lived messenger RNA copy of DNA, which carries instructions from the vault (the cell's nucleus containing the DNA) out to the factory floor, where the ribosomes do their work. Forwarding this process are the transfer RNAs Crick had postulated. Transfer RNAs carry their particular amino acids to the sites of assembly on the ribosomes.[4]

In the same period, Jacob and his colleague Jacques Monod of the Pasteur Institute in Paris uncovered the first genetic brakes, which silenced the expression of the genes by preventing the copying of DNA into messengers. In 1966, Gilbert and Ptashne at Harvard isolated the predicted silencer proteins, called repressors. Others found accelerator "factors" that operated in forward gear.[5]

These feats held more than academic interest. Cancer, of growing importance in an aging population, was a genetic disease. Cancer could be inherited in families, but more often resulted from a succession of mutations in a victim's lifetime that led to an uncontrolled proliferation of particular cells. Viruses, with just a few genes, were known to cause cancer in animals. This fact indicated that only slight changes were necessary to convert cells from orderly citizens of a particular tissue into rogues that could eventually, by the process known as metastasis, enter the bloodstream, settle in distant organs, and kill the victim.

Aspiring to medical payoff for their work, molecular biologists saw the viruses as model triggers for cancer. But even a virus's few genes were, in the late 1960s, a biological black box. Robert Holley's sequence of a tiny transfer RNA was fewer than one hundred letters long. And Seymour Benzer of Purdue mapped, in ever finer detail, mutations of just one gene in a virus. But the much larger sequences of viral or cellular genes, 1,000 letters or more, were shrouded from view. A shower of Nobel Prizes had rewarded the feats of molecular biology, but the machinery of cancer remained hidden.

To open the black box and "read" the sequence of letters, molecular scissors were needed to cut the DNA into manageable pieces. One means to do this had been considered for some years. Arber in Switzerland had begun working on "restriction enzymes," which prevent an invading virus from taking over a bacterium. People began trying to isolate one of these restriction enzymes. Among them was Matthew Meselson at Harvard, who, along with Franklin Stahl, had proved in 1958 that DNA was duplicated as suggested by Watson and Crick. Meselson and Robert Yuan discovered one restriction enzyme, but it was found to cut DNA indiscriminately.[6]

Meselson's search inspired Smith. "With great interest," he gave a seminar on Meselson's work to his colleagues at Johns Hopkins and began working on the problem himself. He and his graduate student Kent Wilcox had started studying rearrangements, or "recombination," of DNA in a microbe called *Haemophilus influenzae*, which caused not the flu but ear infections, meningitis, and pneumonia. They tried injecting into their new microbe radioactive DNA from the virus P22, which Smith had studied for years. P22 normally infects *Salmonella*; would it infect *H. influenzae*? To their surprise it would not. They might have expected a lot of P22 DNA after infection, but they could not "recover" any. It was May 25, 1968.

This was exciting. Had they encountered a restriction enzyme, a way for bacteria to cut up invading P22 DNA? Yes. Smith proceeded to find that their enzyme cut double-stranded DNA only. It recognized a particular six-letter sequence and cut it in the middle. They had found a powerful, "site-specific" scissors for DNA.[7]

A Hopkins colleague, Daniel Nathans (who shared a Nobel Prize in 1978 with Smith and Arber), was then on sabbatical leave in Israel. Smith wrote him about the enzyme, which was dubbed HindII and pronounced "hin D2." Some time later, in a gossipy reply, Nathans jovially wrote that he had "designs on your enzyme" for his planned mapping of genes in the cancer-causing monkey virus SV40. Nathans started using HindII and found other restriction enzymes that also cut SV40.

It was the beginning of a flood. The laboratory of Richard Roberts, a later Nobel Prize winner, at Cold Spring Harbor Laboratory (CSHL) began systematically identifying dozens of restriction enzymes. Soon companies began producing them.

By isolating the first DNA-cutting scalpel, Ham Smith's laboratory had opened up a new level of biology. But this restriction enzyme was only one of the biochemical tools that proved essential to what became the Human Genome Project. Sequencing machine pioneer Leroy Hood noted in 2004 that a battery of enzymes was needed to make the genome project possible. Most important of these, he thought, were the polymerases that built (synthesized) strings of DNA. Also essential were ligases, enzymes that knitted DNA pieces back together after the chemical scalpel had done its work, which were developed in several laboratories in the late 1960s. Reflecting on the genome project, which gave "us something to work on for the next hundred years," Hood said, agreeing with Sydney Brenner, "All of the big revolutions—they are technique driven."[8]

In the early 1970s, California scientists Herbert Boyer and Stanley Cohen made another bacterial enzyme, called Eco RI, part of their system for manipulating DNA. This involved taking pieces of DNA from one organism that had been cut by the restriction enzyme, then stitching the pieces into DNA circles called plasmids with the help of the needle-and-thread ligase enzymes. Bacteria could be induced to take in the doctored plasmids. This meant that animal genes of interest could be moved into bacteria and multiplied fantastically.[9]

This was a matter of high practical importance. A way had been found to produce industrial quantities of human insulin and human growth hormone. But more than the biotechnology industry was born. Fred Sanger in Cambridge had been working for two decades to repeat his feat with the linear sequence of insulin by deciphering the "colinear" sequence of DNA. As the restriction enzymes gradually became available in the 1970s, the production of DNA fragments that could be sequenced and read became more frequent. Sanger invented two techniques in succession and in 1977 used them to complete the first entire genome of an organism, a virus called Phi X 174. His second method became the dominant way to read DNA for the next thirty years.[10]

Smith did not jump into the flood of research that followed. Instead, he continued to study the bacterium that made HindII, *H. influenzae*. The 1978 Nobel Prize he shared with Nathans and Arber came as a shock to Smith, who doubted that he had earned it and felt Meselson had been excluded wrongly. Meselson recalled twenty years later, "Most people are pretty decent and pretty generous. But Ham is very decent and very generous." Smith struggled with the innumerable expectations and interruptions that beset a Nobel Prize winner, and his work suffered. By the late 1980s, his grant applications were being rejected, and he was thinking about closing his Hopkins laboratory.

In the early 1990s, however, a spectacular rehabilitation began. At a conference in Bilbao, Spain, Smith met the rebellious Craig Venter and plunged into a new career that included sequencing in 1995 the first free-living organism, his beloved *H. influenzae*.[11] It was the start of many years' work on stripped-down bacteria that might be tailored for such tasks as producing energy or cleaning up environments. In his sixties and seventies, working like so many biologists on a bridge between nonprofit research and commercial application, he was one of the pioneers of what became called synthetic biology.

Recombinant DNA: Fear and Promise

As the RNA and DNA technologies blossomed, a fierce and long-remembered professional and political battle broke out that threatened the very continuance of the work that had been labeled "recombinant DNA." The struggle, which rose directly out of new technical capabilities, effectively lasted about four years, from 1973 to 1977, and ended, after a succession of cliff-hangers, in the work's scientific and industrial establishment. Although the furor taught biologists lessons they have benefited from ever since and may ultimately have reinforced public confidence in the probity of scientists, it was a darker and riskier episode than some commentators recall.[12]

It began within the community of biologists, who were worried both by their relative naiveté about microbial infection in the laboratory and by remote but imaginable risks from experiments involving cancer, viruses, and antibiotic resistance. They struggled with the issues during a 1971 conference at the Asilomar campgrounds in Pacific Grove, California, the results of which were published by CSHL. The issues intensified in June 1973, when attendees at a Gordon Research Conference in New Hampshire heard a report on the Boyer-Cohen successes. They resolved to send a letter to *Science* alerting colleagues to their worries and to invoke the National Academy of Sciences to assess the situation.

The letter was published, and the National Academy convened a small committee, headed by Paul Berg of Stanford, whose early work had ignited the controversy. The group included such other luminaries as David Baltimore of the Massachusetts Institute of Technology (MIT), Norton Zinder of Rockefeller University, Watson of Harvard, Richard Roblin of Massachusetts General Hospital, Sherman Weissman of Yale, and Nathans. Meeting at Baltimore's MIT office on Wednesday, April 17, 1974, the group decided that an international conference on the subject should convene at Asilomar in February 1975; meanwhile, the committee would sign a letter urging colleagues across the world to suspend three classes of experiments considered risky. It was an experiment in scientific self-governance. The participants thought they were preempting public concerns by bending over backwards to protect safety for lab workers and the public alike.

Such a process necessarily involved talking with the public, politicians, and print and broadcast reporters, whose knowledge of genetics and molecular biology was uneven. Influencing the scientists' comments was their sense of awe at the revolutionary implications of recombinant DNA for all the life sciences. The first coverage of the self-denial manifesto in July 1974 contrasted the scientists' openness about recombinant DNA with the secrecy surrounding the development and use of the atomic bomb in World War II. The result was less an appreciation of the scientists' sense of responsibility than concern about a huge new force in science, one that might call for regulations similar to those for atomic energy.

There was no turning back. There were sixteen invited journalists, including this author and a writer from *Rolling Stone* magazine, when the international conference met on Monday, February 24, 1975, in the architect Julia Morgan's sixty-year-old chapel at the pine-dotted, seaside Asilomar campgrounds. The journalists had all agreed not to report the story until the conference ended. Tensely watching the proceedings were many of the brightest young scientists in the field, who saw their careers hanging in the balance. Over five days, Watson, a signer of the moratorium letter who had changed his mind, was a lonely voice arguing that unquantifiable risks were no basis for regulation. He later reflected ruefully, "We presumably responsible molecular biologists . . . first gave the DNA scare its legitimacy." Day after day, the crowded sessions were followed by anxious after-hours negotiation. On Friday, the worried conferees agreed overwhelmingly on a draft of guidelines that the National Institutes of Health (NIH), the principal patron of biological research in the world, would issue. In June 1976, NIH promulgated its conditions for recombinant DNA work, with restrictions increasing with the level of risk, and set up a board to review these conditions as experience dictated.[13]

The guidelines, however, were far from enough to quiet misgivings about the safety of the work or the possibility that it could lead to a new eugenics of tinkering with human beings. In the wake of the Vietnam

War, public discourse was heavy with environmental concerns and distrust of all authority. Hearings, often noisy and televised, were held in communities with major universities, notably Cambridge, Massachusetts, to consider whether recombinant DNA research would be permitted, banned, or regulated at the local level. Similar processes began in various American states, notably New York. At the federal level, legislators considered setting up a body similar to the Atomic Energy Commission.

Steam built up so strongly in 1976 and 1977 that major scientific organizations resigned themselves to legislation and began bargaining over the details. Some raised the specter of police in the laboratory. But the push for strict regulation fizzled, in part because frantic biologists began learning how to lobby, as physicists had done many years before to assure civilian control of atomic energy and push for nuclear-arms control. After six months of hearings before a special committee, Cambridge did end up adopting the NIH guidelines as a city ordinance. But the governor of New York, Hugh Carey, vetoed a state bill as unwarranted interference with academic freedom. Ill feeling between a House of Representatives committee chairman and the head of the relevant subcommittee derailed the federal bill. In September 1977, Senator Edward Kennedy of Massachusetts, hitherto a major proponent of a federal commission, announced that he was dropping the matter. A few months later, NIH's Recombinant DNA Advisory Committee began a long process of relaxing the regulations. Once-feared risks had begun to appear illusory.

At the same time, momentously, the techniques of recombinant DNA marched out of academic laboratories into industry. Using the methods harnessed by Boyer and Cohen, tiny bacteria could be converted into factories for making hormones, such as pure human insulin for diabetics. Founded in 1976, a biotechnology company, Genentech, became the first of hundreds to follow over the decades since. Upon the heels of a sharp recession, an academic set of procedures was turning into a new industry—a way to create jobs and make money. This hot new business would soon excite the competitive instincts of Japan, already noted for its swift penetration of the consumer electronics and automobile industries. Fancies about human genetic engineering and fears about safety in the lab faded in the bright light of economic growth and international competition.

Reverse Transcriptase: Copying RNA into DNA

When cancer virologist David Baltimore was in the thick of the recombinant DNA fight, passionately arguing that scientists must take "responsible" precautions before outside forces imposed them, he had newly become prominent for helping discover what proved to be a crucial tool of fundamental biology: an enzyme harnessed not only for genomics but also as a weapon in fighting a global epidemic disease not recognized until 1981, acquired immune deficiency syndrome (AIDS).

In the spring of 1970, Baltimore and a fellow cancer virologist, Howard Temin of the University of Wisconsin (who died of lung cancer in 1994, although he never smoked), discovered the enzyme independently. They had long known each other. While in high school in Great Neck, New York, Baltimore spent a summer at the Jackson Laboratory in Bar Harbor, Maine. There, his mentor was Temin, a Philadelphian who graduated from Swarthmore College in 1955. Baltimore graduated from Swarthmore in 1960. Temin received his PhD from the California Institute of Technology, where he was influenced by Max Delbrück, Matt Meselson, and Renato Dulbecco, an early student of cancer-causing viruses. He then settled in Madison, Wisconsin. Baltimore, too, fell under Dulbecco's influence at Caltech, where he went after taking his PhD at Rockefeller University in 1964. In 1968, Baltimore joined the new cancer research group at MIT that was assembling around Luria, who shared a Nobel Prize in 1969 with Delbrück and Hershey. In 1975, Temin and Baltimore shared the Nobel Prize with Dulbecco.[14]

The enzyme that Baltimore and Temin discovered was named "reverse transcriptase." Its strange-sounding name—Baltimore called it "romantic"—reflected how strange its function seemed to biologists. They were used to the idea of a one-way flow of genetic information, from DNA into messenger RNA and then on to proteins. This enzyme did the reverse. As Temin had stubbornly insisted for years, with what Baltimore called flawless logic, the enzyme was produced by viruses whose genes were composed of single-stranded RNA. Over some hours after they infected the host cell, these RNA viruses used reverse transcriptase to copy the viral genome into DNA. Then, the virus could take over and enforce the making of many copies of itself. Such RNA viruses, with what Baltimore called a "unique style," became known as retroviruses. The discovery also threw new light on the sixty-year-old mystery of how RNA viruses can turn a cell whose genes consist only of DNA into a cancer cell.

Because the AIDS virus, which was discovered in the 1980s, uses reverse transcriptase to invade the cells of the immune system, so-called antiretroviral drugs, often used in cocktails of three, became a major means to keep AIDS sufferers alive, often for decades.

Reverse transcriptase also became a workhorse of genomics. Short-lived RNA transcripts, particularly the messenger RNA copies of genes, could be copied back into more-stable DNA for storage and analysis. Thus, the messenger transcripts that direct the synthesis of proteins could be preserved as complementary DNA (cDNA). Biologists could assemble vast libraries of cDNA as a record of a creature's protein blueprints. Many, including Brenner, advocated around 1990 that just the cDNA should be sequenced to get a quick look at what seemed to be the main active regions of human DNA. Although that advice was not followed, cDNA remained a crucial resource for genomics.

Mirrors for Humans: "Model" Organisms

From its beginnings, molecular biology rested on a battery of tools from chemistry and physics. These had been invented over decades to separate and purify chemicals, identify them precisely, and study their three-dimensional structure. Among these were ultracentrifuges trapping the objects of interest in layers in a test tube, electrophoresis to spread chemicals across two-dimensional sheets, electron microscopy to detect objects far smaller than those that can be seen with "visible" light, the labeling of compounds with radioactive atoms like phosphorus and sulfur, and X-ray crystallography to decipher the three-dimensional structure of nucleic acids like DNA, and proteins.

But not all of the tools of biology and genetics were machines. Researchers also had adopted an array of living creatures of increasing complexity to do genetic experiments that could not be done on people. These included the viruses that preyed on bacteria, many types of bacteria, living cells grown in tissue culture, fruit flies that had been used for decades to crack open the field of genetics, and highly inbred strains of mice still used in almost innumerable medical experiments.

During the 1970s, two more "model" organisms became tools for the molecular geneticist: baker's yeast, *Saccharomyces cerevisiae*, and a tiny nematode with the formal name of *Caenorhabditis elegans*, or *C. elegans*, and the universal scientific nickname "the worm." The yeast, one of the family of microorganisms that have been mankind's partners for thousands of years in making bread, beer, wine, and cheese, was attractive because it occupied a niche between single-cell bacteria, the simplest of cells, and the much more complex cells of plants and animals, including humans. Yeasts are the simplest eukaryotes, the type of cells that have a nucleus to enclose their DNA. The shimmering, transparent, hermaphroditic nematode is only a millimeter long, with a lifetime measured in days and an average of 300 offspring. It is a relative of the fantastically abundant creatures that do much of the recycling in the earth's biosphere and one of the simplest multicell organisms known.

To become useful to science, each creature had to acquire its community of humans to study it. Under the leadership of charismatic personalities, the standard tools of committed research centers, mentoring, conferences, short courses, journals, struggles to find a common terminology, and above all a vow of openness and sharing of data and insights had to be developed for each organism.

Yeast, long a subject of biological research, became a candidate for molecular genetics in the 1970s in part because of a looming, ultimately overblown sense that bacterial viruses had provided most of their important insights. One description of this transition comes from David Botstein, a burly, highly vocal figure we will encounter several times in the history of genomics. An interviewer described him in 2006 as "no shrinking violet; he has a deeply held opinion about everything, and he isn't afraid to voice it"; she added, "His zeal is legendary, and his delight in people and ideas is palpable."

Botstein was born in Switzerland in 1942, a son of two Jewish physicians whose family origins lay in Lithuania and the Ukraine. Restrictive laws enacted in the 1920s, imposing strict quotas for each country of origin, delayed their migration to the United States for many years. Botstein came to America in 1949 when his parents, after fourteen years on the list of would-be Polish immigrants, finally received visas almost by accident. After graduating from Harvard College in 1963, he went on to the PhD program at the University of Michigan, a dozen years at MIT, a longer stretch at Stanford, a stint as vice president for research at the biotechnology company Genentech, and, in 2003, directorship of a new institute at Princeton. There, his primary mission has been to develop an alternate introductory curriculum for students who aim at a career in science.

As the 1970s began, many researchers were convinced that they would have to move beyond the simplicity of bacteria and their viruses (phages) to see if their basic principles applied to more complex, "higher" cells with nuclei. Although he had been studying the P22 virus (as Ham Smith had) to learn how it carried chunks of bacterial DNA from one bacterium to another, Botstein concluded, "It's time to get eukaryotic." But how? He recalled, "I determined to find a minimalist solution. I sought a field that would allow me to do sophisticated genetics in real time, as with phage; provide access to biochemical tests of genetic insights, as with phage; be in an open and interactive community, as with phage; be relatively cheap to set up and bootleg if grants were not forthcoming; and finally, be compatible with continuing work with phage and bacteria in the same laboratory."[15]

Would it be animal virology? No, animal tissue culture was expensive. So, with encouragement from Gerald Fink, then at Cornell, Botstein's choice fell on yeast. In 1971 he attended the yeast course at CSHL taught by Fink and Fred Sherman, which, in Botstein's opinion, made "the yeast molecular biology community what it is today." He and other "immigrants" could begin exploring how the genetics of yeast resembled, and differed from, that of bacteria and phages. Three years later, Botstein, Fink, and their Berkeley colleague John Roth gathered at CSHL for a sabbatical year that Botstein recalled as "possibly the most significant year of my career." Although they did some experiments in a newly winterized lab, the three spent most of their time debating the tools and ideas of genetics, thinking hard, "nonstop," about each other's work and where it was headed. He and Fink soon began building recombinant DNA "libraries" with yeast. The way was opening to detailed maps of the location of genes in the yeast genome and eventually sequencing all of yeast's DNA.

The drive to harness the worm *C. elegans* went beyond the ambition to deal with more complex cells than bacteria. In the early 1960s, the hope crystallized among molecular biologists that they could use their techniques to dope out how brains and nervous systems work. Seymour Benzer switched his work to the genetics of behavior in fruit flies. The molecular biologist Gunther Stent started working on the leech. Francis Crick spent most of the last thirty years of his life researching vision and then the sources of consciousness.

But Sydney Brenner, who for twenty years shared a modest office in Cambridge with Crick, looked at the puzzle of the brain and nervous system in a slightly different way and went for the worm. Back in the 1960s, after playing a major role in cracking the genetic code and finding the messenger that carries DNA's information to the protein-making machinery, he sought the wiring diagram of a multicellular organism. He wanted to know how genes influence the building of the nervous system in the universal process that biologists call development. And then he wanted to map the physiology of the nervous system.

In 2003, just after his Nobel Prize, Brenner said, "The brain is the most complicated object in living organisms." How, he asked, do the genes match up with the development, structure, and functioning of the brain? Brenner knew he had to start at a simple level. "Choosing the right organism for one's research is as important as finding the right problems to work on," he reflected jovially at the glittering Nobel banquet in December 2002. He asked himself, "What *does* constitute an explanation that says, 'I understand the nervous system?' What is the *least* we need to know?" Of this minimalist approach, he told an intently listening crowd at MIT, mostly students, "That's the nature of science."[16]

Brenner, born in South Africa in 1927, was the son of a Lithuanian cobbler and a Latvian mother who had each migrated from their native countries. He swept through grade school, high school, university, and medical school about three years ahead of schedule, but he became interested in biological research and published his first solo scientific paper in 1946 at the age of nineteen. In 1952, having determined to study molecular biology, which was impossible without leaving South Africa, he went to Oxford and won his doctorate there. In April 1953, he and his friends, crystallographer Jack Dunitz and chemist Leslie Orgel, drove over to Cambridge to see the Watson-Crick model of DNA just after it was built. "This was the watershed in my scientific life," he recalled nearly fifty years later. The model "was the key to understanding all the problems in biology we had found intractable."[17]

In 1954, a traveling fellowship allowed Brenner to meet a galaxy of molecular biology stars across the United States. He shared the driving with Watson on a transcontinental auto trip that was not without its perils. With characteristic good humor, he recalled decades later, "I couldn't see, I didn't know where we were going, and I was on the wrong side of the road, but we got through."[18]

Two years later, he settled in Cambridge at what became the Laboratory of Molecular Biology. This laboratory is not just any institution. Since the late 1940s, Britain's Medical Research Council has supported it on the theory that the best of the best, bringing in mathematics and physics to join chemistry and biology, should have the money they need to take risks and tackle problems requiring years, even decades, to solve. Brenner summarized the "patient and generous" policy this way: "Innovation comes only from the assault on the unknown." Early fruits of this policy, carried out in cramped rooms scattered around Cambridge, included the first three-dimensional structures of proteins, the first complete amino acid sequence of a protein, and the first model of DNA's structure, which showed how it could function.[19]

Now Brenner took another big risk, one that would occupy him for twenty years. He scoured the literature to find the simple organism he sought and went on to work out its genetics in great detail by 1974. He was disappointed that the available tools, including the computers, were inadequate to convert the wiring diagram into a complete explanation of how the worm's brain worked.

But he began attracting coworkers for the task of making the worm the best-described multicellular creature yet. By generating ultrathin "slices" of the worm in an electron microscope, they determined the three-dimensional placement of all the adult's 959 cells, 300 of them neurons. They also began sequencing some of the worm's proteins. John Sulston undertook the monumental task of tracing the ancestry of each of the adult cells back to the 1,090 cells of the immature animal. He found that the 131 cells always died through what became known as programmed cell death. Joining Brenner's group from the United States, H. Robert Horvitz began identifying the genes involved in cell death. The worm was beginning to qualify as one of the earliest model organisms to be sequenced in the 1990s. Sulston and Horvitz shared the Nobel Prize in physiology or medicine with Brenner in 2002.[20]

Sequencing Methods Leap Forward

The two principal pioneers of DNA sequencing in the 1970s, Fred Sanger of Cambridge, England, and Walter Gilbert of Cambridge, Massachusetts, had already made major contributions to molecular biology, Sanger with protein sequencing and Gilbert with studies of genetic control, such as the RNA messenger and the repressor. They came from different backgrounds and followed different paths to the discoveries that gave them a share in the Nobel Prize for chemistry in 1980.

Sanger, born in 1918 in Rendcome in Gloucestershire, where his father, Frederick, was a doctor, expected in school that he would become a doctor too. He accompanied his father on his rounds by pony and trap in the 1920s and 1930s, but as he told two journalists in the 1990s, he "concluded that medicine was a scrappy sort of life." So, before he went to St. John's College, Cambridge, and already interested in biology, he made a decision: "I would be better suited to a career in which I could concentrate my activities and interests on a single goal than appeared to be possible in my father's profession." He was on his way to some forty-five years of notably focused research. After he retired in 1983 to enjoy his family, his garden, and his boat, he observed that he had rigorously, if selfishly, avoided both teaching and administration, neither of which he liked or felt he would be good at. And he was happiest at the laboratory bench. He wrote, "Of the three main activities involved in scientific research—thinking, talking, and doing—I much prefer the last."

In the 1960s, after what he described as "lean years" of experiments to label proteins with radioactive forms of the elements sulfur and phosphorus, Sanger turned to sequencing DNA's close cousin RNA. He did not recall breakthroughs or "Eureka!" moments. He wrote, "For me, however, progress does not seem to go like that." A succession of small advances kept enjoyment at a more constant level than a long wait for "sudden big leaps forward."

As he moved from RNA to DNA, he worked closely with his two assistants, Bart Barrell and Alan Coulson. The hope with DNA was to sequence long stretches in a reasonably short time. They tried the indirect method of using an enzyme to copy DNA into complementary RNA and then decipher the RNA, a mirror image of the DNA. Then they moved on to what they called the "plus-minus" method. This included sorting DNA fragments according to size on a gel made of acrylamide. It speeded things up dramatically. He recalled this approach to DNA sequencing as "the best idea I ever have had."

But Sanger and his colleagues did not stop there. They soon developed what was called the "chain-termination" method. Normally, DNA subunits, called nucleotides, lack one of the oxygen atoms that are in RNA subunits. But the Sanger group developed synthetic A, T, G, and C nucleotides that lacked two oxygen atoms. DNA fragments of different lengths could be endowed with these "dideoxy" nucleotides. At the points where a dideoxy A or T or G or C showed up in the DNA, the DNA-synthesizing polymerase would stop. "Sanger sequencing" had been born.

When Sanger and his colleagues deciphered the Phi X 174 virus, whose DNA could be regarded as a circle, they began finding surprises of the sort typical of molecular biology since its earliest days. The circle was not just a chain of genes; the genes overlapped. Some stretches of DNA were used for more than one gene. Sequencing was more than a technical feat; it opened up tantalizing new knowledge about how an organism worked.[21]

Gilbert was born in 1932, fourteen years after Sanger, in Boston, where his economist father, Richard, taught at Harvard. In 1939, his father began working in New Deal agencies before moving on to the embattled wartime Office of Price Administration. The family settled in Arlington, Virginia. His mother, Emma Cohen, was a child psychologist and practiced giving intelligence tests to her son and daughter. For the first few years, she schooled both children at home, "to keep us amused," Gilbert recalled. He began visiting the adult section of the public library. In Washington, D.C., he went to public grade school and, because his father wanted him to learn Latin, high school at Sidwell Friends. There he often cut classes to go to the Library of Congress to read up on mineralogy, astronomy, and later, atom smashers.

People at Sidwell Friends were not amused. In 1980, he told a reporter, "The school and I finally came to an understanding. I promised to stay in class most of the time if they would let me get down to the library as often as possible." He impressed his English teacher, H. Hall Katzenbach, who had assigned a research paper to each of his students. Gilbert turned in a paper on the inheritance of characteristics in fruit flies. Katzenbach found Gilbert equally fluent in writing and mathematics. "He was an excellent writer. He wrote lots of poetry, but it was science that captivated him." At Sidwell Friends, Gilbert met Celia Stone, a poet and painter. She was the daughter of the noted journalist I. F. Stone, who had become a family friend. "Wally" and Celia married in 1953, when he graduated from Harvard College.

Drawn to theoretical physics, Gilbert spent a year of graduate study at Harvard before going to the University of Cambridge to finish his doctorate in 1957. While there, at an evening meeting of a science club, he met James Watson, fresh from his and Crick's triumph with DNA, and found they liked talking with each other. This bore fruit a few years later when Gilbert, by that time a Harvard assistant professor of theoretical physics, had found his interests shifting "from the mathematical formulations of theoretical physics to an experimental field."

Watson's Harvard lab was struggling to elucidate several steps in the making of proteins, a subject involving both RNA and DNA. Watson enticed Gilbert to come over and see an "exciting" experiment aimed at finding messenger RNA. It set him on the path to molecular biology, which led him not only to a new method of sequencing DNA by chemical separation but also into the suddenly emerging biotechnology industry. He was one of the founders of Biogen and even served for a time in the early 1980s as its chief executive.

Gilbert did not set out, as Sanger had, to sequence both of the macromolecules of biology—nucleic acids and proteins. Instead, he focused on how genes were controlled. This is a central question in understanding the development of a fertilized egg into a full-fledged organism. Every cell, except for eggs and sperm, has exactly the same two endowments of DNA, one from each parent. To create organs and the complete organism, many genes have to be shut down permanently. And for a cell to live, it must switch many other genes on when they are needed and off when they are not. Gilbert's repressor work investigated a simple system in a bacterium, in which the gene for a sugar-splitting enzyme was shut off when there was no sugar nearby and turned on when there was. He proved that his repressor, like the repressor in a virus called lambda, which Mark Ptashne was studying, actually bound itself directly to DNA. And, over two years in the early 1970s, he worked out the sequence of some twenty-two nucleotides that the repressor recognized and bound to.

Gilbert had sequenced the DNA he wanted to study. But, on two visits, the Russian biologist Andrei Mirzabekov had suggested an experiment. Gilbert shrugged off the idea at first. But then Gilbert and his assistant, Alan Maxam, had lunch with Mirzabekov, and Gilbert thought of a related experiment. It would lead to the chemical-separation sequencing method that could be used widely and was faster than Sanger's plus-minus technique. The method indeed worked out, and soon one of Gilbert's graduate students used it to sequence another complete virus genome. Only later was the technique eclipsed because it lost out to Sanger's chain-termination method as the basis for automating the sequencing of DNA.[22]

Robots for Faster and Faster Sequencing

Early sequencing, even with the rapid advances in technique in the 1970s, remained a laborious business. A graduate student could do an entire PhD thesis on the demanding technical task of sequencing a few genes. Impatient spirits asked if the process could be speeded up and made cheaper so that the whole living realm could be explored more completely. This would free young biologists to do more of the rebellious thinking from which new knowledge sprang.

A pioneer in that direction was the confident and visionary Leroy Hood of the California Institute of Technology. Michael Hunkapiller from Oklahoma was one of those he attracted to building a quartet of machines, two to synthesize proteins and nucleic acids and two more to take them apart. Hunkapiller recalled, "His contribution was recognizing early on that technology development was going to be a key to the advancement of biology." Hood's "best stuff came when he had a biology problem and he realized he needed tools to solve it." According to Hunkapiller, Hood was stubborn enough to stand up to opposition and eloquent enough to sell his ideas. His work in immunology won him a prestigious Lasker Award in 1987. His work in technology has been honored down to the present day.

Born in Missoula, Montana, in 1938, Hood was a son of a telephone company engineer who rose into management and a homemaker mother. Despite offers to go east to Bell Telephone Laboratories, his father chose to stay in Montana. Hood recalled, "He chose to stay and raise his children in what he perceived to be a better environment." Hood had one sister and two brothers, the younger of whom suffered from Down syndrome. This was an early introduction to the challenges of genetic disease.

Even before Hood entered grade school, he began hiking and camping. Although his mother encouraged him "to be and think independently from a very young age," she was "implacable" in insisting that he learn to play an instrument, and so he learned to play the piano, the clarinet, and in high school in Shelby, Montana, the oboe. (In later years, with little time to practice, he took up the soprano recorder.) Although, along with music, he enjoyed mountain climbing, skiing, football—and debating—Hood remembered liking school and excelling in science and mathematics. He delighted his father by attending, and doing well in, a technical course his father taught. On his grandfather's ranch in the Beartooth Mountain Range, he attended summer courses in geology with Ivy League professors and students. He made his first trip out of his home state as one of forty finalists in the Westinghouse (now Intel) Science Talent Search.

Hood went to Caltech and earned an MD at Johns Hopkins in three years. Returning to Caltech for a PhD, he became fascinated by a central problem of the immune system. That system handled the torrent of different "antigens" assaulting the body by making highly specific antibodies. How did genes help make this possible? The problem continued to preoccupy him during three years spent at NIH and after he returned to Caltech as a professor in 1970.

As developments in molecular biology rushed forward, Hood began more than twenty years on the Caltech faculty. William Dreyer, his mentor there, had "a deep interest in technology." He instilled in Hood the principle that "if you really want to change biology, develop a new technology for pushing back the frontiers of biological knowledge." Despite some disapproval from university administrators, Hood divided his time between the genetics of the immune system and his laboratory's development of equipment. Practical-minded colleagues like Hunkapiller, who was charmed by Caltech's "collegial atmosphere," turned his ideas into reality—and qualified them for commercialization.

Throughout his career, in search of an environment for finding "totally enabling" technology, an "environment where biology drove the technology," Hood bumped against limitations imposed by established institutions like universities. At Caltech, these included lab space. With the help of software billionaire Bill Gates, he moved in 1992 to a new institute of molecular biotechnology at the University of Washington in Seattle. In Seattle, he again found a need for more space and was told he must wait ten years. So he moved a short distance away and founded the fast-growing Institute for Systems Biology in 2000. He had concluded that a center devoted to biology as an integrated system could not succeed within a university. It was "very hard to do it in classic academia," Hood said, "because [the] bureaucracy is really honed to do small science and not integrative science."

Hood was equally disappointed by the inflexibility of large companies and eventually became involved in a string of biology-related start-ups. In 1981, after being turned down by nineteen potential backers, Hood got a telephone call from San Francisco venture capitalist William Bowes, who offered $2 million to get things going. The result was a company named Applied Biosystems, Inc. (ABI), one of whose missions was to make and market devices from the Caltech work. Hood reflected, "It was extremely fortunate that all nineteen companies turned me down because I believe none of them could (or would) have attracted the talent, had the focus, or committed the resources necessary to commercialize these technically challenging instruments effectively."[23]

ABI quickly escaped orphan status by attracting a range of blue-ribbon investors, including Lord Rothschild in Britain. It attracted "free ink" in the financial press as a supplier of "picks and shovels" for biotechnology. Hood crowed, "The giants really missed out. Biotechnology equipment will be one of the biggest opportunities in instrumentation for the next twenty years." After an early success with a "gene machine" for synthesizing stretches of DNA for recombinant DNA research and another robot for sequencing protein subunits, the company began generating the cash needed to develop new generations of existing devices. It gained the years needed to push the development of novel machines, such as the automated sequencer using dye labels, before its commercialization in the late 1980s. And so, ABI threaded its way deftly through the usual high-technology thicket of patent lawsuits, licensing inventions, private venture capital funding, initial public offering (IPO), strategic alliances with other firms, and acquisitions. The IPO, which raised nearly $18 million, occurred when ABI began making money in 1983, only two years after the company's founding. By the end of 1987, ABI had already sold fifty of its sequencers, priced at $90,000. One of the first customers was Craig Venter at NIH, who received his machine in February 1987.

The tall, soft-spoken Hunkapiller, lionized in 2000 as one of the "unsung heroes" of the Human Genome Project, led the company for more than a decade before becoming a venture capitalist. Hunkapiller perceived that a company can achieve things a university laboratory cannot. Going beyond building a single machine for one laboratory, a company can make it available to 50 or 1,000 others. "You really need people whose interaction is more than on a simple collaborative basis. You really want a lot of people who have a collective, unified goal of getting something done in order to make the technology available." He viewed his role as bringing people's talents and skills together. Thinking of the space program of the 1960s, in which two Oklahomans were astronauts, he said it was amazing "what you can do when you put great people together with a problem to solve."

Work on what became the ABI nucleic acid sequencer began in 1979, when Hood brought together what he recalled as "an engineer, a computer scientist, a chemist and a couple of good biologists." But the central idea did not take shape until the fall of 1982. This involved a crucial change from Sanger's methods of labeling each of the four DNA letters with radioactive atoms. The new labels were four distinct dyes. To make the design work, Hood said, involved an additional three years of solving "an enormous number of small technical problems." When Hood's group sought grants from NIH, the idea was rejected as impossible. The agency's reviewers never grasped that "there [were] never any technological limits. It was just figuring out how to get around all the incremental challenges of chemistry and challenges of engineering and challenges of the biology and enzymology and stuff. None of those were insoluble." The successive generations of machines based on dyes attached to DNA subunits came to dominate the world market for many years. As 2010 approached, ABI began marketing one of the "next-generation" sequencers of vastly greater speed.[24]

Polymerase Chain Reaction: Brainstorm Followed by Hard Work

Although the first generation of sequencing machines broke DNA down to decipher it, there had to be a second machine to build up, to amplify, the DNA of interest so that the sequencer had enough DNA to work on. These machines use a process called polymerase chain reaction (PCR). In the living cell, the chains of both RNA and DNA are strung together by the workhorse enzymes called polymerases. One job of RNA polymerase is to assemble working copies of a gene, so that the blueprint of a protein can be carried to where that protein is stitched together. A task of DNA polymerase is to make a faithful copy of all a cell's DNA, its genome, just before the cell divides into two daughter cells. A vital further task is to make a new stretch of DNA when the cell's machinery detects an error. Early in the 1980s, a laboratory machine to mimic the DNA polymerization in a cell—and produce a million or more copies overnight—was invented and brought to the marketplace.

In the brightly lighted sequencing centers of the twenty-first century, dotted from China to Europe, the visitor's attention focuses on the sequence-reading machines rather than the increasingly automated preparation of samples in neighboring rooms, or the computer analysis in yet other rooms (sometimes scattered in several places to distribute the computers' weight). In an alcove or room near the reading devices are rows of humble machines that duplicate the DNA of interest a million or even a billion times in a day or so. These are the PCR machines, whose wide adoption around 1990 enormously boosted confidence that a human genome project was feasible.

The technique was not an academic invention. It emerged in the 1980s from the pressure cooker of a young biotechnology company called Cetus in Emeryville on the east side of San Francisco Bay. The company sought profitability by driving for commercial tests for genetic disease. In the midst of this drive, the balky, easily bored, combative, but competent chemist Kary Mullis found the revolutionary way to duplicate DNA fragments of interest far faster than had been possible before. A key to the process was a pair of "primer" stretches of DNA that flanked the sequence to be duplicated by DNA polymerase. Now, there was a chemical method to avoid the widely used recombinant DNA process of the 1970s, in which DNA was "cloned" in rapidly multiplying bacteria. But there was a major challenge: Each stage of this "amplification" process involved heating DNA so that its two strands would come apart for duplication. Thus, at each cycle, the process had to be interrupted so that fresh polymerase could be added to batches of duplicated DNA. To get around this time-wasting and labor-intensive obstacle, Mullis made a second invention, which involved using polymerase from a heat-tolerant bacterium (it grew in hot springs). Now the duplication could proceed unattended. After the bacterium that made it, the "thermostable" enzyme was nicknamed "Taq."

Raised in Columbia, South Carolina, Mullis was trained at Georgia Tech and the University of California, Berkeley. As a boy, he and his brothers could explore "an undeveloped wooded area with a creek, possums, raccoons, poisonous snakes, dragons, and a railroad track." And if that bored them, they "could descend into the network of storm drains under the city. We learned our way around that dark, subterranean labyrinth. It always frightened us. And we always loved it." Not long after, Mullis was confecting rockets that carried a tiny frog in a film can a mile into the sky and parachuted him back to earth. During summer breaks from Georgia Tech, he made "noxious or explosive" organic chemicals for sale.

At Berkeley his thesis committee included Nobel Prize-winning physicist Don Glaser (later a founder of Cetus), evolutionary biologist Allan Wilson, and protein chemist Daniel Koshland, later the editor of *Science*. After earning his chemistry PhD, he knocked around several jobs before joining Cetus in 1979. There he began to play with short stretches of DNA called oligonucleotides to see what they could do. He wondered whether one of these "oligos," bound to a spot on the DNA, could induce a polymerase to synthesize a much longer string of As, Ts, Gs, and Cs.

According to Mullis's romantic, often-repeated account, the idea of using not one but two of the oligos as primers to create a polymerase chain reaction struck him in 1983 during a long Friday night drive with his then girlfriend up from the Bay Area along California Route 128 to his weekend cottage in the woods near Mendocino. The air was moist and cool and redolent of buckeyes. Served well by ignorance, Mullis recalled, "I kept on thinking about my experiment without realizing that it would never work. And it turned into PCR." As he drove along, thinking of the details of the idea, it hit him that he could "do it over and over again." Thirty doublings would give him a billion copies. It would be "an immense signal." After a feverish weekend of writing notes on every scrap of paper in the cottage, he raced to the Cetus library and found no indication in the technical literature that anyone had thought before of this seemingly obvious idea.

After more than a year of quiet, often frustrating experiments, Mullis began telling his colleagues what he was up to. Many at Cetus were certain that it would not work. Mullis was constantly coming up with ideas that didn't pan out. But managers were intrigued enough to give him a chance to prove his assertions and to forgive a physical fight with colleagues. There also were doubts about whether there was a market for PCR. Which scientists would want so much DNA? And would Mullis move fast to develop his idea? Cetus assigned a determined group, including Randall Saiki, Stephen Scharf, Glenn Horn, Henry Erlich, and Norman Arnheim, to push development. Often uneasily, they worked with Mullis and his lab associate Fred Faloona to find uses for PCR that would induce research scientists to buy PCR machines. They focused on the gene for one subunit, called beta globin, of the oxygen-carrying blood protein hemoglobin. In beta globin, the minute, but widespread, mutation causing sickle-cell anemia can occur. In December 1985, Saiki and the others could report a test for sickle cell anemia that combined PCR and restriction enzymes. They wrote, "The beta globin genotype can be determined in less than one day on samples containing significantly less than 1 microgram of genomic DNA."

About this time, Mullis submitted his own paper about his breakthrough methods to *Nature* in London, but it was rejected; nor did it pass muster at *Science*. Instead, after receiving a bonus of $10,000 from Cetus, Mullis was invited to speak a few months later at the high-exposure annual Symposium at Cold Spring Harbor Laboratory. He recalled that his talk ended with an ovation. He was taking on the mantle of the inventor. But Mullis was slow in purifying his Taq enzyme, and so David Gelfand, Susanne Stoffel, and others went ahead with the purification. The paper announcing success with the Taq enzyme in January 1988 again listed Saiki as the lead author. By then, Mullis was long gone from Cetus. In September 1986, after an argument with research director Tom White, Mullis left with five months' severance and became a consultant. He settled in La Jolla near San Diego.

With PCR, samples that had been far too small to analyze could now be amplified for accurate analysis. Paleontologists could use it to probe the ancestry of ancient Native Americans. Because of PCR's ability to detect bacterial and viral infections, the technique entered the frightening world of AIDS. Infants born to infected mothers needed therapy immediately, but antibody tests took six months. PCR needed only tiny samples and could spot an infection immediately. Antibody tests often failed to detect the AIDS virus in infected people for years, whereas PCR could give early warning. PCR proved particularly attractive in "DNA fingerprinting" associated with crimes. It also helped identify the children of parents murdered and "disappeared" in Argentina's "Dirty War." PCR soon became legend to a wider public for its role in the fictional revival of dinosaurs in "Jurassic Park," Michael Crichton's 1990 novel and Steven Spielberg's 1993 film.

Automation made Mullis's process easer and easier for researchers to use. As a medical researcher enthusiastically observed a few years later,

"The repetitive cycle is . . . self-contained and fully automated." In 1989, as PCR's applications multiplied, *Science* named the Cetus polymerase "Molecule of the Year." In 1990, *US News & World Report* wrote of an "amazing gene machine." *Science* magazine said the bonanza of PCR-created genetic sequences was "transforming molecular biology"; its headline was "Democratizing the DNA Sequence." The *New Scientist* put it even more succinctly: "Genes Unlimited." In 1993 a scientist in La Jolla exclaimed, "Pick a stock superlative. Virtually everyone [in molecular biology] uses PCR in some way."

The patents on PCR belonged to Cetus, and as the technique spread to hundreds of laboratories across the world, their value soared. When Cetus was sold in 1991 to another Bay Area biotech firm, Chiron, the patents—newly upheld in court after a titanic battle with DuPont—were transferred to the pharmaceutical giant Hoffman-LaRoche at a valuation of about $300 million. One estimate placed the annual market for PCR at $500 million.

Only three years later, Mullis received a $450,000 Japan Prize and shared equally an $825,000 Nobel Prize in chemistry. Mullis's reaction was characteristically different from that of most Nobel winners. Instead of emphasizing the shock of the early-morning telephone call from Stockholm, he said he had expected the honor. "I figured they had to give it to me eventually." Then he went surfing. At a news conference in San Diego, he said, "I was excited, happy, exhilarated. So I did what I usually do in the morning. I went out on the waves."[25]

"We Can Give You Markers"

Not all the devices that led to the genome project were enzymes and instruments and chemicals. The necessary tools included insights from the world of mathematics and statistics. One of the most crucial of these advances crystallized the search for medically important genes that became increasingly intense through the 1980s. It involved the fact that the length of DNA fragments cut by restriction enzymes differed from one individual to the next. Observing where these differences were found opened a way to find disease markers spread evenly across the human genome. The finding of such markers had been relatively easy in model organisms like yeast and nematodes and in fruit flies, where red eyes or extra wings are obvious. But in humans, it had not been so easy. There were fewer obvious markers. And the matings could not be controlled.

Early in 1978, geneticist Mark Skolnick held his annual gathering of his University of Utah graduate students at the Alta ski resort in the Wasatch Mountains southeast of Salt Lake City. There to comment on the students' presentations were David Botstein and his colleague Ronald Davis of Stanford. At Alta, they also were finishing the syllabus of their genetics course at Cold Spring Harbor that summer. During one graduate student's presentation, Botstein had a brainstorm. Kerry Kravitz was describing his study of 250 Mormon relatives of ten patients suffering from the often-devastating hereditary iron-excess disorder known as hemochromatosis. Was the gene involved recessive—so that the victim had to have a copy from both parents? Kravitz said he was increasingly sure that it was. It lay on chromosome 6, near the genes for a dozen cell-surface molecules called human leukocyte antigens, or HLAs, which had been discovered in France only three years earlier. Skolnick remarked how helpful these antigen genes were in marking the location of the hemochromatosis gene.

Botstein looked silently and intently toward Davis. Then Botstein started talking, with occasional interjections from the soft-spoken Davis. He outlined the possibility that analogous markers were scattered throughout the DNA of human chromosomes, near many other disease genes. Botstein looked at Skolnick and said, "Theoretically we can solve your problems; we can give you markers, probably markers spread all over the genome." At lunch after the session, Botstein recalled in 2007, he and Davis started to get "very excited."

Their confidence was based first of all on the restriction enzymes, the "molecular scissors" found by Hamilton Smith and his successors, who subsequently discovered hundreds more. Biologists had learned how to "clone" restriction fragments in microbes and to make these into radioactively labeled "probes" that would bind to complementary strips of a person's DNA. Some of the enzymes found their cutting points in a million places along a genome, but the specific places varied from one person to the next. Some fragments from one person would be longer than those from another. These fragment-length differences were also called polymorphisms. Similar ever-so-slight polymorphisms in DNA, many involving a single A or T or G or C, play a central role in human disease. The change of one DNA letter can change a single amino acid, predisposing a person to a particular illness. The differences in fragment length received in 1980 the awkward name restriction fragment length polymorphisms (RFLPs, pronounced "riflips"). These RFLPs, Botstein and Davis saw, could open a royal road into understanding human variation, providing milestones to simplify the search for mutations linked to disease.

With enough markers, one could use what Botstein called the "completely standard paradigm" of logarithmic odds (LOD) scores to find genes. Unlike Botstein, few molecular biologists working on yeast and other nonhuman organisms knew anything about LOD scores, and human geneticists knew very little about molecular biology. But, by "an accident of the history of my education," Botstein had taken a degree in human genetics. "And so I was probably the only one at the very beginning who understood how the whole thing might work." More accurate mapping of genetic differences between people was dramatically enabled and soon harnessed for pinning down the causes of genetic diseases like Huntington's.

The RFLP concept began to take shape. Ray White and Arlene Wyman of the University of Massachusetts Medical School succeeded in 1980 in finding the first RFLP. This happened at about the same time that Botstein and Davis published the RFLP method with Skolnick (who by then had taken the Cold Spring Harbor course) and White. The article appeared, with the blessing of Victor McKusick of Johns Hopkins University School of Medicine, in the *American Journal of Human Genetics.* At first, however, the idea drew little attention, Botstein recalled, probably because it was methodological and theoretical and fell into a gap between highly quantitative "model organism" genetics and the largely medical focus of human genetics. Undeterred, Botstein began years of crusading for the methods.

Methods were also a central concern for Davis from the start of his scientific career in the 1960s. Coming from a farming background in central Illinois, he recalled, "My family had no money. And my father said, 'Well, you can go to college if you want to, but I think it's a total waste of time. You should learn a trade. But if you want to you can, but you have to pay for all of this.'" Davis went to Eastern Illinois University in his hometown of Charleston, then on to a scientific career at Caltech, Harvard, and Stanford, in which he constantly bumped up against the inadequacy of the tools available to explore questions in biology. He began to agree with professors who told him he could "make a bigger impact on the rate of progress in microbiology by building tools." But in his preoccupation with techniques, Davis confronted scientists' fierce preference for a science of testing hypotheses using existing tools. Why, scientists asked, do you need new tools? This question was a central dilemma for one of the most significant tool-building enterprises in the history of science: the genome project.

One of the tools that engaged Davis in the 1970s was the bacterial virus called lambda, which provided a relatively simple system to study how genes are turned off and on. This work brought him into the controversies of the mid-1970s over the risks from recombinant DNA. He was a member of NIH's Recombinant DNA Advisory Committee, or RAC, which oversaw the rules governing the research. He was among those who perceived "almost overnight" that the rules had "become too burdensome and unnecessary" and spent several years dismantling them. Davis eventually started a genome technology center at Stanford in 1993, soon after the genome project got rolling. This nursery of new tools found backing for projects that filled a 50,000-square-foot building and produced a string of patents.[26]

The Bits and Bytes of A, T, G, and C

The influence of mathematical insights soon became pervasive in the new biology of the genome. Although DNA is finite in size, the combinations of events that DNA specifies are immense. As biology penetrates ever-more-complex living processes, the amount of information to be handled keeps mushrooming. It must be stored and displayed in a form in which researchers can obtain it at maximum speed, and it must be moved through ever-wider electronic channels. To analyze and understand the information, the statistical and mathematical tradition, dominant since Gregor Mendel's experiments in the 1850s and 1860s and crucial to the science of genetics from the moment it acquired its name early in the twentieth century, must intensify.

Very fortunately, the technologies of electronic data storage and manipulation had already begun the explosion that pervades modern life to this day. The sovereign device has been the silicon semiconductor chip, on which ever-larger numbers of ever-tinier transistors can be built up by an exotic combination of lithography, baking, and chemical washes. The chips for holding or processing information have evolved relentlessly, far beyond the dreams of their originators, demanding ever-faster data transmission. The data traveled not only as electrons in tiny wires, or as beams to and from satellites stationed in orbit, but also as waves of laser light traveling through ultrathin "optical fibers," even across ocean floors. These devices and the "software" to guide them were accelerating just as fundamental biology was taking off. The result was an ever-denser network of millions of computers across the world able to "talk" instantly to each other.

Born at the end of World War II, electronic computers began as huge machines confined to air-conditioned basements that devoted themselves to billing phone and power company customers, breaking enemy codes, and modeling weather systems and hydrogen bomb explosions. But a drastic democratization, central to the genome project, had begun. The changes were already dramatic by the early 1980s. From the beginning, both intellectual and technical trends had been at war with the queen bee model of computing. Determined pioneers pushed relentlessly to make computers ever smaller and cheaper, so that they could operate on a user's desktop or lap and be linked almost seamlessly to each other.

The minicomputers that sprang up by the thousands in the late 1960s were followed only a dozen years later by millions of personal computers, using semiconductor chips from companies like Intel. In 1980, the giant computer manufacturer IBM, long the champion of the giant "mainframe," put its first personal computer on the market, with software from a small company called Microsoft. Every year and a half, from the 1960s onward, solid-state devices doubled their capacity to store data and carry out calculations. In parallel, communications companies learned how to carry more and more "bandwidth" of information. This explosion occurred almost simultaneously with the recombinant DNA revolution in biological science and biotechnology. Biologists, particularly the younger ones, struggled to navigate an ocean of numbers.

With tools to decipher complex systems, biologists could tackle the myriad interactions of living processes. An imperative was maturing.

With the ability to cut and paste DNA, multiply it for study, grab and "clone" genes of interest, sequence DNA at greater and greater speeds, compare humans to a range of model organisms, and channel a mounting flood of data, a battery of enabling technologies for an assault on the human genome was now in place. Ahead lay the summoning of the will to bring biology into Big Science.

CHAPTER TWO

Getting Started

The Huntington's Gene

James Gusella of Massachusetts General Hospital (MGH) in Boston was one of many researchers gripped by the dream of using genetics against disease. Hence, he knew he had a significant announcement to make in the scientific journal *Nature* in November 1983. Gusella and his colleagues, including the indomitable Nancy Wexler of New York, had succeeded, with a considerable amount of luck, in locating a region in the DNA of human chromosome 4 involved in one of the most tragic of human genetic diseases: Huntington's chorea, so named for the nineteenth-century Long Island doctor who first described the disease and the jerky movements of its victims, who would eventually die from the progressive death of particular brain cells. Wexler's mother had died of Huntington's. Although the gene itself would not be identified for another ten years, the marker opened the way for a genetic test that could be given prenatally as well as later. According to the report in *Nature*, the achievement marked "the first step in using recombinant DNA technology to identify the primary genetic defect in this disorder." Looking back twenty years later, the influential population geneticist Neil Risch wrote, "Linkage analysis in humans was no longer an arcane, low-yield dalliance but rather the first step in a realistic effort to identify mutations underlying human genetic disease."

Gusella and his colleagues were so excited by their finding that they spoke about it ahead of schedule at the annual meeting of the American

Society of Human Genetics. And on November 8, an earlier-than-planned press conference made a media splash. Suddenly, "the use of DNA markers had been shown to be effective," *Wall Street Journal* reporters Jerry Bishop and Michael Waldholz later wrote in their classic account, *Genome*.

Afflicting more than 20,000 people in the United States, Huntington's was widely known because it had claimed a famous victim, the folk singer Woodie Guthrie. After a tragic series of misdiagnoses—he was thought to be an alcoholic or schizophrenic—Guthrie died in 1967. Before 1983, Guthrie's son, Arlo, also a folk singer, had no way of knowing whether he or his children were at risk. The public television news program *The McNeil/Lehrer NewsHour* reported, "There's a medical development today which means singer Arlo Guthrie may find out if he, too, will suffer from the same disease that killed his father."

Like many other rare, but devastating, human genetic diseases, Huntington's is agonizing to contemplate. Geneticists call it an autosomal dominant trait: The probability of passing it on to one's own children is 50 percent. Onset often occurs during middle age, after a victim has seen a parent or other close relative die painfully. So, victims often live in dread, knowing that no medical treatment has yet been found. Until development of the test based on Gusella's finding, a potential sufferer could have no idea if, when, or how severely the disease would manifest itself. Often an unwitting victim would fall ill after the birth of children and grandchildren.

The disease generally takes hold slowly, so that a doctor has trouble diagnosing it unless there is a family history. Huntington's thus imposes distressing emotional and ethical issues: Should a person learn his or her fate and if so, when? Should he or she marry or have children?[1]

One Woman's Descendants

Well before the restriction fragment length polymorphism (RFLP) concept was published in 1980, word of it reached Nancy Wexler, who had set up a foundation to pursue the cause of Huntington's after her mother was diagnosed with the disease in the late 1960s. Leading biologists had advised her to focus on locating the best and brightest people to help. She invited David Botstein to one of the foundation's periodic workshops, staged in 1979 at the National Institutes of Health (NIH) in Bethesda, Maryland. There, MGH researchers, already searching for a Huntington's gene in samples from families in Indiana and Venezuela, could hear about RFLPs as a tool to search for a needle in a genetic haystack. Days of often-raucous debate showed that the new tool could help—especially if the families were large.

Seeing that there was actually some chance of finding "this thing," Wexler and her colleagues went back to Venezuela to track down hundreds of additional family members. The family turned out to number in the thousands; one hundred members were afflicted, and many hundreds more were at risk. Dwelling in and around Venezuela's Lake Maracaibo, including in Laguneta, a village on stilts, they proved to be descendants of a female victim, perhaps the daughter of a European sailor carrying the disease, in the early nineteenth century.

With the help of her credibility as the daughter of a victim, Wexler and her colleagues overcame huge logistical and cultural obstacles, convincing the family members to give blood on specified, chaotic "draw days" and devising a practical way to refrigerate and send samples to Boston by air.

At MGH, vast quantities of the lake dwellers' DNA could be manufactured by "immortalizing" lymph cells. Botstein's MIT colleague David Housman persuaded Gusella to abandon plans for a postdoctoral fellowship at Caltech and to lead the work instead. He began four years of searching randomly, contrary to Botstein's advice, through hundreds of clones. Luckily, however, he found what he was looking for—a marker that was present in Huntington's patients and absent in others—on the twelfth marker, called G8. Later searches for other genes were not so easy.[2]

Disease Genes

Like everyone in a modern society, biologists lived in the expectation of scientific and engineering miracles that would save and lengthen lives and lessen both hunger and the pain of illness. Alongside the industrialization of the world over the previous 150 years had come a torrent of discoveries and inventions, including great advances in sanitation and agriculture, that had improved human health. In twentieth-century medicine, the expectation of finding cures—and the willingness to spend tax dollars in search of them—kept growing with every advance. It became increasingly evident that the quest must include a far deeper understanding of fundamental biology. With a tiny fraction of all biomedical research budgets—themselves a tiny fraction of world spending on medical care—molecular biologists rapidly sketched a picture of how the genetic information within the DNA double helix was used in health and disease. The focus on inherited illness increased. But where were the preventives and the cures?

Many intellectual and practical threads led to Gusella's finding, which was universally acclaimed as a major advance. One of these involved identification and classification of human genetic diseases. From the early 1970s, a growing international club of human geneticists, interested in the role of genes in both rare and common illnesses, had systematically built a catalog of "disease genes" and methodically mapped hundreds of diseases to particular parts of the twenty-two pairs of autosomal human chromosomes and the two pairs of sex chromosomes. The collaboration was led by Victor McKusick, a native of Maine and one of the founders of human medical genetics in the United States. As in all collaborations, the participants found the need to meet, face to face, at intervals. In 1973, McKusick and such colleagues as Frank Ruddle of Yale began a series of biennial conferences to collaborate on the latest human genetic "map" and issue ever-fatter volumes under the title *Mendelian Inheritance in Man* (whose publication eventually migrated to the Internet).

McKusick—not shy about bringing the new perspective on the genetic origins of disease to the attention of interested journalists—emphasized the potential benefits of locating genes for disastrous illnesses, allowing diagnosis in fetuses, newborns, and their families, and discovering the proteins specified by the genes (a prerequisite for developing drugs). Beginning in 1964, McKusick and other geneticists met reporters attending "Press Week" during an annual scientific conference at the Jackson Laboratory in Bar Harbor, Maine. The conference was sponsored for many years by the March of Dimes, which had shifted its focus to birth defects after its historic success with the polio vaccine. Journalists' accounts from the annual conferences appeared in many major outlets. Of his work with journalists, McKusick said, "It never entered my mind that I was acting as a salesman." His relentless message was, "This is the future. This is the way it is."[3]

A poster child for advances in medical genetics arrived in 1973. Michael Brown and Joseph Goldstein at the University of Texas Medical Center in Dallas were studying a form of tragically early heart disease that was known to be passed on in families. The disease involved huge excesses of cholesterol in the blood. Studying an afflicted family, Brown and Goldstein discovered a particular enzyme that maintained the delicate balance between too much and too little cholesterol in the body. Mutations in the enzyme degraded its function. This finding helped open the way to designing drugs to control cholesterol levels—the famous statins now used by tens of millions of people every day. Experiments on these drugs were under way in the 1980s.[4]

If Only the Diagnosis Had Been Made Sooner!

Huntington's was by no means the only genetic disease receiving new attention in the early 1980s. Across Boston at Children's Hospital, Louis Kunkel was working on the muscular dystrophy that afflicted many of the patients he saw in the course of his research. He later recalled for a group of journalists a mother who brought in her three-year-old son, who was having noticeable trouble getting around. She also brought along her younger son. The three-year-old's trouble was a sign of Duchenne muscular dystrophy (DMD), which progressively cripples and kills its victims, usually before they reach age twenty. Located on the X chromosome, the disease usually afflicts boys. One of a spectrum of similar diseases of varying severity, DMD causes so much suffering that the parents never wish to have another muscular-dystrophic child. The boy tested positive for the newly discovered DMD gene. More testing determined that his mother was a carrier. Her younger son had the gene too.

Kunkel, a human geneticist at Children's, was horrified. Incidents like this fed his passion to find the cause of such suffering and ways to prevent or alleviate it. At Children's, Kunkel led the effort that not only located the muscular dystrophy gene but then found the huge protein specified by that gene, called dystrophin, which proved to have a crucial place in the structure of muscle-cell walls. When he saw the stricken family, Kunkel reflected, If only the genetic diagnosis had been made sooner![5]

Gusella, Kunkel, and like-minded scientists in many laboratories were engaged in a crusade to find disease genes, in which tiny changes to a single base caused catastrophe. They sought the genes not only for sickle cell anemia, Huntington's, and muscular dystrophy but also for cystic fibrosis (CF), whose gene was located and identified later in the 1980s by Lap-Chee Tsui of the Hospital for Sick Children in Toronto and Francis Collins of the University of Michigan. The excitement over the CF discoveries led to immediate calls for general genetic screening for the disorder, which Collins, Tsui, and the American Society of Human Genetics promptly opposed as premature.[6]

Meanwhile, understanding of the genetic basis of various cancers was opening up. In the mid-1970s Michael Bishop and Harold Varmus of the University of California (UC), San Francisco, who would later win the Nobel Prize, uncovered a gene implicated in cancer in chickens. Soon after, cancer-related genes were identified in people. Michael Wigler of Cold Spring Harbor Laboratory (CSHL), Robert Weinberg of the Whitehead Institute at MIT, and others began finding what were called oncogenes. When normal, these genes specified proteins used in the regular work of the cell; however, even a slight mutation in such a gene could start a cell on the path to dividing uncontrollably and becoming a cancer. Shortly thereafter, biologists began finding another class of genes that normally blocked cancer. When these anti-oncogenes mutated, the path to cancer reopened.[7]

The identification of specific keys to heart disease and cancer further uncovered for geneticists the huge arena of widespread, common, complex diseases involving several or even many genes.

The discoveries of the genes responsible for Huntington's, CF, and some types of breast cancer were heroic achievements of the 1980s. But these relatively easy searches had a darker side. They took years and years; yet, the researchers hunting for disease genes worked almost daily with terribly disabled people, for whom no cures existed, and their agonized families. Despite many years of waging "war" on cancer and achieving successes in early detection and treatment, particularly in children, the age-adjusted death rates from cancer had barely changed. If this were the pattern for the future, the quest for human disease genes would be heartbreakingly long.

Cures seemed maddeningly distant. Years later, James Watson expressed satisfaction that the recombinant DNA techniques, born amid loud public controversy in the 1970s, provided biologists with ways to locate genes as well as grab hold of the specific DNA of those genes. But he told members of the House Appropriations Committee in 1995, "Even with the immense power of recombinant DNA methodologies, the eventual isolation of most disease genes still seemed in the mid-1980s beyond human capability."

It seemed clear that genetics played an even bigger role in both rare and common diseases than had been suspected. Hence, genes implicated in disease should be identified as fast as possible to speed both diagnosis and treatment. This way, many thought, molecular genetics could lead to a medicine that attacked causes, not just symptoms. For the genetics that taxpayer dollars had supported for decades, it was time for the public to see medical payoffs. It was time, Watson told the members of Congress, to stop seeing ourselves "as tiny pawns in a play which we did not write." Could that script be read faster if human DNA were deciphered all at once?[8]

New Awareness, New Industry

In the first half of the 1980s, the world was growing more "genomic," to use a term coined later, and breathing in a bit of a Klondike gold-rush atmosphere. News coverage of issues in genetics increased, as did attention paid by lawmakers and the general public. A flood of the newly discovered oncogenes came to light. With the fanfare about the location of genes related to terrible diseases like Huntington's, somatic cell "gene therapy" was tried in those suffering from the blood-protein disorder sickle-cell anemia, which afflicts African Americans; beta-thalassemia, a disease that affects people of Mediterranean ancestry; and a rare deficiency in the enzyme called adenosine deaminase involved in the survival of the white blood cells of the immune system.[9] Debate continued about "germline gene therapy," that is, whether or when people's germ cells could be altered so that a characteristic would pass down to descendants. A rat gene for growth hormone was transplanted into mice that consequently became "super mice," twice the normal size.[10] The first child conceived by in vitro fertilization in the United States was born in 1981.[11] The increased precision of genetic testing raised questions. Would employers and insurers discriminate if they obtained a person's genetic profile? Would prenatal diagnosis of a genetic disorder lead to societal pressure to abort?

The new field of biotechnology, dotted with start-up companies in the mode of Silicon Valley, drew fevered interest on Wall Street. Major pharmaceutical firms invested millions in recombinant DNA ownership stakes or alliances. In 1982, only five months after being asked, the U.S. Food and Drug Administration (FDA) approved synthetic human insulin grown in cooperative bacteria by California-based Genentech. This first biotechnology product to be cleared by the FDA was a small sign that biotechnology companies might begin making money. Discussion continued as to whether genes could fit the patenting criteria of novel, non-obvious, and useful.[12]

With the waning of the seeming risks posed by recombinant DNA that caused intense concern in the 1970s, the NIH Recombinant DNA Advisory Committee, or RAC, regularly heard petitions to relax regulations.[13] Microbes were harnessed to transfer disease resistance into food crops.[14] Researchers began predicting that a complete picture of the human genetic endowment would exist by the end of the century.[15]

Frustration in Hiroshima

Amid such thoughts, a few minds were beginning to look beyond intense mapping of the genome toward the possibility of spelling out all the subunits of human DNA. The spurs came from widely different sources. One of these was a long-standing problem in looking for mutations among some 12,000 children of survivors of the 1945 atomic bombings of Hiroshima and Nagasaki. This work was important to the

U.S. Department of Energy (DOE), the quintessential "Big Science" enterprise, which not only developed and made nuclear explosives but also pushed the technologies of nuclear power. What genetic risks would be posed by a big expansion of the already large nuclear-power industry, which provides 20 percent of U.S. electricity?

Some of the most respected geneticists in the world, mindful of discoveries they had made many years earlier, had expected that the bomb's radiation would leave a heavy mark on future generations. The number of mutations in the exposed group, it was thought, would significantly exceed those in unaffected populations in Japan. James V. Neel of the University of Michigan was one scientist who gave much of his time looking for the extra mutations—but did not find them. After decades of study by the U.S.-Japan Atomic Bomb Casualty Commission, it was clear that the tools of human genetics were far too crude—about ten times short of the necessary precision to detect an additional hereditary burden.

Meeting in Hiroshima in March 1984, frustrated researchers agreed that developing new tools for analyzing DNA was almost as important as establishing cell lines from the bomb survivors, their children, and the "control" population. As a consequence, DOE cosponsored a conference at Alta in December 1984. Robert Cook-Deegan, author of *The Gene Wars*, called it "the Alta Summit." It was the start of a series of deliberative conferences and workshops that extended over more than three years, building the case for a genome project and thrashing out a strategy that scientists could commit to.

A memorable feature of the meeting, which ranged boldly over a panorama of genetic techniques, was that Alta was snowed in for all five days.

James Neel was among the approximately twenty participants, most of whom were pioneers in methodology. David Botstein of MIT and Ray White, who by this time had moved to the University of Utah and was enlisted as organizer, were there to argue strongly for pushing "to the utmost" their RFLP concept. White described his project to look for changes in the Y chromosome of a single Mormon progenitor from 1850, who now had thousands of male descendants. In Cook-Deegan's account, based on interviews with several participants, Botstein made his points in his usual style, "as always exuding volcanic enthusiasm peppered with sharp humor." But Botstein acknowledged, with everyone else at the meeting, that the method still would not be powerful enough to detect the mutations Neel and his colleagues had vainly sought.

Maynard Olson of Washington University, whom we will encounter again in this narrative, was there to discuss his efforts toward a complete gene map of the model organism, yeast. With the help of "pulsed field gel electrophoresis," codeveloped by another participant, Charles Cantor of Columbia University, Olson was also finding a way to store million-nucleotide pieces of DNA in yeast cells. This technique was used early in the Human Genome Project. Thomas Caskey of Baylor College of Medicine in Houston described the detection of mutations in HPRT, the gene associated with the hereditary Lesch-Nyhan disease, and Sherman Weissman of Yale gave a similar review of the recently discovered human leukocyte antigen (or HLA) region, seat of genes for human immunity and autoimmune diseases. Elbert Branscomb of the Lawrence Livermore National Laboratory in California described a flow cytometry method used in finding altered proteins on the surface of red blood cells.

George Church, a veteran of Walter Gilbert's Harvard lab, was there to discuss his work on a fast "multiplex" method of DNA sequencing. He credited Olson and Richard Myers with helping him to clarify his ideas. Myers was there to describe his method of using an RNA-cutting enzyme to detect polymorphisms at a single DNA letter, A or C. Olson, his roommate at the conference, helped Myers see how the method could be expanded to other mismatches. Edwin Southern of Edinburgh, pioneer of a widely used method of DNA analysis called the Southern blot, gave the closing talk.

Superficially, the 1984 Alta meeting was very discouraging. The sensitivity needed to detect mutations among the atomic bomb survivors could not be reached, as Cook-Deegan put it in 1994, without "an enormously large, complex, and expensive program." As Eric Lander recalled in 2001, "Back in 1984 and 1985, it was an absolutely ludicrous notion to think that we could do this. DNA sequencing was so inefficient, there was this debate: was it even worth it? We had no clue what to do or how to do it." Nonetheless, the participants were excited. For five days they had explored a new world of methods. Many of the technological cards were on the table. Christopher Wills, in his book *Exons, Introns, and Talking Genes*, put it this way: "DNA mapping and sequencing techniques were advancing further than any of them individually had realized." An intellectual corner had been turned.[16]

Walter Gilbert Encounters the Private Sector

The day after the Alta conference closed, Walter Gilbert ended his experiment of serving for nearly three years as chief executive officer of one of the first biotechnology companies, Biogen of Geneva, Switzerland, and Cambridge, Massachusetts. Perhaps under pressure from Biogen's board, he resigned and departed for an extended vacation in the South Pacific.

Given Gilbert's insistent focus on a long-term scientific orientation and financial backers' short-term focus on products hitting the market quickly, it had been a bumpy ride. Gilbert had cofounded the company in 1978, heading its scientific board. Just two years later, in January 1980, the company thrust itself into the spotlight. Biogen-supported researchers led by Charles Weissman of the University of Zurich announced that they had used recombinant DNA methods to induce bacteria to make human interferon. This exciting achievement promised the ability to develop plentiful, relatively cheap supplies of interferon. Discovered in the 1950s in England and Japan, interferon, a signaling protein of the immune system that boosts the body's response to viral infections, appeared to be a potential weapon against cancer as well.[17]

In his years at Biogen and later, Gilbert operated near the epicenter of loud debates on the proper relationship between university research and commerce. These issues resounded in the minds of biologists as they worked painfully over several years toward a commitment to the Human Genome Project.

The pitch loudened considerably in 1980. In June, the U.S. Supreme Court ruled 5–4 in the landmark case *Diamond v. Chakrabarty* that at least some living things, in this case a bacterium genetically engineered to devour oil slicks, could be patented.[18] On October 14, Genentech went public at $35 a share, and before the trading day ended, share prices had flown up to $86 before settling back to $71. It was the hottest new offering on Wall Street in ten years.[19] In December, the U.S. Patent and Trademark Office granted a broad, and eventually very lucrative, patent on the basic recombinant DNA inventions of Herbert Boyer of UC San Francisco and Stanley Cohen of Stanford. The two universities would share the fruits equally.[20]

A few days later, Congress passed, and President Jimmy Carter signed, the Bayh-Dole Act. This major departure in government policy instructed universities to patent inventions emerging from government-sponsored research. The idea was to get the inventions to market and thus help people more quickly than before.[21]

Meanwhile, the new biotechnology companies, such as Biogen, were forming more extensive ties to universities, creating more than a little worry in academia that professors' minds and souls might come to belong more to industry than to their students.

These issues touched Gilbert directly in 1980. Harvard, where he was a professor, had invested indirectly in Biogen through a venture capital firm. But in October of that year, less than two weeks after Gilbert learned of his Nobel Prize, Harvard's president announced at a faculty meeting that the university was considering forming a biotechnology company and directly owning a minority of shares. Leaders in the project were two other Harvard professors, Mark Ptashne, Gilbert's old rival in the repressor work of the 1960s, and Thomas Maniatis, the pioneer in creating "libraries" of DNA fragments. The company was to be called Genetics Institute. The manager of Harvard's investments said publicly that the idea was to garner greater reward for the university than just revenue from license fees. But many scientists said this attitude cut across the academic principle of choosing topics strictly on intellectual merit.

The plan produced an explosion from Gilbert. Interviewed for a front-page article in the *Washington Post*, he denounced the plan as "extraordinarily unwise . . . a fantastic distortion of the purpose of a university." In market competition, his company, Biogen, might have to push the Harvard company "to the wall." He did not like the idea of competing "against Harvard for the best people and the best work." Labs could soon have people working for ten different companies. Would the researcher working for Harvard get preference? It was the same, Gilbert fumed, as the law school forming its own law firm or the English department starting to writing advertising copy. "The idea is completely mad at a certain level."[22]

With stiff opposition from its faculty, Harvard swiftly bowed out of the direct investment, and Ptashne and Maniatis went ahead on their own. They were kept "at arms length," as Phillip Sharp of MIT, also on Biogen's science board, put it.

Not long afterward, Gilbert found the sword of university principles cutting the other way. It seemed imperative for him to assume the chief executive's role at Biogen. But Harvard said he would have to choose between that job and remaining on the faculty. Gilbert resigned his professorship.[23]

Robert Sinsheimer Reflects, "I Had to Have Some Validation"

Freshly back from the South Pacific, with one adventure completed, Gilbert quickly undertook another: advocating for a complete sequence of human DNA. In March 1985, after weeks of trying, Robert Edgar of the University of California, Santa Cruz (UCSC), a pioneer in studying the nematode *C. elegans*, managed to contact Gilbert and urged his attendance at a conference at the university specifically focused on whether and how a genome project would operate. Although the conference was scheduled only two months hence, Gilbert accepted.

The origins and effects of the Santa Cruz meeting, which carried the discussion well beyond that of the meeting at Alta, illustrate that unpredictable contingency operates as strongly in genomics as in any other movement that ends up making history. To be sure, the idea of a complete sequence of human DNA bases was in the air. But no one could have forecast how the first serious conference on the matter would come to be held or how much the skepticism it began with would evaporate over the weekend's discussion.

The path to the meeting began with the ambitions of UCSC chancellor Robert Sinsheimer. A leading student of Phi X 174 (the virus that Frederick Sanger sequenced) and a former head of Caltech's famous

Division of Biology, Sinsheimer was determined to put his campus "on the scientific map," not just in physics and astronomy but in biology. He was familiar with, and unafraid of, Big Science. As it happened, UCSC had a lead gift of $36 million toward a $70 million, ten-meter optical telescope to be run jointly by Caltech and UCSC. But then the gift went glimmering when another donor said he would give the whole $70 million for a telescope named for him and run by Caltech alone.

So now what? Sinsheimer already "wanted to do something in biology." He turned to a life science project that would match the telescope in scope and scientific interest. This would be a brand-new institute in Santa Cruz to lead an effort to sequence the entire human genome. Estimating its start-up cost at $25 million, Sinsheimer sought to steer the lost money from the first gift to that project. But the president of the University of California "didn't see it; he didn't understand it," and he said no in the fall of 1984.

Who else could provide the money? Sinsheimer knew that NIH, focused on grants to individual researchers, was not accustomed to building entire university institutes. He recalled, "NIH, I think, was pretty reluctant. They don't work in that mode; they work much more with investigator-originated projects, [not] with large, managed endeavors." Although NIH director James Wyngaarden (later a crucial backer) was sympathetic to the idea and wished Sinsheimer luck, local congressman Leon Panetta made it clear that federal funds for such a big project would require a formal approach from the University of California. And there would be a national competition. NIH "would undoubtedly have to go out for bids," Sinsheimer reflected ruefully, "and there was no way Santa Cruz could compete with Caltech or Harvard or Berkeley at that stage."

Now he had to think of how to justify his plan to other donors. "In order to do that, I had to have more than just the idea; I had to have some validation that this was really a feasible project." He needed the blessing of leading people in the field. So, with the help of Santa Cruz colleagues like Robert Edgar, Harry Noller, and Robert Ludwig, he invited a dozen of the leaders to Santa Cruz to decide whether a complete human genome sequence was feasible.

Sinsheimer knew that the pace of sequencing was quickening. In 1977, Sanger had used what Sinsheimer described as "his elegant technique" to decipher Phi X 174, with its 5,400 DNA bases. But complete sequences had been getting longer. Despite the slowness of the methods of those days, the viruses called T7, with 40,000 DNA nucleotides, and lambda, with 49,000, had already been done. In 1984, the much longer sequence of the Epstein-Barr virus (172,000), implicated in human cancer, was achieved. People were beginning the long struggle to sequence the *E. coli* bacterium, with 4.5 million bases, and the nematode *C. elegans*, with 80 million.

"For a number of years," Sinsheimer wrote later, "I had been simultaneously excited, frustrated, and concerned about the potential application of genetic knowledge to the human condition." From the work of McKusick and his collaborators, he knew that human development, growth, and aging were largely genetically based, as was human evolution. But to get at the processes, including those of disease, required a much clearer picture of the human genome. As he told a reporter in 1990, physicists and astronomers did not hesitate to ask for big money. He began to wonder "if there were scientific opportunities in biology that were being overlooked, simply because we were not thinking on an adequate scale."

He asked himself, "What if a project could be undertaken to sequence, once and for all, the entire human genome? This would be Big Science, but the product would be an invaluable resource for all biology and medicine. Was it feasible? What would it cost? How long would it take? How might it be organized?"[24]

The Mood Shifts from Why? to Why Not?

Many of the key people who would be needed to make a genome project work were invited to the Santa Cruz conference. Noller wrote Fred Sanger, the sequencing pioneer, and Sanger, already retired, wrote back, "It seems to me to be the ultimate in sequencing and will probably need to be done eventually, so why not start on it now? It's difficult to be certain, but I think the time is ripe." In Sanger's place came one of his principal Cambridge associates, Bart Barrell, head of large-scale sequencing at the Laboratory of Molecular Biology. He wrote later, "Everybody accepted that the then state of the art was inadequate for the task but I do not think it was realized that this state of the art had just reached its pinnacle in the techniques of shotgun DNA sequencing and was being continually refined."

Also focused on DNA sequencing, besides Gilbert, were Leroy Hood of Caltech and George Church of Harvard, who had also been at Alta five months before. The participants were aware of Hood's progress in using dyes instead of radioactive atoms to label each DNA subunit and thus greatly accelerate sequencing. Barrell regarded this as an "exciting" new development, as were Church's ideas on "a way of sequencing many pooled samples at once, which he called multiplexing." But rival Hood was skeptical of this approach. He recalled in 2004 his hunch that "this multiplex kind of sequencing approach was doomed from day one because of enormously high error rates."[25] And he proved right.

The field of genetic mapping was represented by David Botstein, who had also attended the Alta workshop, Ronald Davis, and Helen Donis-Keller. When Edgar first invited him, Botstein was originally unenthusiastic about attending. "I wasn't that interested in sequencing the human genome; I thought it was premature." But, he recalled in 2007, "they basically made it personal, and so I agreed to go. And partly we were still at that time concerned about getting the genetic map done and using the genetic map to find actual genes. Sequencing was relevant." Botstein showed up with "pretty much the idea that, yes, eventually we're obviously going to do this, but maybe tomorrow wasn't in the cards."[26]

To tell about work on the physical mapping of the nematode *C. elegans* was John Sulston, who had begun collaborating with Robert Waterston when they both worked in Sydney Brenner's lab in Cambridge, England. Leonard Lerman, from the biotech company Genetics Institute, another Alta participant, focused on technology. David Schwartz of Columbia University, in partnership with Charles Cantor, who attended the Alta meeting, was working to separate and handle megabase DNA fragments. Others included Michael Waterman of the University of Southern California, who focused on the mathematics of compiling databases and analyzing them, and Hans Lehrach of the European Molecular Biology Laboratory in Heidelberg, Germany, where some of the earliest thinking on the genome was done.

As the conference opened, "extreme skepticism" was rife, Sinsheimer recalled. Hood later said, "I thought Bob Sinsheimer was crazy. . . . It seemed to me to be a very Big Science project with marginal value to the science community."[27] Participants talked both about whether such a project should be done and about whether it could be. As they argued about injecting Big Science into biology, Sinsheimer said the project might not need a big telescope or accelerator, but it definitely required "a massive information base" (of the sort DOE had been handling for decades).

Another topic that arose again and again, before and after the Human Genome Project started, was whether the DNA that did not code for proteins—what people had taken to calling "junk DNA"—was worth sequencing. In a premonition of the ultimate policy, Sinsheimer wrote, "I had little patience" with this argument. "It is only by examining sequence that one can actually decide what is or is not junk. Further, 'one man's garbage is another's treasure.' Analyses of the putative nonspecific DNA regions could be of considerable value to evolutionary and anthropological research."[28]

The mood began to shift, Sinsheimer felt, toward "confidence in the feasibility of such a program. . . . Feasibility was no longer the issue." He looked forward to a complete human DNA sequence, "once and for all time . . . a landmark in human history . . . the basis for all human biology and medicine of the future." The thinking had shifted from why? to why not?

To be sure, participants foresaw difficulties. Perhaps the greatest was that the possession of a complete human sequence, by itself, would actually tell scientists or clinicians very little. At Santa Cruz, Botstein warned that the sequence, without supplementary information, gleaned through comparisons with model systems, would be uninterpretable. The sequence would be like a complete set of Egyptian hieroglyphics with no accompanying text in a known language—no Rosetta Stone.[29]

Over a weekend of intense talk, the select group of guests stayed at a rustic inn in Santa Cruz called the Babbling Brook. Their rooms were scattered across a cluster of wooden pavilions that descended a hillside on both sides of the audible brook. The deliberations boiled down to "reasonable goals." The first, advocated by Botstein, was a whole-genome genetic-linkage map, with genetic markers "at reasonable spacings," locating genes with respect to each other. The second was a corresponding physical map in which overlapping stretches of about 30,000 to 40,000 nucleotides would be placed in order. The third was further development of sequencing tools. Fourth, it was desirable to make a centralized, correlated effort to bring together new knowledge in human genetics, development, and physiology. Because the total sequence was then out of reach, the conferees agreed, efforts should concentrate on polymorphic regions and functional genes (about 1 percent of the genome).

Sinsheimer summarized the conclusions: "First we had to develop automatic techniques for sequencing. While this was being done, the gene mapping could go on, and then comes the actual sequencing. You'd also have to get started very early on developing computer techniques to handle all this information." He looked back on the exercise proudly: "We anticipated remarkably well what needed to be done and what sort of scale it needed to be done at." Indeed, the conclusions drawn at Santa Cruz foreshadowed the eventual biologists' compact of 1988 on how to proceed.[30]

To Hood, Santa Cruz was "the first meeting [at which the genome project] got discussed by a bunch of good people. It was clear that there was going to be a big split [among biologists between] those who thought it should be done and those who didn't think it should be done. But one thing that meeting did agree upon, everyone agreed . . . (a) it could be done, the genome could be done; but (b) a really key part was to do a lot of technology development before you got into large scale sequencing."[31]

For Botstein, "The meeting substantially changed everybody's view about the feasibility. . . . It went both ways. The idea that it's not crazy came from Sulston's success in making a restriction map of the worm, which followed Olson's success in making a restriction map of yeast. For those of us who are the hardball geneticist nucleic acid folks, it made it clear that it wouldn't be a fantasy; that there were plenty of reality checks in the form of the genetic and physical maps that you could actually tell that the sequence was more or less right or wrong." He recalled Hood and Gilbert as the warmest advocates. "They said, 'You gotta do this. You gotta go ahead. And it's not going to be so bad.' And

Wally even made the calculation it was going to be $1.00 a base, which was (surprisingly) close."[32]

Imagining a Sequencing Center

Gilbert left the meeting a confirmed enthusiast for a full-bore, industrial-scale genome program. A day later, he excitedly sent Edgar a sketch of his ideas for a genomics institute. He favored starting with an intense effort at mapping the genome. Using present rates of sequencing per researcher, to find the correct order of all 3 billion DNA bases would involve 30,000 to 100,000 person-years and cost between $3 and $10 billion. "However, I expect that improvements in the technology . . . over the next decade will make sequencing one hundred or one thousand times more rapid." He estimated that a group of about 200 people, developing the technology as well as using it, could do the job in twenty to thirty years.

Gilbert repeated these arguments in a February 1986 memorandum that was a background document for the next genome conclave in Santa Fe, New Mexico. He foresaw an organization divided equally between two groups. In the first, some ten scientists and forty technicians would carry out "mission-oriented" mapping and sequencing, pushing the techniques and building and maintaining electronic databases. In the second group, some twenty researchers on renewable five-year appointments, who would be encouraged to seek additional funds outside, would conduct "a highly innovative program of basic research on human genetics and biology." Their work, also employing twenty technicians and ten postdoctoral fellows, would use such model systems as mice, yeast, and nematodes. With an additional thirty support staff, the institute, using leased space, would have an annual budget of $10 to $12 million. This would require an endowment of $240 million, built up over five years. Support could come from private individuals, foundations, or state governments.

The work, Gilbert held, should begin with the 1 percent of the human genome that would produce "the most rewarding information scientifically." The annual output of DNA subunits at that time, Gilbert estimated, was 100,000 per person. At that rate, thirty people could do the most interesting 1 percent of the human genome in ten years at a cost of $1 per base. In a second phase of perhaps ten years, with more automation, productivity would increase tenfold to embrace the next most interesting 10 percent. In a possibly faster third phase, productivity could go up another tenfold to decipher the remaining 90 percent.

A salient benefit would be the sequences of all the genes associated with some 3,000 inherited human diseases. Comparing genomes—say, human with mouse—would uncover the genes involved in development from egg to adult and in the formation of the brain. Gilbert wrote, "The total human sequence is the Grail of human genetics—all possible information about the human structure is revealed (but not understood). It would be an incomparable tool for the investigation of every aspect of human function."[33]

In the coming months, such ideas would spread widely and spur intense debate. That summer, the topic was discussed at the annual meeting on molecular genetics during the influential series called the Gordon Research Conferences.

Bruce Alberts Bemoans "Big Science"

As thoughts of big biology percolated, an eloquent protest was issued against labs too big for close interaction between principal investigators and younger associates. It came from Bruce Alberts, a professor at UC San Francisco. A specialist in DNA replication and cell duplication, Alberts was the principal author of a major biology textbook, *Molecular Biology of the Cell*; he later became president of the National Academy of Sciences, then editor in chief of *Science*.

He proposed rules that would keep labs at a size compatible with bright ideas and teach young researchers the habits of independent work. Federal grants to a principal investigator's lab should be limited "to the amount of research with which one investigator could be closely involved in on a day-to-day basis." Thus, the focus would be on the challenges of actual science. "It is crucial to recognize that many important research results start as surprises whose implications can easily be missed, and that money is no substitute for careful observation, thoughtful analysis, and scientific skill. Moreover, a single innovative and original publication is worth much more than ten obvious ones."

Alberts wrote, “Science is not a business, and bigger is not better”; good science is different from producing Wonder Bread. Yet, in large labs, the chief was too often bound up in getting money, giving lectures, and processing manuscripts for publication, unable to keep up with new techniques and their strengths and weaknesses. Too many labs had grown so big that the younger researchers felt like factory workers focused “strictly [on] their own part of the production line.” Alberts said, “This does not prepare them to function as independent scientists and may even impede their development, by preventing the acquisition of habits of independent research at a crucial point in their careers.”[34]

The essay stuck in many people’s minds as a strong defense of the interests of the young researchers who represented the future of biology.

Pep Rally in Santa Fe

The deliberative process continued. In the fall of 1985, Charles DeLisi, a veteran of both NIH and the Los Alamos National Laboratory, was serving as director of health and environmental research in DOE. He read a draft report of the Congressional Office of Technology Assessment (OTA) entitled *Technologies for Detecting Heritable Mutations in Human Beings*. The study was led by Michael Gough, an enthusiastic participant in the Alta meeting the preceding December. As he read, DeLisi began to envisage a vast genome project led by DOE. After all, the department’s National Laboratories were used to big science and big computing. They had already started major biology databases. With its vast financial resources, the department could take on the genome easily.

Accordingly, Mark Bitensky, a division leader at the department’s Los Alamos National Laboratory, organized a conference in Santa Fe on March 3 and 4, 1986. The noted human geneticist Frank Ruddle of Yale was chairman. The stated purpose was to generate the information DOE needed to go ahead with a genome project. Bitensky wrote to invitees, “We are attempting to generate a wise and compelling recommendation,” adding that the conference was to consider two questions: “Can we, through a concerted and dedicated effort, sequence the entire human genome in ten or twelve years? Should we?”

A notable participant was Sydney Brenner, apostle of research with the nematode *C. elegans*, which was already so well explored genetically as to be a candidate for DNA sequencing. Restriction enzyme pioneer Hamilton Smith was present, as were Alta attendees Cantor of Columbia, Caskey of Baylor College of Medicine, Church of Harvard, and Weissman of Yale. Gilbert and Lehrach had both attended the Santa Cruz meeting. But no officials of NIH, the dominant supporter of biological and medical research in the United States, came to Santa Fe.

The Santa Fe meeting spurred DOE's excitement and served as an implicit warning to other agencies, particularly NIH, that they risked being pushed aside on the roadway to the biology of the future.[35]

Although the meeting had some of the flavor of a pep rally, many issues that would dominate later discussion were aired in detail. Should the project focus on the 1 percent of DNA that coded for proteins and leave the rest aside? Was the human genome the only genome of interest? What should be done first—mapping the genome or sequencing? What were the best methods for each? What would the project cost? Would there be assurance that funding would involve "new money," so that existing biology research would not be harmed? Should the project be centered at one institute, as Gilbert suggested, or distributed among regional centers, as Weissman urged? How could Congress and the public be persuaded to back the program? What would the medical benefits be? Did those benefits outweigh the possibilities of misuse of genetic information?

One of the most trenchant discussions concerned how to generate "public acceptance—and especially acceptance in Congress and [at] higher levels of government." David Comings of the City of Hope Medical Center in California advised, "It would be essential to emphasize [the genome project's] broad appeal, not only for the usual genetic diseases (such as Huntington's, cystic fibrosis or Duchenne muscular dystrophy), but for the large number of common disorders that have a very strong genetic component to them." He mentioned arthritis, cancer, birth defects, heart disease, diabetes, Alzheimer's, and mental disorders like depression. The brain was of special interest, he said, because it apparently expresses about a quarter of all human genes, far more than the liver or kidneys.

Advocates, Comings said, would have to emphasize that "improved health would come more quickly with this knowledge." Better understanding of human genetic diseases would save "health dollars." And a once-and-for-all human DNA sequence would save science dollars as well. Individual investigators would no longer have to do "extensive sequencing" because a central facility would probably do the job at least ten times faster. Comings was looking ahead anxiously to the inevitable conversations with politicians.[36]

In the closing session, Brenner, wise in the ways of starting and running scientific programs, advised focusing on how to get the project started, rather than trying to spell out the entire design. Expressing the consensus, Hamilton Smith said the program should have two phases, a small-scale and low-cost mapping program, followed by a high-cost sequencing effort that would dictate multiple centers. An early goal, Ruddle said, should be making an ordered library of DNA fragments of modest length, about 30,000 to 40,000 nucleotides, called cosmids. The idea was first proposed in 1981 by Edwin Southern. Such a library, Ruddle said, "would be of tremendous value and would stimulate a great deal of research in human genetics." In other words, like a railroad under construction, the project would begin to produce benefits even before the sequencing phase that all endorsed.[37]

Reporting to DeLisi a month later, Bitensky was very upbeat. He said the meeting "was distinguished by a rare and impassioned esprit." He reported much optimism that the cost of sequencing, which had fallen tenfold during the previous decade, would keep on plunging. There was excitement about the potential of both Church's multiplexing methods and the new automated sequencer that Applied Biosystems was readying for the market. The estimated cost of sequencing more than 3 billion nucleotides over seven to twelve years would reach $1.8 billion, or even twice that. The project would need a lot of computer power and central coordination.

The conferees did not neglect problems. One concerned the highly repetitive stretches where restriction enzymes had few opportunities to cut the DNA. Another involved worry that researchers would focus only on particular areas of interest, leading to "neglect of 'uninteresting' domains" containing the controls for the operations at all phases in the lives of cells.

But the picture of expected benefits was rosy. Bitensky wrote, "The gain in knowledge . . . will markedly improve the quality of human life, expand the duration of an individual's productive years, and substantially reduce loss of productivity associated with acute or catastrophic and chronic or debilitating illnesses. . . . The time is opportune to begin to assemble the funding and personnel that will carry forward the sequencing of the entire human genome." Bitensky reported that "widespread enthusiasm for the activity is beginning to mount in the molecular biology community."[38]

In bureaucratic style, Bitensky was preaching to the converted. By May 6, DeLisi was writing his boss at DOE that the project should be divided into two phases. The first phase would develop three products: a complete physical map of the genome so that each gene could be located; high-speed, fully automated sequencing methods; and computer capacity. The second phase, relying heavily on the first, would focus on the actual sequencing of human DNA.

DOE quickly formed the requisite advisory committee to prepare an endorsement of a genome program and, with the help of New Mexico senator Pete Domenici, pushed money for it into the U.S. government's budget for the fiscal year that began in the fall of 1986.

The eventual design of the project was becoming clearer.

Renato Dulbecco Concludes, "We Are at a Turning Point"

The menace of cancer was a central theme in discussions of a genome project. In the fall of 1985, Nobel Prize winner Renato Dulbecco came east from the Salk Institute in California to CSHL in New York to speak at the dedication of a new laboratory named in honor of Joe Sambrook, a notable associate in his cancer virology lab in the 1960s. Recruited by James Watson in 1969, Sambrook was just completing more than fifteen years as the hard-driving, inventive head of CSHL's tumor virus group. He had also served as assistant director for research and, for a year, as acting director during a Watson sabbatical. Now, he was taking off for the University of Texas Medical Center in Dallas, from which he would later transfer to the Peter McCallum Cancer Institute in Melbourne, Australia.

The dedication of Sambrook Laboratory occurred on a hot September day during the laboratory's annual meeting on DNA tumor viruses. Instead of the usual dedication speech, Dulbecco, unaware of the meeting at Santa Cruz, issued an independent call for a complete sequencing of the human genome. The reason, he said, was an urgent need for more progress on cancer. To be sure, there had been many advances in knowledge, including unraveling the role of viruses in starting cells on the road to uncontrolled division and the discovery of cancer-related oncogenes. But crucial details remained unclear about how cells evolve into tumors able to invade neighboring tissue and then to escape, or metastasize, to distant places in the body.

The cancer problem, in Dulbecco's mind, was hopelessly confused. It resembled the multiheaded monster of Greek legend, the Hydra, which grew new heads to replace any that were cut off. In an interview a few years later, he compared cancer research to investigating the crash of a completely unknown type of airplane, picking through scattered parts, and not knowing whether a part was normal or damaged or whether it had caused the accident. With a complete human DNA sequence, one could compare each gene with the genes of a cancer cell and tell which changes had caused the disease.

Dulbecco told his audience at Cold Spring Harbor, "We are at a turning point in the study of tumor virology and cancer in general. If we wish to learn more about cancer, we must now concentrate on the cellular genome." Of course, new cancer genes could be discovered piecemeal—but at vast expense. A better way would be "to sequence the whole genome of a selected animal species." Humans, he thought, should be the target. The effort should be national—or, even better, international—comparable to the exploration of space "and carried out in the same spirit." If the rate of DNA sequencing could be increased fiftyfold, he said, the job might be done in five years.[39]

Listening skeptically was Watson, who in the late 1940s had been a colleague of Dulbecco's in the lab of Salvador Luria at Indiana University. In 2008, he said that Richard Roberts of CSHL had been working on a sequence of the adenovirus so heavily used there. Watson knew of the Santa Cruz meeting but could not understand the relevance of the complete genome to Dulbecco's field. Dulbecco's argument, Watson recalled in 2007 as he received a copy of his own that DNA sequence, "did not at once convince me." Dulbecco had maintained that the full genome was needed to understand cancer, but that a number of cancer genes had already been discovered. So, said Watson, "I didn't take it too seriously."[40]

By the time of his talk, Dulbecco had prepared a paper for *Science*, which the journal soon sent to Gilbert for his comments. He replied to *Science* editor Dan Koshland that Dulbecco was offering "a stimulating brief" on the understanding needed "to solve the problem of progression of tumors." Gilbert suggested that Koshland organize a special issue on a genome project, with help from Edgar or Sinsheimer. A human genome project, Gilbert wrote, would produce "tremendous" benefits, not only for "genetic and molecular understanding" but also for medicine.

"I Was Not a Very Good Monk"

Meanwhile, the problem of studying common afflictions, such as heart disease and cancer, forced geneticists to confront frustrating problems of mathematical analysis. Surely more than a single factor, genetic or environmental, caused these diseases. In 1985, David Botstein had found a collaborator trained in mathematics who could help wrestle with the multifactor problems. He was Eric Lander.

Born in 1957 and brought up in Flatbush in Brooklyn, Lander had the highest grades in his class at Stuyvesant High School in Manhattan, which, along with Brooklyn Tech and Bronx High School of Science, drew the top math and science students in New York's vast public school system. While he was at Stuyvesant, Lander's math project won the Westinghouse Science Talent Search for 1974. He went on to become valedictorian at Princeton, where he met his future wife and did a summer internship as a journalist. During a Rhodes Scholarship at Oxford, he earned a PhD in mathematics and spent an extra year there writing a book based on his thesis.

But he found himself wondering what he really wanted to do. He recalled, "Really, in my heart, I didn't want to be a true mathematician." It was a monastic career, and "I was not a very good monk." He visited Ed Tufte of Yale, famous as the author of *The Visual Display of Information*.

Tufte sent him to Frederick Mosteller at the Harvard School of Public Health, who, noting that Lander did not know any statistics, sent him on to Howard Raiffa, who had been a mathematician before joining the faculty at the Harvard Business School. After a lunchtime talk with Raiffa, Lander was hired at the business school to teach "managerial economics" in the fall semester. But, though he was a popular teacher, the course, he said, "didn't captivate me." He recalled, "Most of economics, at that time, was the study of rational expectations. Man is the ultra-rational creature solving complex equilibria in his head to figure out what to do. And, it didn't seem right. It seemed like all the interesting things in economics were in the assumptions, not the theorems." He had left the priesthood of mathematics and felt correspondingly guilty, but what would take its place?

He mentioned his dilemma to his brother, Arthur, a cell biologist, who recommended that he look at some papers on the architecture of the cerebellum in the brain. The theory in the papers struck him as "pretty hokey." However, he "thought the biology of it was fascinating. So, I started learning about the biology of the brain." Making sense of it, however, was impossible. His book finished, he had a free summer to start reading, and that fall he took a course in biology. Some might have regarded this as "career suicide," but, he said, "I had nobody around to give me any good career advice."

He went on to various Harvard labs and learned a few techniques, becoming very impressed with the manual dexterity of the researchers who gave him little projects. Something else impressed him even more, "the community of the lab." He loved "the lab meetings and tea, and journal clubs, and beach trips." Lander recalled, "I realized that there was a social and communal nature to modern biology—that was pretty magical."

Harvard biology professor William Gelbart, who was delighted with Lander as a "landsman" from Flatbush, taught him genetics with the help of classic papers. Lander was permitted to sit in on "an amazingly deep and heartfelt genetics course on all classical bacterial genetics" at Harvard Medical School taught by Jonathan Beckwith (a notable critic of the potential misuse of genetics). He also took a neuroanatomy course, to which later Nobel Prize winner Robert Horvitz of MIT, a student of the nervous system of *C. elegans*, gave a talk. Lander recalls screwing up the courage to ask if he could work in Horvitz's lab during a sabbatical.

He still had his day job at the Harvard Business School, where he co-taught a course in artificial intelligence and realized that a year at MIT might teach him more about that subject. He also taught a second-year elective course on bidding, bargaining, and contracts. It taught him, he said, to ask a question that he often used during the Human Genome Project: How do you create win-win situations?

Although Harvard Business School generously granted him leave, he worried that he would never receive tenure there. "I was getting pretty depressed about what my future career prospects were." But while at Horvitz's lab, Lander got to know Barbara J. Meyer, an alumna of Mark Ptashne's Harvard lab, who specialized in *C. elegans*. One day, when Lander was leaving after a seminar at MIT, he encountered Meyer, who was talking with Botstein. Meyer said to Botstein, "This is Eric Lander. He's a mathematician who's learning genetics." Knowing that most biologists bristled at the idea of bringing mathematics into biology, Lander cringed. But Botstein had already heard about Lander from Michael Waterman, his fellow attendee at Santa Cruz. In a short time, the mathematically oriented Botstein discovered that Lander knew quite a bit about genetics.

Because Botstein was from the Bronx and Lander from Brooklyn, they quickly fell into a familiar mode of conversation. They started arguing—about what you could do with genetic markers. Botstein said you could not map a heterogeneous inherited trait, and Lander replied, "Nonsense." They were off, and Lander was hooked. Shortly afterward, Botstein took Lander with him to the 1985 international gene-mapping conference in Helsinki, Finland, and they began working on papers about diseases associated with multiple genetic factors.[41]

"A Palpable Unease" at Cold Spring Harbor

Stunning progress in human genetics created an atmosphere of high expectancy at the June 1986 Cold Spring Harbor Symposium, which proved to be the next stop in the deliberative process. The fifty-first in the Symposium series, it had the ebullient title "Molecular Biology of *Homo sapiens*," chosen by its organizer, James Watson. The only previous Cold Spring Harbor symposium on human genetics had occurred twenty-two years earlier. Now, the participants would hold sessions on the status of the human gene map, cancer genes, genetic diagnosis, human evolution, recombinant DNA drugs to combat blood clots and cancer, the science of cell-surface receptors, and gene therapy.

The following year Watson wrote, "An air of excitement was palpable throughout the meeting; there was a satisfying feeling that DNA science had at last come of age. Contrary to past Symposia, when results were most often of no immediate concern to the 'average person on the street,' research team after research team presented results with significant implications for human health." Watson had taken trouble to encourage journalists to attend so that they might hear these reports. Richard Saltus of the *Boston Globe*, Robert Cooke of *Newsday*, Roger Lewin of *Science*, and Miranda Robinson of *Nature* filed stories.[42]

Discoveries in human genetics were flowing rapidly, and the techniques were gathering. At the Symposium, Kary Mullis gave his first talk on DNA amplification by polymerase chain reaction, or PCR. Leroy Hood of the California Institute of Technology reported the imminent commercial introduction of DNA sequencing by machines that automated the lab process developed by Fred Sanger in England. Early in the first session, Lander expounded on the work he and Botstein had been doing on the mathematical tools to tackle multifactor diseases.

The risk was taken to broach the idea of a human genome project to the community of hundreds of biologists from around the world gathered at the Symposium. Many were hearing of the idea for the first time. The enthusiasm of DOE, reinforced at Santa Fe, overlay the discussion like an ominous cloud. The biggest of Big Science institutions wanted to carry out the project in giant laboratories, such as those in Berkeley and Livermore, California, and Los Alamos, New Mexico. The department was confident that it had the tools and experience to operate at the required scale.

Some biologists, among them the entrepreneurial Watson, had an "uh-oh" reaction that they discussed in the conference sessions and outside, at meals and walking up and down the laboratory's main street, Bungtown Road. One reporter covering the meeting referred to "a palpable unease." The Human Genome Project, if it was judged worth doing and the money could be found, would be crucial to the entire future of biological and medical research. It would bring together a battery of tools to build an even bigger tool—a universal library of genetic information. Could the community of biologists, chiefly supported by NIH through a judgment system called peer review, surrender their futures to a big-lab agency with little tradition of scrutiny by juries, or Study Sections, of scientific colleagues? To be sure, DOE had sponsored important work in biology and would continue vigorously backing new technologies and work on genomic issues of potential interest to energy industries. But should it take the lead in the Human Genome Project?

If not DOE, it seemed logical that NIH must act as leader, even though most of the directors of its disease-focused institutes were unenthusiastic about a centralized program for doing something their grantees were already working on, research grant by research grant. Still, the moving finger was pointing toward NIH.

Before the meeting, Paul Berg called Watson to suggest a special afternoon forum on the Human Genome Project. Berg, who had shared the 1980 Nobel Prize in chemistry with Sanger and Gilbert, had been a leader of the 1970s effort by biologists to keep control of recombinant DNA work through self-regulation. He was chosen as cochairman with Walter Gilbert. To Cold Spring Harbor, Gilbert brought his enthusiasm for a single sequencing factory that would end the mind-numbing work now being done by thousands of biology graduate students around the world. He was sure that methods he and George Church were working on could be industrialized rapidly to do the whole job quickly and cheaply.

Just as the Symposium opened on Wednesday, May 28, James and Elizabeth Watson spoke openly to arriving participants about a personal ordeal. That day, United Press International reported that their sixteen-year-old son Rufus had run away the day before from a psychiatric hospital in White Plains, as he had threatened to do. The previous October, he had withdrawn from Phillips Exeter Academy in New Hampshire, which had recommended he receive counseling. He had been committed after he tried to break a window in order to jump off the World Trade Center. Elizabeth Watson said she and her husband were going public in hopes of finding him and to alert other families to the need to note signs of trouble in their children. She said plaintively, "We were just parents trying to do the best for our child, not exactly knowing what to do." Two days later, the Associated Press reported that Rufus had fled into some woods near the White Plains railroad station, from which Watson had picked him up and brought him home to Cold Spring Harbor.

In 2008, Watson recalled, "It was a tough personal time for our family," one with a direct, immediate, and lifelong influence on his commitment to proceeding with the genome project. A possible diagnosis for Rufus was schizophrenia. So, asking himself, "Will we ever understand schizophrenia?" Watson reflected that "it was pretty obvious schizophrenia had a genetic cause." The only way to give Rufus a life, he reflected, "was to understand why he was sick." This meant that the genes for the illness must be found, "and the only way we could do that was to get the genome." With that, "my son might have some of his life at a normal level."[43]

His son's trouble ended Watson's plans to relinquish the directorship of the laboratory after eighteen frenetic years and to take up a teaching post in England. A probable successor—none other than David Botstein—had been identified, and Symposium participants were discussing Botstein as heir apparent. The two men talked. With Watson now staying around as president, conflict was inevitable. Botstein reflected later that if he had taken over, he would have sought to commit the laboratory heavily to genomics. Watson instead chose to give his major efforts to building up neuroscience there.[44]

On the afternoon of June 3, 1986, Gilbert started the session provocatively. Berg stood at the podium with its symbol of a double helix circling the letters *CSHL*. Wearing a big microphone around his neck, Gilbert strode across the polished floor of newly dedicated Grace Auditorium to the blackboard and chalked a figure: "$3 billion." The audience stiffened. Soon Botstein leapt forward for a verbal duel with Gilbert, a fellow participant in the Santa Cruz meeting. Speaking on behalf of the shocked younger scientists crowding the room, Botstein expressed his profound skepticism of the project. It could spell the ruin of the immensely fruitful leading edge of biology. He said biologists could become indentured "to mindlessness." He exclaimed, "It endangers all of us, especially the young researchers."

Botstein asserted, "In one sense, the sequence is trivial. What we really want is a physical map of the genome." He spoke, as had Ruddle and others at Santa Fe, of an ordered series of overlapping segments some 40,000 bases long. Such a map would have much higher resolution, enabling researchers to pick out a region of interest, say the CF gene, and locate it within a single segment.

Afterward Botstein told Saltus of the *Boston Globe*, "The nucleotide [base] sequence is only the bare beginning. It's as if we were a primitive tribe that collected all the pages of Shakespeare on a beach and figured out how to put them in order. It's Hamlet, but we don't even know it, because we can't read English."[45]

Perched on the back of an auditorium seat, Lander also expressed doubts. Biologists, he said, were skilled at small-scale projects that were not "goal directed." He admitted that "the ability to sequence the human genome has just arrived" and that he expected the job to be done somehow by 2000. But he observed nervously, "The structures necessary to cope with the expenditure of three billion dollars could be inimical for biology. It could create immovable structures." In 2000, Lander recalled, "At the time, it was madness to think we could collect three billion letters of information from the whole genome."[46]

There it was, on the table. A vast project supported by many of the world's most competent, experienced, and ambitious molecular biologists confronted the deep uneasiness of the younger scientists, who would do the project's heavy lifting. Watson was struck that enthusiasm for the big new initiative seemed confined to researchers over the age of fifty. Younger scientists were skeptical to the point of rebellion. Many years later, in the same auditorium, Watson reflected with a trace of bitterness that Botstein "doesn't think straight. . . . Bright people can be wrong."[47] But the audience focused on the price tag and rebuffed Berg's attempts to start discussion of the scientific merits and demerits. A year later Berg recalled, "We could hardly get to the science because of the ominous views people had about the project."[48]

The Redesign Gets Under Way

The enthusiasts could see that they were in big trouble. Robert Waterston hoped, "At the very least, we wanted some of us to still be alive when it was finished." But the idea might not fly. The advocates would have to confront the objections and craft a program design that would bring doubters aboard. As it happened, a heaven-sent forum was opening up. The National Research Council (NRC), an arm of the National Academy of Sciences, had already been discussing whether it could help assess the genome project. Further, an academy group called the Board on Basic Biology of the Commission on Life Sciences, headed by Francisco Ayala of UC Davis, was planning a meeting for August 5, 1986. The place was the National Academy's summer house in Woods Hole, Massachusetts, on the shores of Buzzards Bay. In April 1986, John Burris, the NRC staff officer on the issue, wrote to his boss, "Ayala has some doubts about the value of a concerted effort to map the human genome but would like to discuss in greater detail" with his board. "Ayala is personally quite interested in this issue and posed numerous questions on the feasibility and use of the final product." Ayala noted that the letters of the completed sequence would take up a million pages of text.[49]

Frank Press, formerly science advisor to President Carter and now president of the National Academy, followed up in mid-June by writing to NIH, the National Science Foundation, and DOE. He told Wyngaarden at NIH, "The mapping of the human genome is an idea that has captured the imagination of a large number of scientists." He offered the NRC's help in elucidating questions about the technology and organization of such a project. Wyngaarden quickly replied, agreeing that "the idea is an exciting one" and reporting that he had named Ruth Kirschstein, head of the National Institute of General Medical Sciences, to lead an NIH study of the matter.[50]

Meanwhile popular attention was ratcheting up. The project for total mapping and sequencing was a major topic at the annual Press Week in Bar Harbor. A reporter quoted McKusick, a supporter, as saying, "In many ways this is equivalent to exploring the inner environment in the same way the space program was aimed at exploring the universe.

There's more to be derived from exploring this internal universe than came out of the moon landing." Ruddle said, "It seems to touch every aspect of our existence. That's why it's captured everyone's imagination. It's extremely fundamental."[51]

At Buzzards Bay, a collection of leaders met around a big conference table to thrash the subject out once again and decide whether the NRC should play a role. Kirschstein, a skeptic, was there, taking notes, as were David Smith from DOE, David Kingsbury of the National Science Foundation, and Alan Rabson of the National Cancer Institute. Speakers included participants in previous meetings: Cantor, Gilbert, Hood, Ruddle, Watson, and White. No fewer than twelve members of the Board on Basic Biology attended, along with some new faces—Russell Doolittle of UC San Diego, Alexander Rich of MIT, and Melvin Simon of Caltech—all of whom spoke. Representing the academy were Frank Press and John Burris.[52]

This discussion led to a decision in September to convene a study-and-bargaining committee headed by Bruce Alberts, author of the well-regarded denunciation of Big Science. The organizers picked him for his identification with the travails of the young scientist and his skepticism of gigantism in biology. The committee's work was supported by a grant of $150,000 from the James S. McDonnell Foundation of St. Louis, whose president just happened to have attended the Cold Spring Harbor meeting.

The committee included skeptics like Botstein and Shirley Tilghman, later president of Princeton, vocal proponents like Gilbert, and quiet backers like Watson and his close advisor (and experienced committee hand) Norton Zinder. Other members were Brenner, Cantor, Doolittle, Hood, McKusick, and Ruddle, along with Stuart Orkin of Harvard Medical School, Leon Rosenberg of Yale, John Tooze of the European Molecular Biology Laboratory in Heidelberg, and Daniel Nathans, who shared the Nobel Prize with Hamilton Smith in 1978.

Under Alberts's leadership, the committee steered clear of votes, focusing instead on the testimony of scientists brought in to spell out the current status of a field that would become known as genomics. Besides representatives of several government agencies, the witnesses included Lander, Maynard Olson, George Church, Kay Davies of Oxford University, Ronald Davis of Stanford, Helen Donis-Keller, James Gusella, David Patterson of the Eleanor Roosevelt Institute for Cancer Research, Allan Spradling of the Carnegie Institution of Washington, Jean Weissenbach of the Pasteur Institute, and Ray White.[53]

Maynard Olson Advocates "a Middle Course"

A key figure on the committee was Maynard Olson, then of Washington University in St. Louis, Missouri, who would figure on several occasions as a kind of consensus philosopher of genomics. As one of the few scientists actually practicing genomics—his lab had developed a method for storing big chunks of DNA in what were called yeast artificial chromosomes (YACs)—he appeared first as a witness. He later joined the committee when Gilbert left to try—and fail—to start a private company to do the human sequence (he called it Genome Corp). McKusick recalled that he was "a valuable member," noting that Lander referred to Olson as "the sage of the human genome project." McKusick remembered, "He was very level-headed."

Olson was in the ninth grade when the Russian Sputnik went into orbit in 1957, and he soon felt the benefit. A shocked nation invested in more equipment to teach science and gave science overall heightened prestige. His first steps into science did not concern biology. Instead, he went into inorganic chemistry, first as an undergraduate at Caltech. After earning a PhD at Stanford, he started teaching chemistry at Dartmouth College. But during his five years there, he formed an interest in genetics and molecular biology. He read books that inspired him, such as the history of the Delbrück and Luria bacteriophage group at Cold Spring Harbor and James Watson's highly influential textbook, *Molecular Biology of the Gene.*

One chapter of Watson's book describing a "skeptical chemist's" view of the bacterial cell struck him. Olson thought of himself as a skeptical chemist and wondered how much biological detail would ever be grasped. The skeptical chemist, Watson indicated, would say, "Instead of there being one biochemical pathway to metabolize glucose, how do we know there are not 1,000 pathways?" Why was biology not "infinitely complicated from a chemical standpoint?" Olson was worried: "I'm that class of scientists that needs to get solid answers to questions. I don't do well in areas where everything is murky." But then Watson offered reassurance: "There can't be thousands of pathways—because the genome is finite." A living cell has only so much DNA, enough to fill a 750 megabyte disk. There can only be so many enzymes and pathways to do the cell's work. So, there is "a finite bound" to the problem of mapping all the DNA of a cell, "and we can do this."

Although there had been many triumphs in molecular biology since the 1950s, Olson reflected, there would be "another round . . . a molecular biology of complex organisms," that he could get into. On sabbatical at the University of Washington in the laboratory of Ben Hall, Olson got caught up in the feverish field of recombinant DNA and was involved in "a very nice project" that sequenced a mutant gene in yeast, the smallest cell with a nucleus.

Taking up his first official job as a geneticist at Washington University, he started working on a long slog toward a complete physical map of the yeast genome. Despite the project's novelty and complexity, Olson was sure it could be done, and the usually skeptical grant-review committee agreed with him. Soon his next-door neighbor, Robert Waterston, a veteran of the work on the worm in Cambridge, England, told him about John Sulston's projects and told Sulston about Olson's. On a project "far below the radar," Olson labored for seven years without publishing, comparing notes with the researchers in England. Then, in 1986, Sydney Brenner, the father of the worm studies, decreed, "You've got to publish this now," and sent the papers from St. Louis and Cambridge, stating plans for physical mapping in yeast and the worm, to the *Proceedings of the U.S. National Academy of Sciences*, of which he was a foreign member. The experimental and analytical ideas were "well developed," Olson recalled, "and it was clearly working."

Spurred by the new "pulsed field gel" techniques, Olson went on to think about handling much bigger chunks of DNA than most researchers thought possible, and just as the debate over the genome project crystallized, he and lab colleagues David Burke and Georges Carle developed YACs to hold the DNA for genetic studies. For most molecular biologists and geneticists, the topic was unfamiliar. "Not inappropriately," Olson recalled, fellow scientists "viewed me as somewhat eccentric." But he and Sulston were then the principal researchers who had practical experience of what came to be called genomics. He concluded that most of the advocates of rushing into big-scale DNA sequencing were seduced by an idealized "Platonic form" of what to do, not mindful of "the tremendous messiness of data errors and unstable clones and just sample mix-ups and so forth." With his chemist's background, Olson saw himself as charting "a middle course."

Testifying to Alberts's committee, Olson said that "there was a lot to go on," but there was also "potential for a debacle" because the ideas of "the more energetic cheerleaders" were unworkable. He told the committee it would not be easy: "You don't need a big international program to do something that's going to be so easy." He was convinced that "a lot of trial and error" lay ahead, and the project had to be big enough and last long enough to go through it. Immediately consolidating the work in one center would have been "a disaster." Alberts asked Olson to summarize his views in a letter, and the committee judged him a suitable replacement for Gilbert.[54]

"A Multiple-Genome Project"

Having heard the testimony, Alberts and the committee spent a year hammering out the text of a deal that would get the project going and yet square the doubters. The project had to be financed with "new money" that would not be subtracted from existing funding for biological research grants. Gilbert's proposal of an immediate industrialization of sequencing was set aside in favor of a more measured approach. While sequencing technology was pressed forward for several years, attention would focus on mapping the location of genes, not only in people but also in an array of model organisms—classical lab workhorses of genetics like *E. coli* bacteria, baker's yeast, the gleaming "worm" called *C. elegans*, and the fruit fly, *Drosophila melanogaster*. Detailed mapping of entire genomes would create mileposts for ordering the sequences of thousands of carefully warehoused and amplified segments of DNA. The deal included a place in the program for many research centers—in effect carving up the genome into pieces of scientific turf.

Soon after the report was published in February 1988, Alberts summarized it for a House subcommittee. The foremost conclusions, he said, were "that a special effort is justified in this case, that this project does make both technological and intellectual sense, that the benefits are important, [and that] the information obtained would be important for biological research and for medicine."

The scope of the project would go well beyond human beings. To understand the human sequence, Alberts said, it was essential to have "a multiple-genome project," focusing on model organisms like the mouse, the fruit fly, and yeast. All of these, he said, are "very similar to ourselves in their basic workings."

Research for the project would have to focus on developing techniques of mapping and sequencing the human genome. But sequencing immediately would "distort other biological research." Big centers were not needed at present and would only be worthwhile when new, more efficient, and cheaper methods had been developed. But "the way forward is not clear: we don't know exactly, we can't predict what technologies will win out." So there would have to be "a bunch of competitive projects . . . [of] modest scale." To decide on these, the time-tested process of review by panels of scientific peers should be used.

Premium tasks, he said, would be developing central data banks equipped to handle the flood of data to be expected and building a biological "stock center" to store samples to distribute to researchers. "It is important," he observed, "that the materials and the information gained from this project be made immediately available to everybody else in the project." Alberts was foreshadowing an unprecedented scale of cooperation among highly competitive people.

Recommending an annual spending level of $200 million, Alberts said that "this project should be done with some enthusiasm and focus . . . to give the project visibility." Thus, it would be essential "to recruit and train outstanding people who will be really involved in developing these technologies [and also] to provide the necessary coordination, ensuring quality control."[55]

The discussions of three years had been codified in an extensively rethought scientific program. Although the committee left it to others to determine the precise roles of NIH and DOE, it recommended a leading agency, probably NIH, rather than a cumbersome interagency board. Watson told a journalist, "It has got to go ahead, it is so obvious. The only question now is the rate and under whose auspices."[56]

"Can It Be Done? Yes. Will It Be Done? Yes."

Besides DOE, two other organizations were debating that question: the Howard Hughes Medical Institute (HHMI) and NIH. Each held deliberative conferences on the subject in July and October 1986, respectively.

Interest in the HHMI meeting on July 23, 1986, was heightened by the institute's new status as the largest private charity in the United States. Until a year earlier, it had been proprietor of Hughes Aircraft, once the prized possession of the eccentric aviation pioneer Howard Hughes, who died in 1975. After ten years of tangled, highly public maneuvering over Hughes's will, the huge enterprise was sold to General Motors, yielding a $5 billion endowment for medical research. Those interested in a genome project swiftly contacted HHMI, which began thinking what it could do in the field. HHMI already supported the major gene-mapping project at Yale and McKusick's catalogue of disease-related human genes.

The meeting drew many of the people who had attended or spoken at previous sessions. Among newcomers to the list were McKusick, Kenneth Kidd of the Yale genetics project, Francis Collins of the University of Michigan (noted for his part in finding the genetic marker for CF the year before), Lloyd Smith (a Hood collaborator and pioneer of robotic DNA sequencing), and two representatives of the University of Texas Health Science Center in Dallas, noted cholesterol researcher Joseph Goldstein and Joe Sambrook.

Donald Frederickson, the institute's president, opened the conference by saying, "Can it be done? Yes. Will it be done? Yes." But he added that there were many unresolved questions about how, when, and by whom the task would be accomplished.

Lander, favoring an early emphasis on mapping, noted that the completed human sequence was not a goal but a "tool." The sequence itself, he said, would tell little until more was known about the structure and function of proteins. He thought data storage and networking were easier than working out useful mathematical theory.

One journalistic report of the conference asserted that "all-out sequencing" was "slipping into the background," a conclusion based on Leroy Hood's cautionary view that sequencing techniques needed much more development. Hood said, "It would be a serious mistake to jump into a full-scale sequencing effort with the cottage-industry techniques we have at the moment. If we make the proper investments in developing technology, then in 5 years' time we will be able to do the job more effectively, both technically and financially." Almost certainly thinking in terms of evolution of the equipment he had unveiled in June, Hood foresaw a sequencing system that would be "10 times faster, 100 times more accurate and 100 times less expensive."

Hood's comments reflected a consensus: Map the genome first, develop sequencing techniques, and then do big-scale sequencing. Frederickson said the HHMI trustees would decide whether they could help coordinate the project. Ultimately, HHMI decided not to participate.[57]

NIH took the opposite action. Stung by the news of DOE's ambitious plans, NIH director James Wyngaarden quickly began working to create conditions for genome work despite skepticism, even opposition, from the directors of the disease-related institutes. They argued that a genome program was either superfluous or disruptive.

Treading carefully, Wyngaarden organized a conference of his own on October 16 and 17, 1986, in the same Bethesda auditorium where HHMI had held its forum. Along with Cantor, Botstein was present to discuss mapping and sequencing. Botstein, who would soon face these issues on the Alberts committee, observed, "The central endeavor of human molecular biology is to understand the relationship among genes, proteins, and function." He added, "Most of what is known about many human genes comes from the study of lower organisms, as it is easier to obtain mutants from these organisms and they have less DNA to sequence." Noting that the human sequence was 1,000 times as long as *E. coli* and a hundred times as long as *C. elegans*, he urged attention to picking regions of greatest interest, such as genes that are turned on—or expressed—and genes that have been conserved over long periods of evolution (a sign that their function is not dispensable).

Representatives of government agencies, Kirschstein for NIH, DeLisi for DOE, and Kingsbury for the National Science Foundation, outlined their present efforts in the field. Discussing data management were Russell Doolittle and Allan Maxam, Gilbert's collaborator of the 1970s on sequencing. With Ayala of the Board on Basic Biology as chairman, three topics were discussed: improving physical mapping (Cantor), cloning of million-base fragments (Ronald Davis), and somatic cell genetics (Ruddle). After Wyngaarden discussed the pros and cons of NIH involvement, comments came from Botstein, Alex Rich, and Renato Dulbecco.

As he closed the session, Wyngaarden said there were two options for NIH: to "remain on our present course" or to "embark on a coordinated endeavor to develop a map of the entire human genome in the shortest possible time." As a sign that he was keeping an open mind, he threw out the idea of a pilot program to sequence one or two chromosomes.[58]

During the wait of more than a year for the publication of the Alberts committee's report, Wyngaarden received largely negative, often strident letters. He recalled, "The reservations were many. Some were primarily financial. Others were philosophical and involved fears that NIH was moving toward centrally managed science." Still others feared disruption in graduate students' experience as the trend shifted toward "mindless sequencing."[59]

While the Alberts committee deliberated, support solidified, and journalists paid still more attention to both the plan and the controversy over it. In May 1987, a biological science lobbyist named Brady Metheny escorted David Baltimore and Watson to Capitol Hill to meet with leading members of Congress—six, Watson recalled—with purview over NIH's budget. One of these was Rep. William Natcher of Virginia, chairman of the powerful House Appropriations subcommittee with purview over health research. Baltimore was seeking additional funds for AIDS research. Watson asked if the budget contained money for the genome project, and when Natcher asked how much new, specific funding was needed, Watson said $30 million. But it was clear that Wyngaarden would have to say he wanted that money for NIH. Soon after, Metheny recalled, he and Watson met with Wyngaarden. Luckily, Watson reflected many years later, Wyngaarden was a human geneticist. With typical directness, Watson said that if NIH was not prepared to take the lead, he would throw his support to DOE. Even though most NIH officials were "violently opposed," Wyngaarden replied that NIH was ready.[60]

A Start-up Leader Is Chosen

With the necessary new money committed for the start-up, the ducks were in a row when the Alberts committee's report came out. About two weeks later, Wyngaarden assembled a large group in Reston, Virginia, to consider how the recommended NIH program would be organized and who would lead it. Baltimore, still a vocal skeptic, was enlisted as the meeting's chairman. It was decided that the program would start as a small unit in Wyngaarden's own office and that he would offer Watson the inaugural directorship. Watson shared the widespread disapproval of Big Science. His view was, "We are all small scientists. Big science is no good. We have to give the money to bright people. The program has to be run by scientists, not by NIH administrators." After some months of negotiations, Watson took office and immediately announced that a percentage of the genome project funds would go to studies of ethical, legal, and social implications.[61]

The choice of Watson to lead the start-up could be considered odd. To be sure, he was perhaps the best-known biologist in the world, made famous by his role in discovering the structure of DNA. But he had also published in 1968 the most indiscreet—also the most widely read—memoir in the history of science, *The Double Helix.* Its casually brutal tone reminded some of J. D. Salinger's *Catcher in the Rye.* Many scientists were scandalized by the raw spirit of competitiveness revealed in the book and indignant at its disparaging portrayal of Rosalind Franklin, the X-ray crystallographer whose work contributed greatly to the discovery (and whose fame Watson greatly increased).

In his career, Watson had often proved himself very hard to get along with. His path was strewn with explosions and disputes. He appeared to be unappeasably restless, taking on project after project, from architecture and horticulture to botched attempts to found biotechnology companies. His personality was the antithesis of the usual concept of the leader, the sunny, confident team captain. How could such a weird person fly down for a few days each week and function in the large NIH bureaucracy in Bethesda?

On the other side of Watson's ledger were his achievements. For more than thirty years, he had been an extraordinarily observant, relentlessly gossipy scientific entrepreneur, spotting young researchers strong enough to take the pressure and keep their self-respect as he drove their talents to the limit. Watson's intellect was sufficiently cool, inquisitive, and commanding to put him at the center of a complex network of researchers striving impatiently at the highest level. His motto, as well as the title of a memoir he wrote many years later was, "Avoid boring people."

At Harvard, starting in the 1950s, he assembled an exciting laboratory that figured prominently in DNA research, while building a molecular biology department with a minimum of politeness. Then, in 1968, he took on a huge second task at the decrepit but famous CSHL. For decades, the laboratory had been a world center for the exchange of the latest techniques, discoveries, and research plans in molecular biology, but its buildings were rickety and year-round research staff needed to be expanded. With the money he planned to raise, Watson aimed to create research groups, larger than those accepted at Harvard and varied enough to tackle vast and complex problems like the induction of cancer by viruses. The venture was so risky that Watson continued on his Harvard salary, oscillating back and forth for eight years.

By this time he had won a unique position as a pied piper of his field. In mumbled lectures at Harvard, filled with that day's gossip from the laboratory, he caught the imagination of people looking for a grand challenge, people like Walter Gilbert, David Botstein, Mark Ptashne, and Mario Capecchi, who won a Nobel Prize in 2007. Before his scandalous memoir, he had written the canonical textbook of the new field, *Molecular Biology of the Gene*, which has gone through six editions in forty years. Its brightly written, assertive style had quickened the imaginations of some of the most intelligent and willful researchers, drawing them into the new frontier of biology.

At Cold Spring Harbor particularly, he had demonstrated an uncanny ability to strike the right themes to attract needed funds from governments, foundations, and private individuals. In Bethesda, he would be called on again to impress leading members of Congress to keep the new money flowing, while scouring the country for the best talent to join the genome project. Watson had the horsepower to get the project off the ground, but would he last?

CHAPTER THREE

Scaling Up

When the Human Genome Project got under way in 1989 and 1990, its planners were certain that big-scale sequencing would have to wait some years until new technology, dramatically better than what was available, had emerged. Few suspected that the course of events would be quite different. Incremental developments, based on dramatically streamlined current technology, led to a system that was good enough to start climbing up the mountain of 3 billion nucleotides.

Congressional Hearing: "The Time to Act Is Now"

From the start, public interest was high. So was the confidence of those leading the National Institutes of Health (NIH) genome work. Late in April 1988, at a four-hour House hearing, the Congressional Office of Technology Assessment (OTA) unveiled its report about possible ways to coordinate the various federally supported genome projects. It was an occasion to spell out the principal players' intentions at the outset. Testimony made it clear that a once-reluctant NIH now had the bit in its teeth and was determined to be foremost, to be the "lead agency." Others in the field, such as the Department of Energy (DOE), the Department of Agriculture, and the National Science Foundation, would devise their own programs. This remarkable hearing also made clear that NIH saw the project pointing toward more and more powerful tools of biological research. It was a technological enterprise.

Ron Wyden, then a Democratic representative from Oregon and later a U.S. senator, chaired the hearing and set its upbeat, even urgent tone in his opening remarks. Citing a decade of "tremendous advances" in medicine, Wyden observed, "The human genome holds seemingly infinite knowledge about the human body. Precise knowledge of the genome could create a remarkable biological tool." Besides speeding up the identification of genetic diseases, he said, "it could help explain why some people get certain diseases and others don't. It may offer clues to growth and the process of aging." He added, "We are going to have to recognize that the pre-eminence of this country in the life sciences is at stake." Later in the hearing, Rep. Doug Walgren of Pennsylvania asked, "Would anybody dispute that the watershed that is being crossed by this technology has almost no parallel in human history or science?"

Under the bright lights of the hearing room, the coast was evidently clear for a maximum effort to convince Congress that this new, complex, arcane project would need—and deserved—robust budgets.

Excellence, Bruce Alberts argued, was essential. Outlining his National Research Council committee's report of two months earlier, he said, "We don't want to waste money." Young participants would need a first-class experience: "When you set up something like this it acts as a training ground for young scientists and the future of this country." The technology would depend on "training scientists well. That can only be done if we support only the highest-quality research."

Wyden asked whether the concerted genome effort would accelerate progress in biology and medicine. Without the genome project, Alberts replied, progress would be much slower. A key to greater speed lay in the effort's technological focus. Emphasizing the development of tools designed to operate at a scale as much as a hundred times greater than before would challenge the culture of biology and medicine. "It's important to realize that biological scientists have sort of an inherent bias . . . against people who develop technology," Alberts said. "Journals do not like to publish 'methods' papers. NIH [peer review] study sections want to know what biology you're going to accomplish, not what methods you're going to develop." Hence, genomic research would call for "a special program that will give credence to technological developments, and to people—chemists, engineers, and biologists—who want to work together to develop automated machines or whatever else is needed to improve the science of working with large genomes."

Could Congress measure progress by such means as counting the number of human disease genes found each year? Alberts refused to be pinned down to a numerical measure, making a frank confession of the uncertainties ahead. "I think it's very dangerous to make firm predictions. . . . We really don't know what the methods are going to be or how rapidly they will develop." He noted that mapping genes to particular locations in the genome was further along than sequencing, "but there's still great room for improvement." So, the focus should be on "those efforts that promise to make it easier" to map and sequence. Robert Cook-Deegan, staff director of the OTA study, concurred: "The whole point of genome projects is to systematize the work."

The project was not a Big Science enterprise like the huge Superconducting Supercollider for advanced physics, then under construction in Texas but killed by Congress five years later. Instead, Alberts said, the model was California's Silicon Valley, "a bunch of small, modest-size groups competing with each other to find the best technologies." The genome effort would be "a coordinated series of Little Science projects."

Also emphasizing genomics as a tool-building enterprise was Maynard Olson, a member of Alberts's committee. Looking forward to the "trickle-down benefits" of the effort, he said, "The means of genomic analysis are as important as its ends. As presently formulated, the essence of the genome project is not just the acquisition of a huge archival data base, but rather a sustained effort to improve the efficiency with which genetic material can be mapped and sequenced." Through "major advances in automation and the training and motivation of highly skilled technical personnel, and in multi-disciplinary integration and in quality control," efficiency would increase by ten times or even a hundred. He added, "We are talking here about training a whole generation of young scientists in the type of experimental activity which barely exists at the present time. If this project prospers [it] will grow to a major activity in American science."

Like Alberts, Olson was conscious of difficulties. When Wyden asked about the economic payoff—would the accuracy of sequencing affect its usefulness in the marketplace?—Olson exclaimed, "I think it's a bit of a scandal that we don't know what the accuracy of current DNA sequence is." Millions had been spent creating a data bank of 20 million bases, but "not one cent has been spent in estimating what the error rate is."

The subcommittee heard further testimony about payoffs. Mark Pearson of the DuPont chemical company said the work would result in "rapid, accurate, and inexpensive diagnostic tests for human genetic diseases." After George Cahill described the Howard Hughes Medical Institute's support of Victor McKusick's international gene-finding program, McKusick said that Congress would see payoffs on the way to the goal line: "One doesn't wait until one has the complete map and the complete sequence in hand to derive benefit." The project, he was sure, would lead to a detailed genetic diagnosis of cancer. Equipped with "a new human anatomy," he predicted, "we will go directly to the DNA."

Then, as now, ethical questions were prominent. Thomas Murray of Case Western Reserve University took a distinctly nonalarmist view. The project would not subtract from human "moral status," he said, because humanity has known for centuries that "human traits have a heritable component." The work of Copernicus, Galileo, Darwin, and Freud has "not deprived us of moral significance," he said, even though each set of revelations made huge changes in how humans place themselves in the universe—changes far bigger than would flow from incremental mapping and sequencing of the genome. The process was like "a symphony." Indeed, in his opinion, "we've had our revolution in genetics. The human genome initiative is in a sense merely the filling in of an outline of what we have known for decades." But he admitted that "knowing more about all of our genetics is going to change the way we think about ourselves."

An important concern for Congress was how the project would be organized under coherent leadership. After some two years of debate within NIH, Director James Wyngaarden was convinced that "NIH is a logical home for a good deal of this effort." To be sure, NIH already supported a large amount of genome-related work that was making "good progress" in tracking down individual genes. But this was not enough. "There are thousands and thousands of genes that would not be sequenced, at least during anyone's current lifetime. . . . A systematic, coordinated effort toward mapping and sequencing is necessary."

NIH and DOE had each begun allocating money specifically to genome work. In the federal fiscal year that began on October 1, 1987, NIH was spending $17 million and DOE $11 million. The DOE program included drives toward mapping, sequencing, and computing, said David B. Nelson of DOE's Office of Energy Research. Coordination between NIH and DOE was "working well," he said, although if directed to do so by Congress, the two agencies could agree to a formal "Memorandum of Understanding." Congress did direct it.

As was usual for him, James Watson took a more categorical, outspoken approach. Already negotiating with Wyngaarden about leading the institute's genome work, he asserted, "Now, I sense a general consensus that we have already talked and argued enough. We should get on with the program" while creating the techniques for "cost-effective" mapping and sequencing. He saw several immediate tasks for an "active scientist" leader, backed by a "strong" advisory board. These included starting "exploratory research" on new technology, launching pilot projects to fix the cost and accuracy of the technologies, moving to larger-scale contracts for mapping where the technology was ready, and increasing the number of experts "in the new DNA technologies." Without these, Watson said, a project that should not cost more than $3 billion would cost $10 billion. Currently the costs of sequencing were "out of control."

Talent in the field was so scarce, Watson said, that there was little room for two efforts of the same size. NIH should take the lead with a dedicated "line-item" budget from Congress. Counseling an immediate decision on this rather than years of dithering, he opposed what he called a "disastrous" interagency setup—management by committee—which would prevent "long-term stable leadership." Instead the enterprise would be "effectively in the hands of an ever-changing, unpredictable bureaucracy." While not excluding a major role for DOE's big national laboratories, Watson said that NIH was best qualified to lead because of its "long experience in supporting first-class DNA research in human genetics" and its "strong cadre" of experienced administrators who had shown during the problematic "War on Cancer" of the 1970s that they could get a grip on a big program.

Noting that Congress would now have to decide "whether you have liked what we have said," and settle the question of a lead agency,

Watson concluded, "I see an extraordinary potential for human betterment ahead of us. We have at our disposal the ultimate tool for understanding ourselves at the molecular level and for fighting the genetic diseases that diminish the quality of so many of our lives. The time to act is now."[1]

The Bomb, the Moon, a Telescope, a Railroad?

A complete map and sequence of human DNA had such vast implications, both inside and outside biology, that supporters and detractors often resorted to comparing it with other grand-scale, focused efforts like the atomic bomb project of World War II and the moon program of the 1960s and early 1970s. Because it involved human health, nutrition, and energy, there was a tendency to describe the genome project in even grander terms than its putative predecessors. Indeed, continued government support for genomics might prove easier to sustain than in other technical fields.

The genome project, however, differed significantly from both the atomic and space projects in both cost and scale. The bomb project cost $2 billion in the 1940s, equivalent to more than $20 billion today. The American moon program cost $20 billion in the 1960s, equivalent to more than $130 billion now. The genome project, which spanned fifteen years—instead of about five for the bomb and ten for getting to the moon and back—used far smaller equipment and employed fewer scientists and engineers than either the bomb or moon projects. The $3 billion genome project represented less than 2 percent of the total NIH budget, which in turn equals less than 2 percent of the annual American medical bill of some $2.5 trillion.

Motivation and politics differed among the three enterprises. The secret bomb project aimed to develop a weapon before Germany; in the event, a pair of nuclear bombs was used to end the war with Japan. The highly public space program strove to place humans on the moon before Soviet Russia. Both programs can be considered command decisions by Presidents Franklin D. Roosevelt and John F. Kennedy, respectively. The genome project bubbled up from the scientific community and relied principally on patronage from Congress. And perhaps its purposes were grander. Saying that neither analogy "rings true," Eric Lander wrote in 1996 that "the new genomics" sought to create a "periodic table" for biology similar to the one invented in the nineteenth century for chemistry.

To be sure, all three efforts created very large industries and institutions still in use decades later. The bomb project gave birth to a raft of companies making radioactive isotopes for industry and medicine and to nuclear power plants for submarines and civilian electricity. But there were wider effects. The prestige earned by their work on the bomb, radar, and the proximity fuse in World War II enabled the physicists to win support for programs at the frontier of their field. Considering this a way to refresh the knowledge that they had drawn on so heavily during the war, they launched dramatic efforts to probe ever deeper into the heart of the atom, using larger and larger, ever-more expensive machines called accelerators. They went beyond the electron, the proton, and the neutron into the realm of quarks and even more exotic constituents of matter. Despite the passage of decades and some major setbacks, government support for this work has continued to the present day.

The space program created the rockets and generated the experience used to keep astronauts flying in shuttles and a space station, at a cost of billions of dollars each year. As with nuclear power, however, space technology from the very beginning has been harnessed to a range of practical services that have become so universal as to fade from popular consciousness. Using satellites launched into orbits around the earth, or even to "stations" 22,000 miles above the Equator, humanity can track weapons deployments around the globe, monitor the earth's weather minute by minute, use Global Positioning Systems to determine where a person is within feet, and send electronic communications between continents. As in the nuclear field, sustained political support has backed the vast program of science in space. For the first time in history, spacecraft give the human race close-up views of other worlds, flying by or orbiting them and sending down packages of instruments, such as research carts that migrate on the surface of Mars. Yet other orbiting spacecraft, such as the Hubble Space Telescope, which astronauts repaired for the last time in 2009, survey the heavens at all the major frequencies of light reaching earth from billions of light-years away, creating a picture of events near the birth of the universe.

Genome research, with continued support in Congress, has already spawned an array of large-scale sequencing centers, DOE's in Walnut Creek, California, the Wellcome Trust's outside Cambridge, England, and three supported principally by NIH in Cambridge, Massachusetts; St. Louis, Missouri; and Houston, Texas. As we will see in later chapters, genomics is constantly creating a stream of new enterprises and research institutions.

Although its scale is far smaller, the genome project also shares other challenges with the Erie Canal of 1825; the nineteenth-century U.S., Canadian, and Russian transcontinental railroads; the Panama Canal of 1914; the Tennessee Valley Authority begun in the 1930s; the post-1953 flood-control dike system in the Netherlands; the American nuclear-power industry; and the Channel Tunnel between France and England. Comparable scientific examples include telescopes like Palomar in Southern California or oceangoing research vessels used to look inside the earth and to trace the climate-changing interactions between the planet's air and water.

Almost always these gigantic enterprises appeared improbable or just plain impossible. Each had to be technologically ripe and required wide acceptance of the need for it. On more than one occasion, the spur was a disaster, such as the North Sea floods in 1953 in the Netherlands. But sometimes, as with the American railway to the Pacific, it took decades to secure the upfront, long-term investment in advance of the market, usually from governments. The money had to keep flowing as enthusiasm cooled and costly problems inevitably cropped up. This called for massive, relentless publicity. Even the secret atomic bomb project was immediately and massively publicized after Hiroshima and Nagasaki. Hence, it was crucial to win the loyalty of journalists and politicians and, perhaps most importantly, the immense enthusiasm of thousands, even millions, of hobbyists.

Focused management was essential to all these endeavors. As Watson said of the genome project, someone had to take the blame for problems or failure, requiring one to a few individuals to serve as highly visible symbols and managers. A deadline was indispensable to reinforce the enthusiasm within and outside the project and to assure that, in general, one generation of participants could complete it.

Highly varied resources and widely divergent skills, each with a history and culture, had to be focused on a common object. The required collaborations among many people at many sites were complex. The massive infrastructure not only had to meet the goal but must allow success to be exploited. A prominent feature that surprised genome project leaders was the early and intense involvement of private industry in providing money and equipment.

Like the quest for the complete human genome, the big projects were often built step by step. With canals or railroads, one section could open for settlement and traffic before the whole route was finished. Hence, the enterprise could earn some money and respect along the way, while bringing in supplies for the next section. The genome work progressed from mapping to sequencing and from smaller model organisms to the ultimate human target. The byproduct of scientific information found wider and wider use in medical diagnosis and in biology generally.

Quite soon, the genome project spurred such related efforts as the worldwide breeding of ever-more productive food plants and the development of genetically engineered microbes to produce renewable biofuels or clean up pollution. Meanwhile, the massive introduction of new medical diagnoses and treatments continued.

In many of these great projects, the true scale of the ultimate product's uses was unforeseen, as was the case with the almost fantastically rapid evolution of computers, communication channels, and the World Wide Web (so vital to genomics). Each project faced decisions about waiting for the better technology or going ahead with the good—the genome project was no exception, as we will see. These decisions involved a triangle of specifications, time, and costs. Sometimes, as in the Manhattan Project, prodigality was thought essential to beating the Germans to a bomb. Sometimes great savings were found, as with development of a lunar mission plan that started off with one huge Saturn rocket rather than two. In the drive to the genome, processes were continually streamlined.[2]

Watson as Pied Piper

Until he was dumped after about three and a half years, James Watson did everything expected of him as director of NIH's genome project. He charmed Congress with a vision of better health for humanity and adroitly stayed continually in the public eye.

He had forceful views on the vital subject of recruiting. At one meeting, a leading scientist advised Watson, "You put lots of cash on the table, and you'll get people to come." According to Ronald Davis, Watson replied, "Yes, but you'll get the wrong people to come. The people who will come to the table with a lot of cash there are the ones who need it. And those are the ones you do not want. You want the people who are too busy and there's lots of things going on for them. Those are the ones you want." Davis, whom Watson had known and admired for nearly twenty years, was one of the targets. Taking Davis to dinner in Palo Alto, Watson exclaimed, "You gotta go sequence yeast." Davis answered, "I do?" And Watson said, "Yes."[3]

He pushed successfully for more money and fought to keep the program running at the speed he favored. He quickly managed to have the program converted into an autonomous NIH center that could make its own research grants. He led a resolute campaign, including direct conflict with critics at public meetings, to beat back opposition to the project as a misapplication of money and a distortion of intellectual focus.[4]

True to form, he also plunged into hot water when he used highly emotional terms to denounce Japan for hanging back on genomics (to be sure, some members of Congress took delight in his unvarnished references to Pearl Harbor). In an angry letter to Kenichi Matsubara of Osaka University, head of Japan's new genome program, Watson demanded a minimum Japanese commitment to the international project of $300,000 a year. Anything less, he said, would be "halfhearted" and excite "continuing ill will and resentment." The chairman of his scientific advisors, Norton Zinder, admitted it was "a very bad letter" but added, "You should have seen the first draft." Soon after, McKusick was dispatched to Japan to mend fences. Watson, however, was unrepentant. He told Leslie Roberts of *Science*, "I've found you never get anywhere in the world by being a wimp."[5]

Just two years later, he crossed swords with Bernadine Healy, the new director of NIH, when he termed a plan she endorsed to patent EST gene labels as "sheer lunacy." He did not stop there. He moved so aggressively to forestall recruitment of two of his best scientists by a private company in Seattle that its founder wrote Healy in anger, accusing Watson of conflicts of interest in holding pharmaceutical and biotechnology stocks—and insufficient respect for the free enterprise system. Healy seized the classic Washington, D.C., conflict-of-interest crowbar to push him out. Perhaps suggesting Watson was a has-been, she praised him as a "historic figure in the annals of molecular biology."

Watson's three and a half years of directing the genome project exemplify the trials experienced by the founders of many a start-up enterprise that went on to thrive, as the genome project does to this day. Probably more than ready to get out when the job threatened to evolve into micromanagement, Watson was delighted to claim that he had been fired. He sent in his resignation via a curt facsimile from Cold Spring Harbor Laboratory (CSHL) and immediately turned to the task of turning the lab over to his successor. In after years, he continued as a potent lobbyist for a program that could not have begun without him.[6]

Getting Down to Specifics

In January 1989, the project's twelve-member council of advisers met for the first time in an "obscure corner" of the vast NIH campus, Conference Room 6 of the C Wing of Building 31. It was time to set directions and resolutely move on from "cottage industry" biology. A prime assignment was drafting a formal plan demanded by Congress. Watson told the advisers, "I will push people probably harder than they want. I am impatient." Chairing the meeting was Watson's sometime rival and longtime ally, Norton Zinder of Rockefeller University. "Today, we begin," Zinder said. "We are initiating an unending study of human biology. Whatever it's going to be, it will be an adventure, a priceless endeavor. And when it's done, someone else will sit down and say, 'It's time to begin.'"

For Zinder, the finished genome would be a Rosetta Stone. Sharp-tongued and peppery, Zinder had the knack of both public and behind-the-scenes work in protecting biological research from derailment. In the 1970s, he had written a quiet, but devastating, evaluation of some of the more bloated aspects of the "war" on cancer declared by President Richard Nixon. He had also pushed for what became the famous

1975 Asilomar conference on the risks of recombinant DNA. Amid the furor that followed, Zinder lobbied tirelessly against moves to enshrine regulations in local, state, or federal laws and investigative commissions. Along the way, he chaired a U.S. Army committee on a highly contentious subject: how to dispose of large stocks of old chemical weapons at eight sites around the country. And he was a veteran member of the CSHL board of trustees. So he was an old hand at "explaining Jim Watson to the world."

Sitting at an oval table thirty feet long were computer scientists and biologists. Several came from industry, which had already grown into a major force that would intervene more than once to speed and steer the project. Also around the table were veterans of the earlier conferences and committees on genomics, such as Maynard Olson and David Botstein. One member, Nancy Wexler, heroine of the drive to find the Huntington's gene, focused on the ethical challenges of knowing a great deal about people's genetic backgrounds. Wexler headed a panel to fulfill Watson's pledge, mentioned in Chapter 2, to allocate a percentage of the project's budget to studying the social and moral risks of genomics. Ever since, the focus on ethical challenges has been a prominent theme in maintaining public confidence in the program.

Watson declared that the fifteen-year project aimed "to find out what being human is." Unlike the giant accelerators that physicists built at great cost in money and time, the genome project would "generate important results in five years. We don't have to wait until the end."

More prosaically, NIH director James Wyngaarden said the committee's biggest task was "to define the boundary between what would be going on anyway and what is different." Olson, similarly down-to-earth, extracted the committee's agreement to the near-term goal of an improvement of three- to fivefold in knowledge or technology—more modest than the five- to tenfold goal he had pushed on the Alberts committee. Zinder underscored the big problems that the program would face in his naming of four panels, including Wexler's on ethics, headed by renowned scientists. The training panel would be led by Joseph Goldstein, famous for the studies of genes involved in heart disease. Botstein would chair the panel on databases. The fourth panel would wrestle with the anticipated concentration of mapping and sequencing efforts in a handful of big genome centers, designed for tasks beyond the capacities of a small laboratory.

Nothing symbolized the new approach to biology inherent in the genome project more than the centers. How many would there be and where? Two men oversaw this explosive topic: Phillip Sharp of the Massachusetts Institute of Technology, a discoverer of the fact that genes are "in pieces" with long silent stretches in between, and Olson.

Some would regard the big centers as elitist, Olson admitted, and they might prove difficult to kill off if they turned out to be unproductive. But the centers were vital to the sharing of major technical resources. Realistically, they were "the only way progress is going to be made." The controversial plans for centers would also set the pathway to the ultimate goal of sequencing all 3 billion bases of human DNA.

Anticipating his recruiting efforts, Watson, too, was convinced that centers were essential to reaching the goal line in under a century. As he told hundreds of genome researchers at a meeting in the fall of 1989, "People want to do this with a cottage industry approach, but I don't think it will work" without adding centers that would ultimately spend half the genome drive's money. He said he was convinced that construction money must be held out as bait for universities where the new centers would be started. As it turned out, only a few academic institutions had the flexibility to set up genome centers.

Definitions of success were another topic. How soon would the payoffs be evident? McKusick, the dean of gene searching, said visible results would come quickly from gene finding. Botstein disagreed, making a significant comment: "We are in the tool business, and not the application business." Olson immediately saw a public relations problem. Many of the inherited diseases had vocal constituencies. "It won't do to tell them we are building a tool that may help them in the next century."

The pull between research and applications was played out a few months later. In the summer of 1989, scientists in Toronto and Ann Arbor, Michigan, succeeded after long years of struggle in pinning down the actual gene for cystic fibrosis (CF). But the consequent burst of enthusiasm and public attention created demand for universal testing for the CF mutations. Research leaders Lap-Chee Tsui and Francis Collins

(Watson's eventual successor as head of the genome project) hastened to declare that universal testing was premature.

International cooperation and competition were discussed. Japan had already started a program in the Japanese "science city," Tsukuba, to sequence the DNA of the rice plant, the world's biggest single source of food. And Europeans planned to have some thirty-five laboratories in seventeen countries cooperate to sequence the genome of yeast. Botstein was skeptical, calling the idea "incredibly expensive and time-consuming"—but this cottage industry collaboration went ahead anyway, reaching completion in 1996.[7]

Fateful Choice: *C. elegans*

Humans were not the only species on the table. Which of the nonhuman species, the "model organisms," would be the most attractive test bed for large-scale sequencing when the time came? At its second session in June 1989, the committee made a fateful choice. Many people, including Watson, began with the notion that it must be the fruit fly, *Drosophila melanogaster*, the classic target of gene mapping from early in the twentieth century. But advocates of the smaller "worm," *C. elegans*, refused to be brushed aside. After all, they had not only worked out the genetics of their organism but determined the exact location of every cell and even traced its lineage.

Stimulated by Watson's apparent skepticism, John Sulston of Cambridge, England, and Robert Waterston of Washington University in St. Louis, heads of two of the earliest and largest genome centers, developed maps of the worm's genes so detailed that they stretched along an entire wall at the annual worm meeting at Cold Spring Harbor. Watson was impressed. Although Waterston's close colleagues told him he was "nuts," the worm-sequencing got going. At the Cambridge and St. Louis centers, the genes of *C. elegans* were mapped by 1992, and the worm's 97-million-letter genome was completely sequenced in 1998. To Waterston, the worm was "this marvelous tool kit with all the basic information for making an animal." Just two years later, the fly sequence was done—as a dry run during the frantic race between industry and academia to be first with the human genome.[8]

Opposition: Do Not "Underestimate the Extent of Unhappiness"

Even after the genome project got rolling, the surrounding temperature remained hot. Opposition among biologists had not been dispelled. During a San Francisco conference, Victor McKusick said the finished genome would be "a source book for biology that will be used for all time," but another speaker branded the enterprise as "an expensive tinker toy for a technocratic elite." Supporters fervently claimed the project would ultimately change medicine, while detractors denounced it as premature and unnecessary, charging that it threatened to starve the rest of biology. Even among supporters, fights broke out over whether to push for the entire sequence or, as Sydney Brenner had suggested, to decipher only the "expressed" pieces that carried the blueprints for specific proteins. An outgrowth of the debate over expressed sequences was the dispute over patenting short DNA "tags" for the genes, which influenced Watson's 1992 departure from the project.

As always during these years, the news was full of genomic topics—such as prenatal tests that could lead to abortions, the widening use of in vitro fertilization, possible misuses of genetic information by employers, the prominent use of DNA testing in criminal trials and a fight over the standards needed to make the tests technically accurate, and new discoveries of genes associated with disease. All this attention brightened the spotlight on the genome project.

On the morning of July 11, 1990, microbiologist Bernard D. Davis of the Harvard Medical School appeared by invitation at a Senate hearing to express his doubts about the scale and direction of even the revised human genome project. He spoke last, after a series of witnesses favoring the project, including Watson, and maintained that other lines of biological research would produce faster gains for health. Besides, biology was unsuited for Big Science. Saying that he was expressing a majority view among American microbiologists, Davis urged that at least some of the money should go back into the general funding of biomedical research, which had become unusually tight.

Listening was Senator Pete Domenici, a Republican from New Mexico and a fierce backer of genomics work by DOE laboratories, such as

Los Alamos in his own state. He burst out, "I am thoroughly amazed, Doctor, at how the biomedical community could oppose this project. I am absolutely amazed at how you could do that. . . . I cannot conceive that this small amount of money for this project, $200 million a year for the next five years, could conceivably be opposed by the biomedical community of this country. . . . I cannot believe that you are going to insist on business as usual in this field. It is beyond my comprehension, I repeat, beyond my comprehension." Shaken, Davis replied that Domenici should not "underestimate the extent of unhappiness in the biomedical community."[9]

Davis spoke near the peak of a months-long campaign by disgruntled biologists to overturn the conclusions reached in 1987 and 1988 that the project should go ahead. Two scientists, Martin Rechsteiner of the University of Utah and Michael Syvanen of UC Davis, were writing hundreds of letters decrying the genome project. Rechsteiner said to a *New York Times* reporter, "It's bad science; it's un-thought-out science; it's hyped science." Syvanen denigrated the idea that knowing the whole sequence would speed cures for genetic diseases. Even twenty years after the gene for sickle-cell anemia had been found, he argued, no treatment had emerged. The *Times* reporter wrote that the program "is arousing alarm, derision, and outright fury" among biologists who emphasized that even the complete sequence would generate few clues about how the genes in the genome catalog actually functioned. Predictably, opponents disliked the plans for big centers. Flying into Washington to lobby for the full appropriation, Francis Collins said, "We have a PR problem of major proportions." A reporter for the weekly *Science* wrote, "The honeymoon is over for the genome project."[10]

Watson heatedly responded to the critics, "Our project is something that we can do now, and it's something that we should do now. It's essentially immoral not to get it done as fast as possible." Zinder asked, "Why do they so completely misunderstand what we are about?" Paul Berg of Stanford, who would succeed Zinder as head of the NIH genome advisers, was more contemptuous of the opponents: "How can $60 million or $100 million have that big an impact on the $5 billion NIH budget? It just doesn't make any sense."[11]

The opponents' arguments had special bite because of a crisis in funding at NIH, which could give money to little more than one-fifth of all proposals judged "meritorious," compared to one-third only three years before. The previous month, this problem had led a House appropriations subcommittee to cut the George H. W. Bush administration's request of $108 million for the project to $66 million. This had happened despite furious lobbying by Watson, who said the reduction would cripple his new program of major genome centers. Not for the first or the last time, he threatened resignation. Adding to Watson's difficulties was the fact that the NIH director who appointed him, James Wyngaarden, had been forced out a year earlier and still had not been replaced. The July Senate hearing was an opportunity for Watson to get back the money the House had denied him. Eventually, he recovered a lot. The final figure was $87 million.[12]

It turned out that the criticism had reached a high-water mark. The program now involved hundreds of scientists impatient to get on with genetic and physical mapping and with accelerated development of faster sequencing. They made this clear in October 1990 in a conference at San Diego. As part of the campaign to confront the criticism, Watson and his coorganizer, Charles Cantor of DOE, invited two opponents, Bernard Davis of Harvard and Donald Brown of the Carnegie Institution of Washington, to lecture to what proved a restive audience. Davis said he approved of the project's mapping goals but deprecated the plans to sequence the whole genome, most of which appeared to be "junk." Watson did not respond politely: "Saying that you support mapping without sequencing is like saying, 'I will marry you but there will be no sex.'"

He was equally contentious with Brown, who was less than tactful himself, saying that the "top-down" project was "overbudgeted, overtargeted, overprioritized, overadministered, and micromanaged." Watson answered that the genome was big—so how could they keep the program small and "get it done in a reasonable period of time?" There would be many small grants, he said, but the program was not one of pure discovery. Indeed, it was aimed at a specific target. "It's not whether things are targeted, but whether you have the wrong target." When Brown said the program showed a lack of confidence in the grant system,

Watson shot back, "Don, you're the product of rich-man's science. You have a Carnegie endowment, but you should be honest with the way you exist." The assembled scientists were eager to get on with the scientific part of their program.[13]

Expressed Sequence Tags

A very different fight loomed on the horizon. This time, the dispute concerned a strategy to streamline the project, to make it go faster, not slower. The central combatant was an impatient NIH neurobiology researcher named Craig Venter. Well-spoken, drivingly ambitious, combative, and full of himself, he was destined to be a principal goad to finishing the human genome faster than planned.

Determined not to "waste my life," Venter said, "I reject absolute authority unless it's rational." Eager to sequence the genes of the brain, Venter was already testing a new $92,000 sequencing machine from Applied Biosystems, with the help of money from the Defense Department, not NIH. He had read about the machine in a paper in *Nature* and flew out to meet Michael Hunkapiller to see it. Within a year, Venter had sequenced 100,000 nucleotides, but he felt this was paltry. Early in 1989, he pitched Watson for $5 million to sequence the entire human X chromosome, and at first Watson was enthusiastic. But others in the genome program were appalled at the idea of so much money staying inside NIH and thus being subtracted from the priority effort of gene mapping, largely performed outside. But the deeply stubborn Venter went right on with sequencing the X chromosome.

Venter looked back on this episode regretfully. He had envisioned the Human Genome Project as Walter Gilbert had, as "a single industrial-style centre." Instead, the project became a giant cooperative spreading money to many research groups. "It became a political project with people fighting to control the money. New ideas were shut out."

Still hoping to win Watson's backing, Venter had an inspiration on a flight back from Japan later in 1989. He would pile one sequencing shortcut on top of another. It related to the discovery of "split genes" back in 1977 by Richard Roberts, Phillip Sharp, and others. Not only does the human genome contain huge, dark, often highly repetitive stretches of DNA between genes, but the genes themselves are broken up into "expressed" pieces, called exons, each carrying the design of part of a protein. Intervening between the exons are "noncoding" introns. When a protein is to be made, the cell makes a crude RNA copy of the entire region containing the exons. Then the RNA is edited into a version containing just the exons, the "messenger" that goes out of the nucleus to the places where proteins are assembled. Such messengers are extremely short-lived, existing just long enough for the job at hand.

But, as mentioned in Chapter 1, the exotically named reverse transcriptase can make DNA copies complementary to the messenger. And these complementary DNAs (cDNAs) can be stored in a cDNA library containing just protein blueprints—no intervening "junk." That was the first shortcut, but there was a second. To get round the tedium of completely sequencing a stretch of cDNA, one could use short DNA stretches as labels. Mark Adams in Venter's lab soon found these labels, called expressed sequence tags (ESTs), in abundance in brain tissues. These biochemical catalogue cards might be the way to get off what Adams thought was "a slow boat to China."

In Venter's mind, the goal was finding genetic "targets" for drugs. That meant focusing on the part of the genome—only 1 or 2 percent—that specified proteins. But Venter's strategy cut across Watson's determination to decipher the entire genome. Going for the tags would undercut the big genome budgets that were building up. And Watson did not appear too impressed with the impulsive Venter as a scientist or with his "bargain in comparison to the genome project." As Watson noted later, the ESTs were only a few hundred bases long. There were also two other nasty questions: How many genes give rise to ESTs, and how many ESTs does a gene produce? The ESTs not only inevitably produced a bias against rarely expressed genes but also failed to capture the regulatory stretches of DNA both upstream and downstream of genes that controlled when genes were turned on and off.[14]

The discussion was "bitter," Jane Peterson of the then National Center for Human Genome Research recalled. The EST approach was "very seductive; it was so much cheaper. But it would not have driven the sequencing technology." Venter's proposal did not survive a brutal meeting with the relevant "study section," at which, according to one scientist present, Venter's grasp of technical details appeared weak. But later, Watson did acknowledge that the EST approach had "proved immensely useful, and it should have been encouraged."[15]

Still focused on swifter movement of genomic insights into medicine, Venter went along with a plan for NIH to seek patents on the many ESTs he was finding. It was an idea in tune with recently passed federal laws—such as the Bayh-Dole Act of 1980—inviting, even commanding, nonprofit labs in the government and universities to seek patents on their discoveries so as to encourage private companies to invest sooner to bring the federally funded discoveries to market.

The patenting plan first intrigued NIH's chief of technology licensing and then the new NIH director, cardiologist Bernadine Healy from the Cleveland Clinic, who took the job after a long interregnum during which several prominent candidates said no thanks. A Republican and strong backer of the market approach to innovation, Healy was all for patenting the ESTs. In June 1991, Venter published the first few hundred of his tags, and NIH, the owner, at once applied for patents on them.[16]

Watson exploded almost immediately. Both Venter and Watson attended a Senate meeting called by Domenici and held in a nearly empty hearing room. It was here that Watson denounced the patenting approach as "sheer lunacy." Patenting mere strings of DNA could not pass the legal requirement for utility. Stinging Venter, he exclaimed that his machines "could be run by monkeys." Already convinced that the start-up phase was ending and eager to get back to Cold Spring Harbor, Watson saw ruin for the open, public cooperation of genome researchers around the world that was essential in completing the genome. At a later congressional hearing, he said, "Speaking for myself, and not for NIH or the center, my belief is that these small pieces of DNA should not have been patented. To get a patent, you should have a useful product."

Many scientists were outraged. Olson exclaimed, "I think it's a terrible idea. If the law is interpreted to give intellectual property rights for naked DNA sequences, then the law should be changed. It's like trying to patent the periodic table. To put patent value on cream-skimming is sending entirely the wrong signal. Scientists who advocate such approaches are playing with fire." Britain's Medical Research Council (MRC) was particularly indignant. Tony Vickers, manager of MRC's genome project, declared, "We don't believe that there's anything inherently patentable in one of these fragmentary cDNA pieces." If the U.S. patents went ahead, British scientists would have to follow suit. Vickers also said, "All deals are off. I'd get crucified if I made the sequences publicly available now."[17] The furor was worldwide, and relations between Watson and Healy, never very warm, deteriorated. She began looking for a way to force Watson out, and Watson provided the excuse. He owned shares in a number of biotechnology companies, and when he joined NIH, he had listed them and, by a waiver, not been required to sell.

At the end of 1991, Watson was told of the plan to grab his two star champions of the worm, Waterston in St. Louis and Sulston in Cambridge, England, and bring them to a new company that a Connecticut investor, Frederick Bourke, planned to set up in Seattle. Adding credibility to the scheme was a very tight budget for science in England, which might increase temptations for Sulston. Watson went into crisis mode, talking rudely to Bourke and urging his many contacts in Britain to boost that nation's genome spending to keep Sulston—which was done. Bourke's plan collapsed. He wrote Healy, accusing Watson of insufficient respect for private enterprise and pointing out Watson's shareholdings. Healy refused to sign another waiver on his ownership of the shares, precipitating Watson's noisy departure.

Eventually, Watson and the legions of scientists who agreed with him were victorious. The U.S. Patent and Trademark Office twice refused to grant NIH's applications for EST patents, and in 1994, Harold Varmus, by then NIH director, abandoned any attempt at appeal to the courts.[18]

In 2006, Venter looked back on this episode resentfully. He told Alun Anderson of *Prospect* magazine, "I developed a method for labeling genes quickly and the NIH tried to file patents on hundreds of genes. I was attacked by Watson and others even though the so-called 'Venter patents' were filed by NIH, not me, and I would never have got any money from them."[19]

The genomic rumbling did not stop. Just a few months afterward, Venter and Claire Fraser, then his wife, and a number of coworkers also left the government to found a new private research institute, The Institute for Genomic Research (TIGR). They had $2,000 in their savings account after twenty years in science. The creation of TIGR loudly signaled an unexpectedly prompt Wall Street interest in the genome. An ambitious venture capitalist, Wallace Steinberg, saw the publicity about Watson's public denunciation of Venter's EST shortcut. Steinberg envisioned a key technology being hampered in the United States—to the advantage of other countries like Japan. So he rushed down to NIH to see Venter, who was intensely frustrated by the slow pace of genomics.

This approach was not the only one. Venter recalled,

> The irony is that, thanks to Watson attacking me, everyone heard about my work and I started to get big offers from biotech companies. One of them offered me a five million dollar signing fee, and when I turned it down because I wanted to keep doing science, people saw it as a negotiating ploy. But I eventually did a deal with private investors. I said, 'If you allow me to start my own not-for-profit institute in parallel with a for-profit institute, and they are kept independent of each other except for funding, I will do it.' So I got seventy million dollars over ten years to fund my lab called The Institute for Genomic Research (TIGR) with the intellectual property rights going to the for-profit institute (Human Genome Sciences).

Venter was working his way around a typical pothole in scientists' participation in industry. He recalled, "The scientific founders of biotech companies are mostly just looking for new ways to get their science funded, and when the company comes under pressure to make products, the founder gets the chop. I was trying to avoid that fate by coming up with a model that would allow me to do basic science and reap commercial benefits."[20]

According to the deal, Human Genome Sciences (HGS) would support TIGR's research to the tune of $10 million a year; in return HGS would get six months to evaluate the commercial possibilities of a discovery. As Watson remarked later, the business community saw opportunities in the genome far earlier than he expected. It turned out, he said, that "you can't keep private people out of something as interesting as the genome."[21]

Less than a year after she forced Watson out, the new Clinton administration told Healy it would like to appoint its own person to head NIH: Harold Varmus of UC San Francisco, famed for his discoveries of cancer genes.

A New Leader: Francis Collins

Given the political and financial momentum behind the genome project, the question of who would succeed Watson was not a casual matter. Healy lost no time putting out the word to potential candidates. One of her confidants, Richard Klausner of the National Cancer Institute, called Francis Collins at the University of Michigan, urging him to put his name in. Although reluctant at first, Collins was a natural for a complex job that combined public advocacy, maintaining cooperation among hundreds of scientists, and leading the tricky search for consensus on goals and schedules. He was a central figure in finding genes for cystic fibrosis, neurofibromatosis, and breast cancer. As one journalist put it, Watson spoke of building tools while Collins spoke of saving lives.

Collins also had a certifiably middle-American story. Born in 1950, the youngest of four brothers, he grew up on a farm in Virginia's Shenandoah Valley. The house lacked plumbing. His father, who had been a folk song collector in the 1930s, ran the drama department at a small college. Naturally enough, Francis learned to play the piano and the guitar. His mother, who wrote plays, homeschooled him and his next-oldest brother through the fifth grade. In 1992, he told an interviewer, "My mother decided that the county schools were not a place where a young person would learn to love learning." At the farm, he helped take care of the stock, a task that he remembered as particularly arduous in the winter. Each summer, the Collins family housed half a dozen to a dozen participants in the local summer theater, and Collins began taking part in the plays. At four he played the Cowardly Lion in a stage version of *The Wizard of Oz*. Thus, he received a "good training" in how to "get over the jitters of being in front of an audience."

Soon after, in high school, a chemistry teacher captured Collins for science by demanding that he and his classmates design experiments to figure out what might be hidden in a closed box. "I got caught up in it.

It was the first time, I think, that somebody had challenged me to come up with the ideas." He grew up an atheist in a family "without faith" but found faith later, not only in contemplating the beauty of nature but also in observing close-up the strength of will that belief in God gave to suffering patients. In later years, he would write a book arguing that there is no inherent conflict between God the Creator and Darwinian evolution and take part in a National Prayer Breakfast with President George W. Bush, singing a song to the accompaniment of a guitar.

But guitars were not just for prayer meetings; they also figured at scientific occasions. At the 2003 Cold Spring Harbor Symposium, held to summarize and celebrate what was known then about the human genome, Collins played a black guitar on which many participants in the project then signed their names in gold ink. The guitar now hangs in iconic splendor in a glass case in the basement bar of Blackford Hall at Cold Spring Harbor, surrounded by photographs of researchers who figured in the fifty years of biological revolution leading up to the complete genome.

The idea of leading the genome project struck Collins as odd at first. He had not been a member of the Alberts committee that set the program's strategic direction. He was a full professor at the University of Michigan and a Howard Hughes Medical Institute investigator to boot. He had "great postdocs" working in his laboratory. Despite his lack of enthusiasm, he was interviewed by a search committee that included Shirley Tilghman, who had been a member of the Alberts committee and asked him what the genome project needed next. He answered that the project needed an internal research group on the NIH campus. But Collins didn't think it was "a good fit" for him and withdrew his name. NIH turned to others, including geneticists Mary-Claire King and Frank Ruddle. Thomas Caskey was not interested.

So, for Collins, the arm-twisting began again. Dining with Collins, Lander described a "great opportunity" for a physician to lead the program. Congress was interested in health outcomes. In November 1992, when Collins was in Bethesda for a study-section meeting, Healy demanded to see him. She said, "I don't think I'm getting through to you." Collins recalled his reaction: "She's a smart cookie. She knows what she wants." The idea of Collins starting his own lab and having his own separate institute resurfaced. He reflected, "There's only going to be one genome project, and it might succeed." And so, on April 1, 1993, he took office as director.[22]

Less than two years later, Collins faced a potentially dangerous situation for what was now called the National Human Genome Research Institute. Energized by Newt Gingrich, the Republicans took over both houses of Congress at the beginning of 1995. They soon abolished the Office of Technology Assessment, which had been advising Congress on science and engineering for twenty years. Although Gingrich delighted in high-technology gadgetry, it was not clear what the attitudes of others would be to the long, expensive, and somewhat arcane genome project. Of particular importance was the attitude of Rep. John Porter of Illinois, the new chairman of the health subcommittee of the all-powerful House Committee on Appropriations. Porter had his hands full. The budget committee of the new Republican-dominated Congress wanted to cut the NIH budget by 5 percent per year for the next five years. Porter struck back. He trooped a delegation of researchers and biotechnology executives to see Gingrich, the new Speaker of the House. The near-term upshot was an increase for NIH of nearly 6 percent, and Gingrich became a long-term supporter of health research.

It was to a predecessor of Porter's on the appropriations subcommittee, William Natcher of Virginia, that Watson had appealed in 1987 for the money to start the enterprise. Now it was Collins's turn. Even though Porter's father had contracted polio and Porter thought the Salk polio vaccine was the greatest scientific achievement of the twentieth century (and he had studied at MIT), Collins felt he must make sure of Porter's strong backing. With his heart in his mouth, Collins went to see him in January 1995. With some boldness, Collins said he feared the program might be in jeopardy and that he needed Porter's commitment. Porter, "completely riveted, thoughtful, well-informed," gave his support.[23]

Mapping Speeds Up: "The Academics Were Horrified"

In the genomic community of the early 1990s, the doctrine was that locating genes along the DNA of the human chromosomes—mapping—was easier and cheaper than wholly sequencing the human genome. The techniques for mapping were more developed. Besides, detailed maps would be crucial to putting sequenced stretches in the correct order in a top-quality sequence with few errors. This was a conservative path to the ultimate sequence. The method involved handling chunks of DNA that had been duplicated, or "cloned." Each clone would be sequenced. It was crucial to know just where the clones' DNA was located. To determine this accurately, one had to have genetic maps, with the finest detail attainable. Mapping would open the door to sequencing.

The challenges of mapping turned out to be bigger than expected. They drove a major adventure in automating genomic laboratories, particularly those of Eric Lander in Cambridge, Massachusetts, who not only campaigned against "Balkanizing" chromosomes in particular laboratories but also opposed parceling out particular human diseases. The automation, in Lander's opinion, was necessary to achieving the industrial scale demanded by mapping, let alone sequencing. Most biologists focused single-mindedly on human phenomena and jealously guarded their human DNA territory. But, to Lander's delight, the genome of the humble mouse—an extremely close cousin of humans—lay wide open. On the theory that it is better to ask for forgiveness than permission, Lander's team went ahead with a complete mouse map. As his associate Bruce Birren recalled in 2006, "There weren't enough skilled, smart people in the world to hold all the pipettes needed to do it in the usual way." To help with the problems, Lander brought in mathematicians, engineers, and young people who were free "of all the years of being told the right way." The mouse work, according to Birren, "proved to be incredibly efficient in cost and time; and then, the center was funded to start doing that on humans."

Lander insisted on a brutally simple approach. "Okay, we want to sequence the human genome. What would that take?" Birren would start to say, "Well, here is what I know how to do." Lander would reply, "No. What would it take? Imagine sequencing the human genome in a few years. What would we have to do?" Birren recalled that Lander had the "ability to walk in every day and challenge all prior assumptions, including our own." Lander was "willing to throw out everything, whether it was machines or approaches, when it was clear that they weren't up to the challenge, or there was a new, better challenge."

Lander was ready, Birren said, for a new type of person in biology, someone "who could think about turning previous activities, that were designed for research, into production activities." Birren recalled, "The academics were horrified."[24]

The same problems beset all the major sequencing centers.

Madame Alavi's Bequest

Another accelerating trend in mapping was international competition, particularly from France. Not long after the French immunologist Jean Dausset shared a 1980 Nobel Prize, a woman in her late nineties named Hélene Alavi left a group of paintings, by Balthus and others, to the cancer research laboratories that Dausset directed. She had collected the surrealist works in the 1940s and 1950s while married to a Paris banker. She often bought at La Galérie du Dragon, which Dausset co-owned. When sold at Sotheby's in 1984, the paintings brought over $6 million—more than twice the amount expected.

Dausset was famed for his discovery of human leukocyte antigens (HLAs), vital for matching donors to recipients of transplanted organs—and an ingredient in Botstein's 1978 conversation at Alta, Utah, that led to restriction fragment length polymorphisms (RFLPs). After Madame Alavi's bequest, Dausset turned to an entrepreneurial associate named Daniel Cohen, an engineer who had become a geneticist, and said, "With money, one must do something great. Go and find me a project." Dausset's ideas of something great were influenced by deep enthusiasm about evolution. He wrote that "all living beings, vegetable and animal, whatever they are, are all related, all formed of the same materials, and function in the same way."

The joke-telling, piano-playing Cohen, a man described as able "to swim in any water," had worked with Dausset on HLA. Why not, Cohen asked himself, turn genetic cartographer? Why not make a complete inventory of human variation, of polymorphisms, not just in a single gene but in the entire genome? After all, polymorphism had been the basis of the science of genetics since it came into being in 1900. The 1978 insight about RFLPs had aroused fresh hope of using well-documented, multigeneration families to correlate diseases with genes. The money from the Alavi estate was big, Cohen reflected, far bigger than could be found in government grants. It called for a big research project, concentrated in one place. The way to start was to gather DNA that could be used to find genes, then quickly get into the development of robots for analyzing DNA.

And so, years before mapping reached a big scale in the U.S. project, private philanthropy in France had cranked up an effort that proved very important to the whole enterprise. The Center for Study of Human Polymorphism (CEPH) in Paris became the world custodian of detailed evidence from some forty families, American, French, and Venezuelan, that had volunteered blood samples and data from several generations. About half of these families were in Utah, where, with the cooperation of the Mormon Church, researchers like human geneticist Mark Skolnick of the University of Utah could make use of a rare resource—detailed genealogies of large Mormon families. These had been preserved and gathered in accordance with the Mormon faith to permit descendants to convert their ancestors retroactively. The DNA of such families could show even minor differences between members who suffered from a disease and those who did not.

Adding great force to such efforts was the success of McKusick's international club of scientists in identifying inherited disorders. The markers on chromosomes for such diseases as muscular dystrophy and Huntington's chorea were discovered just as CEPH got going in 1983.[25]

As the years passed, however, just as Maynard Olson had predicted, there was considerable nail biting about whether the goal of producing maps to open the door to sequencing in about five years would be met. As Watson had understood, Congress was impatient for early payoffs—perhaps a gene for Alzheimer's—and he had promised a human map in that time frame. He did not hesitate to drive the scientists to meet or beat that goal.

Reassuringly, new methods of great use for mapping sprang up. These included the yeast artificial chromosomes (YACs) developed in Olson's laboratory and their younger cousins, bacterial artificial chromosomes (BACs), invented in 1992 in the Caltech laboratory of Melvin Simon. Just two years later, the Simon lab also invented artificial chromosomes based on a virus called P1 (PACs). Hopes were raised. But then, as happens routinely in science, after the new methods soared in popularity, new bumps appeared in the road. What had looked like a liberation became a new burden. Additional tricks, including use of more than one method at a time, were needed to keep up the accelerated mapping.[26]

Two of the most promising methods were announced in 1989. One, devised by David Cox and Richard Myers of UC San Francisco, used radiation to sharpen the precision of measuring the distance between genes in particular chromosomes, which could be found in mouse cells fused with human cells.[27]

Another arose in the summer of 1989 in a meeting at the Banbury Conference Center of CSHL. Intense discussion there led Olson, Botstein, Cantor, and others to propose a "common language" in identifying genes with so-called sequence-tagged sites (STSs). These, in turn, lent themselves to the polymerase chain reaction (PCR) method of producing millions of copies of a gene of interest overnight. The information could now be electronically coded and transmitted to different laboratories across the world—and be understood. So now, labs could generate gene sequences they wanted without waiting to receive actual tubes of DNA from other labs or a central repository.[28]

The mapping drive kept heating up. In 1987 two rival U.S. groups, Ray White's at the University of Utah and Helen Donis-Keller's at a private company, Collaborative Genetics, put together "genetic-linkage" maps with markers scattered across the entire human genome. Both used data stored at CEPH. The announcement in 1987 of the two competing maps, each with some 400 markers, came as the members of the Alberts committee were finishing their report endorsing the Human Genome Project.[29]

Soon after, the work of CEPH, which included several European projects to develop robots for genomics, received a big boost from the "scientific marketing" so typical in the United States, where the genome project had been launched with typically loud éclat. Aiming at maps of the entire human genome would go beyond the numerous projects for individual chromosomes. In the fall of 1988, fresh from success in a sequencing-robot project with a private manufacturer, an eager Daniel Cohen went to see Bernard Barataud, head of the French muscular dystrophy association, known by its French initials, AFM. Cohen asked, why not spend something like $40 million and get a map with 5,000 markers as a step to the whole DNA sequence? Barataud was oppressed by the immense number of genetic disorders the AFM was struggling to cure, including an array of muscular dystrophies of varying severity. Seeing the futility of attacking diseases one by one, he said yes after ten minutes and took the idea to his board. The resulting big project, called Généthon, was fed with millions of dollars raised from the public via telethons. The AFM found space for the gene hunters in Evry, an industrial suburb of Paris.

Looking ahead to complex, multigene diseases, CEPH put ads on television, asking for families to volunteer for a genetic study of one type of diabetes, and enlisted more than 500 in France and elsewhere. In less than two years, CEPH uncovered mutations in several genes that affected the level of sugar in the blood.

It remained as clear to the French as it was to the Americans that the technology for sequencing was not ready for the big push. The genome project would have to walk before it could run. Cost-effective, large-scale sequencing remained years away. The project's fifteen-year timetable was fast—but the total budget was not supposed to exceed $3 billion. Besides, mapping was familiar. It was simple and comfortable to intensify a classical activity of geneticists, dating all the way back to the landmark studies with fruit flies in Columbia University's "fly room" after 1910.

Acting as David to the American Goliath, the French aimed to be smaller but smarter. Cohen's colleague Jean Weissenbach led a project to construct a greatly refined map of gene locations based on the genetic data from families. Cohen would lead a group to produce a second kind of map, called a "physical map," starting with human chromosome 21. Charles Auffray would start on the yet longer task of making a map of all the human cDNA (the library of protein blueprints).

Although leading French biomedical researchers told Cohen that the genome was uninteresting, a French government concerned about being left out began supporting CEPH to the tune of $4 million a year, and then, spurred by a dynamic new head of the French genome effort, leading molecular geneticist Piotr Slonimski, the government contributions increased.

Soon, CEPH won a place in major international efforts to develop maps, as with the genetic map published in the fall of 1992 by the NIH/CEPH Collaborative Mapping Group. Cohen's enterprise was setting a fast pace. It found a way to make "mega-YACs," clones 1 million bases long. This cut the number of clones needed for a physical map even further than labs like the Whitehead Institute had done with chromosome Y, site of the maleness gene. Maps the CEPH had led in making would be grist for the cooperating and competing teams that were mapping chromosomes one by one. Cohen delightedly recalled the impact as "an extraordinary shock wave." One British geneticist told the press, "He has the Americans terrified. The speed has frightened all the people who got big grants from the U.S. National Institutes of Health."

During the 1991 annual genome meeting at Cold Spring Harbor, Cohen had shown a video of twenty robots humming away at Evry, doing the work of 400 people. "It was cinema for sure," he reflected, "but what an impact!" Jim Watson turned pale. Later, the maverick Venter told Cohen of "a little meeting" at which Watson exclaimed, "We must stop them! But how?" When Cohen's physical maps began coming out, Watson praised them but said they contained many holes. It was not certain that the fragments were in the right order. Still, Watson had to admit that the French gamble on a genome-wide map was "now paying off." He expressed relief that CEPH, as it had from the beginning, would be sharing its results with the rest of the global genomics community. He was sure the whole-genome human map could be done by the end of 1995, as planned.[30]

The French had taken a lead—and the Americans knew they must collaborate even more intensely than before. By the end of 1995, the Whitehead Institute–MIT genome center and Généthon had combined two types of physical map and one genetic map to produce a whole-genome human map with just over 15,000 markers. To build up DNA supplies for the effort, robots at Whitehead had sped up to perform half a million PCRs. The array of PCR robots was dubbed the Genomatron. Leading the effort, Lander said that the collaboration had produced the equivalent of an atlas of national, state, and city maps. Because it contained enough genetic landmarks to begin large-scale sequencing, this atlas was billed as fulfilling Watson's promise. The French, after a few months' work, published a detailed human genetic map at the same time as the Whitehead map.

A few months later, commercial and academic scientists in the United States, France, and other countries made another leap in gene finding. They used the Cox-Myers radiation hybrid and YAC mapping to sort through hundreds of thousands of the gene tags called ESTs. The resulting map, Francis Collins said, was "a police lineup." The locations of 16,354 genetic markers along the DNA were known. Lander now said the genes no longer just had zip codes but had actual street addresses. Interestingly, many genes clustered in the equivalent of cities, while others were sparsely scattered in an immense countryside, a "gene desert." Some of the genetic neighborhoods had thirty times as many genes as others. Genes for related jobs, such as creating the structure of skin, tended to cluster. This feat was announced with fanfare that included a statement by Vice President Al Gore. It was just weeks before the finding of genes for Parkinson's and prostate cancer was announced.

The mappers and their patrons were in a heady mood. Lander likened the map with 16,000 markers to the thousands of position measurements that Lewis and Clark had recorded on their epochal journey up the Missouri to the Pacific and back. With the precision of the maps, Collins said, a moderate-sized laboratory could shrink the task of finding a disease gene from years to months. He announced that detailed sequencing of the genome could finally begin.[31]

Venter's Coup: Sequencing the First Free-Living Organism

While the nonprofit network glowed with pride, the impatient spirit of Craig Venter did not rest. In the spring of 1995, after less than three years in the private sector, he not only scored a publicity ten-strike but simultaneously vaulted into a grudgingly accorded scientific respectability. He and Hamilton Smith at TIGR announced that they had worked out the entire sequence of the DNA in the microbe called *Haemophilus influenzae*. Naturally it was Smith's favorite because he had found the first restriction enzyme in it. The announcement at the annual meeting of microbiologists in Washington was quite an event. Adding explosiveness to the announcement was the fact that sequencing "H. flu," containing some 1,750 genes, had taken less than a year. Venter and his colleagues had made a significant test of what was called whole-genome shotgun sequencing. This shotgun method brushed aside the methodical clone-by-clone approach of the international, nonprofit program. Instead, the entire DNA of an organism was blasted into smithereens, into tens of thousands of very short stretches of bases. Each piece was sequenced. Then, advanced software put the puzzle back together, "assembling" more than 24,000 fragments of shotgunned DNA. Chillingly for the public genome project, Bad Boy Venter had proved the usefulness of another genomic shortcut. Would it apply to bigger organisms, even humans?

Venter was disappointed in the reaction. He told an English interviewer, "Instead of people saying, 'Wow, here is a faster way to go,' I just heard, 'Oh that could not possibly work on the scale of the human genome.' Everyone wanted to keep going without trying anything new. The reaction was something like: 'Very nice, Mr. Ford, you developed the automobile, but you know our horses and buggies are just fine, thank you.'"

It was the first complete sequence of a microorganism and the beginning of a cascade of "extremophile" sequences from TIGR, including organisms that survive at the sunless bottom of the sea, in pools of boiling water, or in streams of radiation a million times the dose that would kill a human being. The sequence, some ten times the length of an average virus, had been worked out in record time. A second sequence, apparently the smallest genome of a microbe, was only months away from publication in a scientific journal. Indeed, at the end of his lecture to the microbiologists, Venter projected this sequence of *Mycoplasma genitalium* on the screen. A team led by Venter's wife, Claire Fraser, had worked it out in three months.

By comparing the genes of H. flu with genes from other organisms whose tasks were known, Venter and Smith could predict the functions of just over 1,000 genes. The others, equally interesting, would probably have novel jobs. Such comparisons highlighted the usefulness of microbes for tracing the evolution of DNA. Smith said, "All the machinery is right there in front of you." The team at TIGR was already at work on clues to the tricks H. flu employed to survive the attacks of the human body's immune system. The money for this bombshell had not come from NIH: Its reviewers doubted that Venter could close the gaps in his sequence. So, as happened at a number of crucial junctures in the genome project, the Department of Defense supported the work.

Scientists, including rivals and those who had disparaged Venter only a few years before, were impressed. Frederick Blattner of the University of Wisconsin was using NIH money to sequence the classical microbe of molecular biology, *Escherichia coli*. At the microbiology conference, Blattner exclaimed, "This is really an incredible moment in history." He added that "the whole paradigm of genetics is reversed." Now, geneticists were freed from finding genes by looking for the effects of small mutations in DNA. Instead, they could "take the whole sequence of an organism and work down from that to its genes, which is what geneticists have been dreaming of for a long time." The troublesome Venter was heightening the drama.

Collins called the success "a significant milestone." Watson said, "I think it's a great moment in science. With a thousand genes identified, we are beginning to see what a cell is." He wrote that the achievement was "clearly a landmark in the history of biology." Less than two years later, a British observer recalled that the world had been skeptical, but Venter had changed microbial genetics "overnight" and opened the way to many new strategies for attacking infections.

The new enthusiasm quickly spurred new competition. In 1996, a stunned NIH paid for an effort at TIGR to sequence a highly infectious strain of the tuberculosis bacillus that had recently caused an outbreak in a clothing factory in Tennessee. But Britain's Wellcome Trust also jumped in, with money for a sequencing drive on a long-studied different strain of tuberculosis, carried out in Cambridge, England, and Paris. In 1996, the European effort came in slightly ahead of TIGR's.[32]

Of course, the sequencing story was never all Venter, all the time. In the spring of 1996, an international consortium of some 600 scientists in about seventy laboratories in Europe, Canada, the United States, and Japan announced success in a seven-year drive to spell out the entire 12 million letter sequence of baker's yeast. Yeast had some 6,000 genes (of which many resembled human genes) arrayed in sixteen chromosomes. The research had been parceled out according to a "gentleman's agreement" not to compete. This classical tool of genetics became the largest organism yet sequenced, far longer than the laboratory bacterium *E. coli* or the bacterium *Helicobacter pylori* (recently discovered to be the principal cause of ulcers).

Yeast was the first eukaryote, a cell with a nucleus, to have its genome deciphered. Ron Davis, whose Stanford laboratory had done 7 percent of the job, said, "Until now, the yeast genome has been a black box," but now "we know what's there, and what's not there." Collins crowed, "The yeast genome is closer to the human genome than anything completely sequenced so far." If the human genome could be compared to sending Apollo astronauts to the moon, Collins said, the yeast sequence was equivalent to John Glenn's Mercury flight around earth in 1962. Now an intense effort could begin to work out the functions of yeast genes—as a step toward doing the same in humans.

Cooperation sprang up alongside rivalry. TIGR was already taking part in the consortium to sequence a fast-growing weed called *Arabidopsis thaliana* that had proven highly useful for genetics. It was easy to grow, and its time from "seed to seed" was about twenty days. It was a model for the food grains on which humanity depends for sustenance. Also, Steven J. Norris and George M. Weinstock at the University of Texas, who had been studying the spirochete that causes syphilis, asked Fraser of TIGR to join forces to sequence the organism. They worked together not only to find the numerous tricks the spirochete had evolved to get around the human immune system but also to discover proteins that could be targets for drugs. They completed the sequence in 1998.[33]

The impatient Venter had again delivered what he called "a real shock" to the genome establishment, having proved that a controversial method of sequencing could be made to work. Within three years, again with Wall Street intervention, he would get the chance to use the same method on the human genome itself.

"This Isn't a Rocket Science"

There also was impatience within the Human Genome Project. With advanced sequencing still far away, both the headquarters staff in Bethesda and the leaders of the sequencing centers began asking whether the existing technologies could do the job. The conversation came to a head at a meeting of some twenty center directors on December 16, 1994, that focused on next steps. There, Sulston and Waterston, with a confidence buttressed by their experience in sequencing the worm, described their plan to produce a draft human sequence in 2001, years early, at a cost of between $300 and $400 million. The project would then go back and produce a "finished" sequence. Sulston summed up his attitude this way: "Why fiddle around?"

The plan had come up earlier in the year. Having taken his St. Louis group over to Sulston's lab in England for their annual joint meeting, Waterston started talking in a general way about a human sequence. But on the plane ride home, he crafted a plan that he e-mailed to Sulston for his approval, which was given. Waterston then told the St. Louis sequencing team about it. He later recalled wryly, "Now they get scared when I come back from long flights."

In St. Louis, Waterston's team was already processing 15,000 "reads" (stretches of DNA some 400 to 500 bases long) a week, and they thought they could boost the rate more than fivefold, to 84,000 reads a week. If three large sequencing centers could achieve this rate, they would cover 99 percent of the human genome at a cost of about twelve cents a nucleotide. The error rate of 1 part in 1,000 would be ten times less refined than the accuracy they were achieving with the worm, but they felt that the higher standard was not necessary throughout the genome. Although some money would have to be directed away from existing genome research, other money would be saved on gene searches that would no longer be necessary. Olson of the University of Washington took a similar tack. He continued to back a continued push for advanced technology. But even if such a technology appeared right then, shaking it down it for mass production could take longer than would be necessary for existing methods.

Under the spur of what Eliot Marshall, reporting in *Science*, called "a startling message," intense debate followed. The plan was not adopted. Project leaders wanted more proof that they could sequence long stretches of repeated sequence. They were nervous that sequencing technology was evolving rapidly and wanted costs to be driven still lower. Hence, they favored a three-year pilot project from 1996 to 1999, which would demonstrate high-accuracy sequencing (producing a substantial share of the total at an error rate of 1 in 10,000). The program should concentrate on creating clones in BACs, which were more stable than YACs.[34]

A few months later, in October 1995, Maynard Olson published his take on the project in *Science*. The rapidly improving maps, he thought, were "good enough" to support big-scale sequencing. In his role as consensus philosopher of the genome project, Olson noted that about 15 percent of the genome had markers only 100,000 bases apart. For half the genome, the spacing was a still-respectable 300,000 bases. This was encouraging. Furthermore, he saw a chance to combine the final stages of physical mapping with sequencing. "Healthy competition" had sprung up among several ways to make many copies of stretches of DNA, such as BACs. It was getting a lot easier to pick out stretches of DNA suitable for finding the exact order of their DNA bases.

As for sequencing, Olson, like many others, had expected a breakthrough. The extra funding from the project had been intended to enable "powerful new sequencing tools" to thrust aside the methods of the 1980s. These would include the automated sequencing machines from Applied Biosystems, which made such a splash in 1986 and were used in Venter's triumph with H. flu. Instead, as so often happens in the history of technology, the existing Sanger-based techniques survived by adapting and improving. Incrementally, they simply continued to get more efficient and faster. They successfully defended their turf against apparently brilliant new ideas that, in those days, never seemed to get any closer to practicality. Such technological conservatism had precedents, Olson wrote. Gel electrophoresis for separating molecules, for example, had changed little in forty years and yet had gradually grown universal in fundamental biology. Of sequencing, Olson observed, "The basic approach works in any of several well-tested variations." Improvements would continue to come piecemeal. The time had come to cut back funding for blue-sky sequencing and rely on the tools at hand.

Olson's essay showed that the project's leaders were beginning to appreciate the power of what Leroy Hood called "just figuring out how to get around all the incremental challenges." It had taken NIH years to recognize "that they don't need a new technology," Hunkapiller recalled.

He explained at a workshop, "This isn't a rocket science. . . . They didn't understand how far you could take an existing technology."

To be sure, Olson reflected, there would be costs. With sequences in a swelling flood, electronic sorting through the avalanche of data would burn money. Some useful technological innovation in mapping and gene finding might be delayed. The project would have to be careful not to stifle work that would uncover diseases. Ahead lay the vital task of determining the functions of human genes with the help of the "model organisms." So did the key hopes of cataloguing millions of tiny variations between human beings and "tailoring" drugs to individuals.

Still, to Olson, the potential benefits outweighed the risks. The program could turn to mush if it wavered from its goal of a complete sequence that could match up powerfully with the maps that still dominated the program. Focus on the goal of a finished sequence would sharpen international collaboration. And the expense of waiting to start sequencing, itself a $1 billion item, could add up to another billion in just a few years. The events of recent years had shown that Big Science in biology was working. "Dynamic resource allocation" to focus effort was possible in a system based on scientists' projects reviewed by peers. The process had steered clear of permanent institutions tied to "the past rather than the future."

As always, Olson was fascinated by the finite size of the human genome. All the information for reproducing and developing a human being was contained in "just 750 megabytes of digital information" in eggs or sperm. It was time to produce the single CD-ROM that could contain all this.[35]

No Patenting: The Bermuda Principles

Before a full scale-up to mass production sequencing could begin, an essential issue—the total, prompt, free exchange of data—had to be settled. This was an issue of paramount importance to Sulston, the worm scientist who now headed England's major sequencing center, and his counterpart in St. Louis, Robert Waterston. They persuaded England's Wellcome Trust, whose role in genomics kept increasing, to sponsor a strategy conference. This was held in late February 1996 in Bermuda, which was "suitably neutral ground," Sulston recalled, on "British territory but within easy reach of the United States." The genomics leaders would decide "who would do what and how the data should be managed." Scientists from the United States and colleagues from Britain, France, Germany, and Japan met in a windowless room in the pink-walled Princess Hotel in Bermuda's principal town, Hamilton. The participants, Collins recalled, hardly went outdoors in the gray, late February weather. It was the first of a series of annual conclaves.[36]

Upsetting Waterston and Sulston were the plans of a company called Myriad Genetics (of which Walter Gilbert was a founder) in Salt Lake City to patent two genes that predispose women to breast cancer. This step would enable the company to charge royalties every time a test for the genes was carried out. Sulston and Waterston wanted to head off a parceling out of genes among profit-making companies striking deals with each other. Hoarding data on humans and other mammals could lead not only to vast duplication of effort but also delay battles against specific disease outbreaks.[37]

Big pharmaceutical companies were equally worried. Fear of endless patent tangles led them to pay for putting a great deal of genomic information in the public domain. In September 1994, the giant company Merck announced it was bankrolling a public database of cDNA clones and ESTs, Venter's specialty. Clones manufactured at Columbia University would be shipped to Washington University for sequencing and prompt posting on the Internet. Just days later, Venter's TIGR and its commercial partner, William Haseltine's HGS, announced their own plans for a database that would require advance notice of scientific publication and first refusal on commercial applications. Scientific opinion was divided. The Howard Hughes Medical Institute indicated it would allow its grantees to accept the conditions, but Collins at NIH was firmly opposed and Britain's Wellcome Trust inclined the same way. By 1996, the Merck Gene Index had about as many ESTs as Venter's collection. It was a direct attack on the commercial basis of the TIGR-HGS relationship and may have contributed to the "divorce" of TIGR and HGS in 1997. Venter then gave up years of financial security because he had grown tired of the conflicting pressures of pursuing scientific credit and maintaining commercial secrecy. And so TIGR relied on grants from many sources, including NIH, to plunge further into the world of microbes.[38]

Despite qualms about accuracy, the fifty participants at the Hamilton Princess meeting, many of them from the big centers that would do the bulk of the sequencing, agreed to release all sequences longer than 1,000 bases into the public domain—each day. The sequences, raw or "annotated," would be freely available on accessible electronic databases. There were to be no exceptions. There would be no patenting. This ringing declaration became known as the "Bermuda principles," required of all who sought government or foundation grants for sequencing. Donors agreed to enforce them.

A few years later, Waterston thought the open-release policy had paid practical dividends. The information put up on the Web every twenty-four hours was available to any commercial enterprise. Researchers anywhere, on the track of genetic disease and the drugs to fight them, received stimulating information more quickly. Eleven new disease genes were discovered in the next four years. Thus, the scientific bottleneck was no longer just finding genes but the even more complex challenge of understanding how they work. And information about the sequence grew not just linearly but exponentially.[39]

As the participants contemplated the flood of work ahead, they were united in the goal of maintaining the highest quality. The genome, when finished, would not be quick and dirty. So, it was heartening to hear of new computer programs to count errors precisely, devised by Philip Green of the University of Washington. These products of Phil's intellect received the jocular names of Phred and Phrap. It was possible to assign to every single base a probability that it had been identified correctly, thus adding to the accuracy of sequences. Because the assembly of the deciphered segments called for a painstaking process of sequencing each "cloned" stretch of DNA, the tone was not friendly to shortcuts.

But James Weber of the Marshfield Clinic in Wisconsin made a pitch at the Hamilton Princess for just such a shortcut. He had worked it out with Eugene Myers of the University of Arizona. The human program, Weber argued, should scale up the whole-genome shotgun sequencing that Venter had used so successfully on microbes. Green shot the idea down for the public project. Whole-genome shotgun, he felt, was not ready for a standard of one error in every 10,000 letters. The technique would not meet Collins's target of a sequence that would "stand the test of time." The argument did not end there, however. Within a year, Myers and Weber published their idea, and in the same issue of the journal *Genome Research*, Green attacked it.[40]

After acrimonious sessions in which the major sequencing centers made escalating promises of speed and cost—then retracted those promises—center directors agreed at Bermuda on particular targets that each would meet at benchmark times. The way was open for Collins to start the "full-throttle" phase with a three-year, $60 million project of "pilot" grants to six big centers in the United States.

CHAPTER FOUR

Crisis

Throwing Down the Gauntlet

Craig Venter and Mike Hunkapiller bluntly told the leaders of the Human Genome Project that their world was about to change radically. It was Tuesday, May 12, 1998, the day before the eleventh annual genomics conference at Cold Spring Harbor was to begin. This year some 450 researchers would soon check in at the green campus along the harbor shore, and many would claim their rooms in dorms and cabins. They had come ready to look beyond the achievement of a complete human sequence into the immense challenges of interpreting it.[1]

But they were being wrenched back into the present. A couple of days earlier, the tense group of about forty project leaders had seen press accounts of what was coming. Now they heard directly of the dramatic plan for a private firm, as yet unnamed, that would brush past them to sequence the human genome four years earlier than the worldwide, nonprofit project's deadline of 2005. In the private drive, a principal shortcut would be the "whole-genome shotgun" method of dividing up DNA for analysis, which had never been tried on an organism bigger than a microbe. Another key shortcut was $70 million worth of a new generation of sequencing machines called Prism 3700s, with a list price of $300,000 and a capacity ten times that of their predecessors. The 3700s bid fair to streamline the "reading" of DNA samples drastically. The expected price of this quick-look sequence was around $300 million, the sum committed by Perkin-Elmer, parent of Hunkapiller's operation.

At this pivotal moment of the entire genome project, the research chiefs sat around a great expanse of table in the Plimpton conference room of Cold Spring Harbor's Beckman Laboratory. The room had been the setting for many power meetings just before and during the annual genome conferences that had begun in 1988 at the lab. Windows ran along both sides of the room. One side looked out onto a sailboat-studded Cold Spring Harbor and the other onto the courtyard that unites the Beckman Laboratory with its neighboring visitors' dormitory, Dolan Hall, and the Lita Annenberg Hazen bell tower, each of whose four faces bears near the top a single letter: *A*, *T*, *G*, or *C*.

Most of those in the room were still absorbing the shocking implication that a for-profit enterprise might preempt their own. The previous Sunday, the *New York Times* had published an exclusive article by science reporter Nicholas Wade. The next day, a press conference brought together Francis Collins, director of the National Human Genome Research Institute (NHGRI); Craig Venter, director of The Institute for Genomic Research (TIGR); Harold Varmus, director of the National Institutes of Health (NIH) since 1993; and Ari Patrinos, head of the genome effort at the Department of Energy (DOE). Their comments caused an immediate avalanche of press coverage across the United States and in Europe.

At the May 11 press conference, Collins had made clear that the public project would not throw in the towel. The new venture, he acknowledged, was a significant event, but he dismissed the notion of changing course as "vastly premature." He added, "This is not the moment for genome sequencing centers to go on vacation for a year and a half to see what happens."[2]

Now, in the Plimpton conference room, Venter spoke briefly, and then Hunkapiller described the 3700 machines and their related chemical reagents, which were expected to be available for shipment about one year hence. Venter got a hot reception. David Cox of Stanford University thought he was "unusually well behaved," but Collins recalled Venter as "his usual supremely confident self, bombastic, dismissive of the efforts of anyone else."

Particularly indignant was Robert Waterston of the St. Louis sequencing center, who criticized the technology as "very untested." And there was the matter of instant, open disclosure of sequences. He asked whether Venter would observe the Bermuda data-release principles. When Venter said data would be posted every three months, Waterston exploded, "Quarterly! That's a lot different than overnight."

Eric Lander was particularly scornful of Venter's business plan. Venter planned to build an information company, giving away the raw data but charging fees for access to user-friendly ways of analyzing it. His analogy was Bloomberg, purveyor of up-to-the-minute business information to thousands of terminals around the world. "I just stared in disbelief," Lander recalled. On Bloomberg, information "grew stale with a half life of thirty minutes," and so customers were paying for a constant stream of updates. "The human genome sequence was stable over periods of several hundred thousand years," Lander reflected. "They did not understand their business model."[3]

While Hunkapiller talked, Venter ducked outside to enlist the collaboration of Gerald Rubin of the University of California, leader of the fruit-fly sequencing effort. Rubin's project was going slowly. Venter wanted a highly credible, demanding test bed for the much bigger task of determining the human sequence. He needed a stepping-stone—a way of determining if the whole-genome shotgun method was ready for the big time. *Drosophila* was a good deal more complex an animal than the worm, *C. elegans*, which had been the test bed for the public project's drive toward mapping and then sequencing. Publication of the worm sequence was just months away. Rubin agreed to collaborate and was ready to try the whole-genome shotgun method and the new machines to finish the fruit-fly sequence far sooner than anticipated. Rubin set one condition: The fly sequence must be posted immediately on the Web according to the Bermuda principles and thus completely released into the public domain. Venter agreed.[4]

Notably absent from the group was the first director of the NIH genome project, James Watson. He had already heard what Venter was proposing and concluded that the message was, "We should give up; and I really didn't think that was the appropriate response." At one point in that very tense week, Watson was, as usual, in the Blackford dining hall, where he delighted in asking questions, catching the gossip, and goading people. He spotted Collins and, with his usual penchant for the sharp edge, called out that this was Munich and Venter was Hitler. Would Collins be Chamberlain or Churchill?[5]

At the end of the presentation, Venter, Hunkapiller, and Mark Adams left the room, while the leaders moved on to other agenda items. As they were leaving, Collins calmly remarked, "Well, you've certainly given us a lot to think about." But, according to James Shreeve in his 2004 account, *The Genome War*, Cox and Ari Patrinos of the DOE had a sharper reaction: just what we need, something to shake us up.[6]

The Red Carpet Club

Collins was among the few who had heard from Venter in advance. The previous week, Collins was preparing to fly from Newark to Los Angeles, when Venter asked him to break his journey and meet him in the Red Carpet Club, the business-class lounge of United Airlines at Dulles Airport outside Washington, D.C. Shreeve reports that Venter said, "Francis, I think you need to know about something we're about to announce. We have to meet with you right away." Collins was unsurprised when Venter walked in with close associate Mark Adams. But the presence of a "mystery guest," Mike Hunkapiller, gave Collins a clue about what was brewing. By way of giving Collins a heads-up, Venter said, "We don't think people want to wait another seven years [until 2005] for you to finish the genome." Medical researchers, Venter implied, were avid for a mass of DNA data as soon as possible, even if the sequence had gaps and other imperfections.

He told Collins about the plan to harness some 200 of the new Prism 3700s from Applied Biosystems, Inc. (ABI). The new company, which three months later would be christened Celera, connoting speed, would not patent the sequence itself. Instead it would release its data at intervals, say, every three months. Nonetheless, the new venture would insist on having the first chance to develop possible diagnostic tests and drugs and, if it saw the need, to patent some 300 genes with important functions. More menacingly, the private genome enterprise would not just focus on the human genome but would also push aggressively into the key postsequence task of measuring the tiny differences between individuals that make one person more or less susceptible to a disease. The plan could have been interpreted as a bid to own the future of genomics.

But Venter offhandedly suggested a collaboration. In an interview a year later with Douglas Birch of the *Baltimore Sun*, he recalled saying to Collins, "If you want to consider it a co-project that we jointly lead, I would be willing to accept that." The public project could concentrate on its "finished" sequence, filling up the many gaps that an early draft would leave. To keep the scientists of the public project busy, Venter added, his group could concentrate on the human sequence, and Collins's colleagues could go after the mouse, which he asserted was "just as important for research and biomedicine as the human." Collins told Birch that he sat speechless: "I was stunned." Shreeve reports that Collins felt he had been "kicked hard from behind." Venter had a different, more combative recollection: "He just thought that my proposal was the height of arrogance. His view was that he was already the head of the Human Genome Project. He didn't need to share his crown with me."

After Venter left, Collins and Hunkapiller, both imbued with the habits of a long-distance runner, boarded a California flight and had a reassuringly cordial five-hour chat. According to Venter's 2007 autobiography, Hunkapiller agreed to supply the same new sequencers to the nonprofit genome program. A second large market was opening.[7]

"Numerous Mysteries from the Methodological Point of View"

Comments began reverberating around the world. Predicting that the public project would lose the race to sequence the human genome first was Venter's spurned former associate, William Haseltine of Human Genome Sciences, which was doing a lot of its own sequencing. Denouncing the public project as "a gravy train," he said, "This has to feel like a bomb dropped on the head of the Human Genome Project. All of a sudden somebody is going to pull a $3 billion rug out from under you? They must be deeply shocked." Swatting Venter as well, he delightedly pronounced the Perkin-Elmer venture to be not a business but an altruistic "turn-of-the-century gift to science." He told *Business*

Week that the situation was "a tremendous black eye for the government." But, he added, "The Human Genome Project has never been a commercial venture."

Nonetheless, he told *Science*, "I'm sure Craig can do it . . . and it will be a great thing for science." Haseltine went on to write an op-ed piece for the *New York Times* arguing that the federally sponsored project might not add much knowledge to the privately generated sequence. He said the effort of decoding the vast amount of "junk" DNA made "little sense." Because the priority task was finding medical uses for the genes—the nonjunk DNA—NIH should go back to funding individual research grants. He called for the end of "the era of government-sponsored big science."[8]

Another sequencing rival, Incyte Pharmaceuticals of Palo Alto, California, was incensed. Its chief executive denounced Venter's plan as a "publicity stunt" and asserted that being first with the human sequence mattered little compared to linking the DNA to specific diseases and potential drugs. Incyte's president, Randal W. Scott, said, "What they're describing is not a commercial venture. It's really Craig Venter going after the Nobel Prize for sequencing the genome."[9]

Others were skeptical that Venter could pull it off. The warrior of error detection, Phil Green of the University of Washington, remarked gloomily, "It's a million times harder to put the human genome together than a bacterial genome." He reiterated his forecast of three years earlier that the whole-genome shotgun method wouldn't work on the human sequence because of its high proportion—75 percent—of stretches of DNA "repeats." Some of these numerous repeats are two to five bases long, some span six to fifty bases, and others measure several hundred. To be sure, many of the repeats are concentrated in the centers (centromeres) and tips (telomeres) of chromosomes. But those trying to assemble tiny fragments into a true sequence would face serious trouble doping out where a particular fragment was to go.[10]

Jean Weissenbach, the highly respected head of France's principal sequencing center at Evry, near Paris, was impressed by the plan to pull together in one place 230 machines with ten times the capacity of their predecessors. But he doubted that the new commercial enterprise could attain its goals in terms of qualitative and scientific value. He did not believe that the gaps in the sequence would be as short or as few as Venter claimed. There were "numerous mysteries from the methodological point of view," he told *Le Monde* on June 3. Weissenbach pointedly mentioned plans already brewing for nonprofit centers to "modify the initial plan" of finishing in 2005.[11]

Waterston continued his criticism. He told Wade of the *New York Times* that the Venter plan would "significantly" compromise quality, creating "an encyclopedia ripped to shreds and scattered on the floor." He told another reporter, "It strikes me that this is a cream-skimming approach. It's clearly an attempt to short-circuit the hard problems and defer them to the [research] community at a very substantial cost." The private venture, Waterston told a third reporter, posed the threat that lobbying would sap the will in Congress to support the public project and derange the research: "People will do the wrong experiments. Computer programs won't work as well. It's like reading a book with sentences missing and chapters rearranged."[12]

Venter was not shy about firing back. A master of the sound bite, he said that his projected sequence would be more valuable to medicine than the public project's because he was using DNA from only a few individuals (who, he announced later, included him) rather than many. He told the *New York Times* that NIH was "putting humanity in a Waring blender and coming up with a patchwork quilt."[13]

Several news reports fastened onto the notion that the international nonprofit project might now be superfluous. Venter's defiance of a vast scientific establishment evidently fascinated the press, as it had for most of the decade. One reporter wrote, "The audacious bad boy of genetic research is at it again."[14]

Quality and Quantity

The announcement of the new private venture occurred at a moment of both confidence and embarrassment for the international project, whose starting gun had been fired more than seven years earlier. At the beginning of May 1998, in the journal *Genome Research*, published at Cold Spring Harbor, Maynard Olson and Phil Green had enunciated "a 'quality first' credo for the Human Genome Project." They observed that the project was "lurching toward large-scale genomic sequencing" but faced "sobering uncertainties" over costs, rates, and quality.

Alluding to a favorite theme of Olson's, he and Green wrote, "Of course, it is these very uncertainties that make the sequencing of the human genome a scientific and managerial challenge worthy of the committed attention of many hard-working and talented people." But the quality of the sequence had to be good enough for a vast array of biologists and medical researchers to use it. The two stuck by the standard of no more than one error in 10,000 bases. They also demanded validation of the segments as accurate and contiguous and, through rigorous tests, proof of the final "assembly" of the whole sequence. To achieve this, "strong genome centers" would have to "implement well-chosen production methods on a large scale," while controlling costs. And all of this required immediate adoption of "stringent quality standards" in order to generate a sequence that "will meet the test of time."[15]

This essay appeared when it was not yet clear that the 1996 commitment to scaling up the sequencing of the human genome with refined first-generation sequencing technology could be met. The more than half a dozen competing U.S. sequencing centers had been thrust onto a rocky path. In 1997 and 1998, struggling to be included in the final round, the centers failed to sequence their promised numbers of DNA bases and could not cut the price per base as much as hoped. Lovingly adopted robots were not boosting efficiency as much as expected. After two years of confronting the disciplines of industrial assembly lines, the centers ruefully acknowledged to Elizabeth Pennisi, a reporter for *Science* magazine, how steep the slope up to the human genome seemed.[16]

After spending $1.5 billion or more, centers in the United States and overseas had produced a "finished" sequence of only 3 percent of human DNA, 150 million bases, although they also had another 12 percent in draft. They were still some distance from the price of twenty cents per base that Collins had decreed was essential for the assault on the summit. To be sure, the cost had shrunk from as much as $5 down to a two-year average of around $1 and, in some of the most recent runs, to fifty cents. But there was that 2005 deadline for getting a "finished" sequence.

Would the technologies on which the global project was relying meet that target? A sour Craig Venter, who attended part of the Bermuda meeting at which the targets were set, had begun acidly calling the center directors "the Liars Club."[17]

The centers' work was financed through the $60 million pilot program that Collins had launched in 1996. In announcing the program, Collins had said, "Current technology is good enough for starters but needs significant improvement if we are to reach our goal on time." The idea was for the centers to concentrate on different human chromosomes as they streamlined the successive stages of sequencing with such things as robot arms and new chemicals. Pushing to do more and more bases every hour, they would strive to slash the cost per DNA letter well below the current figures, while staying under the error rate of 1 in 10,000 in the data they rapidly released in accordance with the Bermuda principles. The first grants went to Mark Adams of Venter's TIGR, Richard Gibbs of Baylor College of Medicine in Houston, Eric Lander of the Whitehead Institute at MIT, Richard Myers of Stanford University, Maynard Olson of the University of Washington, and Robert Waterston of Washington University in St. Louis.

Collins insisted on weekly conference calls involving the leaders of the five big centers in the consultations they needed to goad each other on—and stay on the same page. He set the agenda and took notes. A few years later, George Weinstock, then of the Baylor sequencing center, described how Collins managed the process. "Francis has a manner of presenting things in a way that makes everybody feel good. There is always some tension, some fear that your turf might be allocated to someone with more capacity. Francis has a very soft touch. He can be very understanding, and he can look at both sides of the situation and make people feel that even if the decision that is being made is not necessarily the best one for them, it's certainly the best one for the project." Weinstock added, "While he may have his own needs in terms of how to move the project along, he handles all that behind the scenes and allows the groups to interact with each other in a way that builds relationships that can keep the project moving forward."[18]

Two months later, in June 1996, $13 million in pilot grants went to nine more researchers, including James Weber of the Marshfield Clinic, six at universities, and two at the companies GeneSys Technologies and CuraGen. The focus was developing technology to miniaturize, accelerate, and automate genome research, not only to achieve the immediate goal of the first human sequence but to help biology and medicine for many years more. The work represented a gigantic effort by biologists, to whom the notion of automation for big-scale processing was unfamiliar, even alien. Collins said briskly, "Current technology is good enough for starters but needs significant improvement if we are to reach our goal on time." His associate, Carol Dahl, added, "We want these investigators to follow a DNA sample from the very beginning of the process to the very end and come up with ways to make the whole process work faster and cheaper."

Eventually, considerable headway would be made. When the pilot projects ended in 1999, Waterston noted a year later, "accurate, gap-free sequence extending over several millions of bases could be generated in a cost-effective manner." He told a radio audience that when his group began processing samples, they were lucky to do forty a week. Now the number was 40,000.[19]

Despite this sunny recollection, the stark fact in the first half of 1998, when Venter laid down his challenge, was that the public project appeared stalled. Observers like Watson and Norton Zinder of Rockefeller University, the first chairman of Watson's genome advisors, wondered with characteristic waspishness if the project schedule was designed to carry Francis Collins to his retirement. Aware of his budget projections, they said Collins wasn't asking Congress for enough money. Watson, a master of gossip, kept in touch with the program and its supporters in Congress. In January 1998, according to reporter Richard Preston, Watson and Zinder got word of a confidential projection in Collins's office of a steady sequencing budget of $60 million a year up to 2005, when perhaps half the sequence would have been determined. Watson hurried to New York to hear Eric Lander deliver the prestigious Harvey Lecture at Rockefeller University, then to meet him and Zinder afterward in the university's faculty club.

Over cocktails, with the impatience of a man who has undergone quadruple bypass surgery and is in treatment for cancer, Zinder (who soon would become an advisor to Venter) pounded on the table and declared, "This thing is potchkeeing along, going nowhere!" The idea was to persuade Lander to get Collins to push more money into the sequencing effort. Lander rejoined that the problem was not money but organization. Perceiving the public genome effort as "too bloody complicated, with too many groups," Lander wanted to lead an elite group of sequencing shock troops. Zinder remained discouraged. He told Preston, "I had essentially given up seeing the human genome in my lifetime."[20]

A second factor in the lumbering progress of the public project arose from an ethical concern: Whose DNA would be used in mapping and sequencing? The earlier stages had involved building vast artificial clone libraries of DNA from workers in the laboratories of Pieter de Jong, then at the Roswell Park Cancer Institute in Buffalo, New York, and Melvin Simon of Caltech. The labs had developed techniques to store DNA in artificial chromosomes made either from a virus called P1 (PACs) or a bacterium (BACs), thereby avoiding some of the problems with using yeast artificial chromosomes (YACs). The researchers, including de Jong and Hiraoki Shizuya, had donated their own DNA. De Jong recalled long afterward that he did not feel like an "elitist." Grant money had to be stretched and the risk of lawsuits averted. But the issue of "whose DNA" crystallized soon after the 1996 Bermuda strategy session. Were the donors capable of giving "informed consent" for the use of their own DNA? Was it advisable to have an identifiable individual as the subject for humanity's first DNA sequence? How would the public react? As center directors like Lander wondered whether the situation was "all right," Collins focused on the issue and convened a meeting on the ethical implications of using the DNA of a few, identifiable people.

The upshot was that the libraries would have to be rebuilt from scratch, at a cost of months at least. The DNA would come from sources whose anonymity was thoroughly protected. And so, an advertisement appeared in the *Buffalo Evening News* seeking anonymous donors. Many responded. DNA was taken from some of these, whose identities were hidden from the researchers and themselves. The DNA clones were rebuilt and laboriously arranged in new libraries. The delay of as much as a year and a half may have opened a window of time during which Venter's impatience intensified. His Maryland lab, TIGR, despite successive triumphs in sequencing disease microbes, was at that time playing only a small role in the Human Genome Project. Perhaps another route to genomic preeminence existed.[21]

A different set of problems that had already emerged during the mapping phase were now becoming even more critical, especially to Lander. The genome centers were still mostly in a highly manual mode of handling DNA samples. The current system of funding genome laboratories discouraged investment in new technology. Furthermore, in what Lander regarded as "an important process failure," the labs were still claiming chromosomal territory, which they intended both to map in great detail and then to sequence. To Lander, "this whole business of signing up for individual territories was madness." If a particular center went too slowly, it could hold up the entire project—just as the speed of the pokiest vessel governs the pace of a wartime convoy of ships. In that situation, "everything's below scale. Everything depends on everything. Bad idea," Lander recalled a decade later. In sometimes heated discussions with colleagues he respected, Lander argued that mapping and sequencing should be separated and that a few centers should operate on a greatly enlarged scale.[22]

The Capillary Sequencer

In the midst of this worrisome situation, the new Prism 3700s reached advanced development. Hunkapiller at ABI and his rivals had kept improving their original machines of the early 1980s. The original 377 model that Venter tested in 1987 evolved into improved machines, with total sales of 6,000 over the next decade to laboratories around the world. It was these first-generation machines that automated Frederick Sanger's "chain-termination" method, substituting chemical dyes, one for each nucleotide, for the radioactive labels Sanger had used in the 1970s. A detector could "read" the dyes to tell each base apart, in order, on gel films pressed between two plates of glass. The DNA fragments distributed themselves by size, over some four hours, under the impulse of a gentle electric current. Despite this smoothing of a toilsome, cumbersome process, the work had to be interrupted after every run to replace the slabs of gel.

Escaping the gel technology demanded a decade of struggle in many laboratories, notably at ABI, which Perkin-Elmer acquired in 1993. It was a classic case of the race between the tortoise, represented by Sanger methods, and the hare, representing those driving for radically different technologies, which would not reach maturity until the middle of the first decade of the twenty-first century. So for more than two decades, it was a matter of enhancing the existing technologies, rapidly pushing down a "learning curve" toward greater speed and lower cost per unit of effort. This process of tweaking capabilities went along at least as fast as the most famous example in the history of technology: the seemingly endless fulfillment of Gordon Moore's prediction in 1965 that the number of electronic components on a chip of silicon would double every eighteen months. The process went on and on, decade after decade, while alternate technologies, such as chips based on gallium arsenide, waited in the wings.

At last the makers of DNA sequencers found an answer: a bundle of polymer-filled microtubes, called capillaries, through which the DNA samples would flow at about twice the pace of the "slab gel" technique. Each capillary was about the diameter of a human hair. The improved machines could operate around the clock. Another sweet feature was that reagent chemicals needed changing only once a day. Now, as the public project struggled forward, the 3700s would be disclosed to a world thrilled with the potential profits of biotechnology and genomics.[23]

Well before Venter's startling announcement in 1998, Hunkapiller had been forced to think about customers. Perhaps driven in part by the relatively low ABI stock price in the dot.com boom environment of the late 1990s, he began discussing this with the dynamic Tony White, who, as chief executive of Perkin-Elmer, was shifting the company entirely into biotechnology. At a business meeting, Hunkapiller and White were joined by the Lebanese-born Noubar Afeyan, whose own company had just been purchased by Perkin-Elmer. According to James Shreeve, Afeyan told his colleagues, "You know, with enough of these machines, we could sequence the whole human genome." White's reaction was, "Let's just do it." Their idea to set up a new Perkin-Elmer subsidiary to buy lots of 3700s and sequence the whole genome was born. At that point, some observers think, there was no thought of a bidding war with the public project. But who would run the new enterprise? Hunkapiller thought of Venter, who had been one of the first purchasers of an ABI machine in the late 1980s.

Hunkapiller phoned to invite Venter out to California to see a machine fifty times faster than any at TIGR. Venter exclaimed, "You guys are certifiable. You're crazy, but I'll come listen." Venter took one look at the new Prism 3700, about the size of a hotel-room minibar, as one newspaper later remarked, and knew it could put him on a much faster track to the genome. He also knew that the computer work needed to reassemble the machine's output of 60 million DNA fragments, each 500 bases long, into a continuous sequence was stupendous. But, as Hamilton Smith remarked, Venter was a character ready to do a high dive into a pool with no water. Zinder, by now an advisor to Venter, went out to California also and exclaimed, "We've got the genome!"[24]

"The Bill Gates of Biology"?

To the leaders of the public project, Venter had all the charm of a pirate ordering a prisoner to walk the plank. They bridled at the suggestion that they could gratify their wish for a gold-standard sequence—with the mouse! Glen Evans, head of the sequencing center at the University of Texas Southwestern Medical Center in Dallas, voiced the general feeling: "We don't want Craig Venter to become the Bill Gates of biology." The public program would just have to produce data faster than Venter. Rick Myers of Stanford agreed, "It would be horrible if [the project] gets derailed," but the new company "will help galvanize us and [make us] work together."[25]

Although he both relished and lamented having a crowd of perceived enemies, Venter did not imagine the fury and determination he would excite. He did not expect that his galumphing opponents would insist on matching him, headline for headline, milestone for milestone—fighting to overtake and surpass the new deadlines Venter had laid down.

Leaders like Lander knew at once that they must hit back—and soon. Indeed, they knew that they would have to fight to retain the support of their patrons in Congress. Among these were two Republicans who headed the relevant appropriations subcommittees, Senator Arlen Specter of Pennsylvania and Rep. John Porter of Illinois. They had joined forces to double the budget of NIH over five years. After all, this was an era of rampant private enterprise and high-tech booms on Wall Street. A market-oriented Congress might love the idea of turning the slow, expensive genome project over to a for-profit company.

There was danger that the information embodied in the DNA sequence, which philosophically belonged to every human being, might be locked up in a horde of little fortresses with walls built of patents. It was to keep the genome in the public domain that the global consortium was posting its new sequence data on the Internet every twenty-four hours. The genome project researchers were determined that the tradition of openness and data sharing, built up in the communities investigating the bacterium and its viruses, yeast, and the worm, would be upheld. Commercialization must not be allowed to inhibit the free exchange of information that would be essential, in years to come, to elucidating the elements of how a living cell works in both health and sickness. Without openness, the insights of genomics would be hindered in providing the insights needed to prevent, delay, palliate, or cure diseases.

"Arguing This Thing Through"

These dark reflections took a dramatic turn a couple of days after the big Cold Spring Harbor conference began. John Sulston, director of the British sequencing center outside Cambridge and an apostle of a public-domain genome, and Michael Morgan, an executive of the huge Wellcome Trust, made an emergency flight across the Atlantic to the genome conference. Both were indignant: Sulston denounced Venter for "cherry-picking." He added, "I really don't see this as being any great advance whatever. We are going to provide the complete archival product and not an intermediate, transitory version of it." Morgan fumed, "To leave this to a private company, which has to make money, seems to me completely and utterly stupid."

He and Sulston brought news of an extremely aggressive and immediate response to Venter's initiative: The Wellcome Trust, perhaps the largest private foundation in the world at the time, would double its commitment to the work at Sulston's center, to the equivalent of $325 million, thus enabling it to take on one-third of the entire public project's sequencing. They dropped hints that they would, if they had to, finance half the project. It seemed clear that the Wellcome Trust intended to stiffen the spine of the U.S. program and to stifle any tendency to leave the human genome to Venter. When this commitment was announced at the conference, the shaken audience, mostly consisting of scientists who had bet their careers on genomics, rose in cheers.[26]

As the heads of major genome centers began thrashing out what to do next, it became evident at once that there were at least two schools of thought: Respond to Venter promptly or stay the course. Waterston from St. Louis and Gibbs from Houston joined Sulston, Lander, and Collins for dinner in an Italian restaurant in downtown Huntington a few miles from the Cold Spring Harbor lab. With typical insistence, Lander began months of lobbying for a riposte to Venter that would recapture the headlines. But to Sulston and others, including Maynard Olson, this plan looked like a diversion. The energy required to generate a finished, highest-quality human DNA sequence could be dissipated. The argument would last for months.

Furious at the threat of a patent-fractured genome, Sulston told *Nature*, "What the new company is proposing to do is an incomplete job. We know that the sequence it plans to produce will have gaps in it; they may be deferring some of the costs involved in obtaining a complete sequence, but the result is that what they will produce has less power of interpretation."

He wanted to press ahead with the complete sequence—Venter be damned. But Lander reminded Sulston of a fundamental difference between the British effort, supported by a private charity able to make long-term commitments, and the American program, dependent on a Congress acutely sensitive to public opinion and only able to appropriate funds annually. Staying the course was not politically viable. "We had to do that which was going to be understandable and simple," Lander said later.

Lander remembered these conversations as "a bunch of intellectually honest but emotionally passionate people arguing this thing through."

In such a face-off between the leaders of what Lander recalled as "the hot-headed places," a mediator was needed. Often, it was Waterston who "played the sensible peacemaker and intellectual broker."

Influenced by his experience with both journalism and business, Lander was worried that the public project would not "survive in the sound-bite world of media." As in a political campaign, the project would face a barrage of claims of success, "news flow," that would have to be countered immediately to show that "we weren't bumbling fools here." And the key to the strategy was to "rearrange our priorities," doing a draft early and finishing later. Lander argued, "That was going to be politically viable. And, it didn't hurt the science in the long run, because we had the commitment and the stamina to stay the course."

In effect leaders from both sides of the Atlantic were moving toward an oath in blood to do a draft and then go right on to finish. In Lander's recollection, "We were not abandoning our principles."[27]

Varmus Fires Back

Soon after the Cold Spring Harbor meeting, these urgent matters sharpened the focus of a two-day meeting in Virginia to go over the NHGRI's next five-year plan. According to a report in the scientific journal *Nature*, the idea of a "rough draft" by 2001, while still finishing in 2005 as anticipated, was taking hold. The time had come for rapid feasibility studies of scaling up sequencing tenfold to 500 million bases a year in five years. The project would continue using its cautious "clone-by-clone" method of sequencing. A draft, Collins said at the Virginia meeting, would speed the placing of "very useful" sequence in the hands of "the people who want it." Yale researcher Robert Lifton said, "More [sequence] would be better, sooner rather than later." An attendee, Lee Hood, exclaimed, "From the point of view of identifying genes, [the rough draft] would be absolutely fantastic."

There was talk of how the public and private efforts could complement each other so that "everybody wins." Hunkapiller and Mark Adams were both at the meeting. After close questions about Celera's data-release and patenting policies, Hunkapiller said it would be "absurd" for the rival projects not to collaborate.

Collins, however, expressed worry that the project was "massively undercapitalized." Would Congress come up with the money for a swift doubling of funds? After all, there were open calls to end federal funding.[28]

Work to counter such ideas had already begun. Wade's articles had not only suggested that the public project might be superfluous but that its NIH sponsors had waffled about staying in the human field. In reply, NIH director Harold Varmus fired an early round. To the notion that "this task is a *fait accompli* and the publicly financed Human Genome Project may not be needed," Varmus wrote to the *New York Times*, "This is not the case." He said it would not be known for eighteen months whether Venter's approach would work; nor was it established how much access would be granted to the new venture's data. "Sequencing experts have significant reservations about whether the 'rough draft' this private group aims to produce will meet the needs of the scientific community and the public." He added that the public research, which involved far more than the human genome, had met its deadlines from the beginning and, indeed, had made Venter's effort possible.[29]

Congress Hears the Rivals Out

Friendly voices in Congress surfaced almost immediately, such as a spokesman for Rep. Porter: "I don't think that there are many people in Congress who are looking at this and saying, 'Gee, this is great. The private sector is going to do all the work on unlocking the genetic code. Let's move on.'" Support from Porter and colleagues, the spokesman said, remained "very, very strong. There continues to be a great deal of interest in having federally supported research in this area."[30]

Arrangements began early for a hearing in the House of Representatives that would air the views of supporters and detractors of the government-sponsored genome project—and demonstrate that Congress was still behind the work they had been supporting. As happens often in big technology projects, a hearing before Congress does not make policy but instead spells out a consensus. It serves as a benchmark of facts and what people think about them.

As Rep. Ken Calvert of Riverside County, California, the Republican chairman of the energy subcommittee of the House Science Committee, opened the hearing on June 16, 1998, he made it clear that the principal subject was "just how" the Perkin-Elmer venture should affect U.S. government genome efforts, on which $1.9 billion had already been spent. Evidently reflecting considerable preparation, Calvert asked four penetrating questions: "Are the goals of this initiative realistic or just an optimistic vision? Will this private sector initiative duplicate the federal program and make it redundant or is it another approach that can complement the federal program and make it stronger? Is the pace and the cost of the federal program increased by the bureaucratic nature of any federal program or does the timetable and cost reflect what is necessary to do a thorough job? And will the federal program utilize the latest technology described in the private sector announcement?"

The big guns rolled up as witnesses: Venter, Collins, the present and former heads of DOE's genome program, and, showing up as usual at a critical time, Maynard Olson. It would be a confrontation, admixed with lip-biting cordiality—but it also showed how fast things were moving.

Venter testified first and immediately set a conciliatory tone. Of the notion that private-sector efforts would make the Human Genome Project irrelevant, he said, "Nothing could be further from the truth. The Human Genome Project is truly a success that both the scientific community and the federal government can look upon with pride, [and] which will continue to generate important information." The genome project, he added, was "a huge boost to the scientific community" and was leading to radical change in the way drug companies searched for new medications.

He reiterated his proposal of a collaboration: "I hope after today you will recognize the success of the program that you have funded, and also recognize the vast potential to improve human health that lies just around the corner by linking both the federally-funded initiative and our new private-sector venture." He said his "sincere hope [was] that this [private] program complements the broader scientific efforts." He urged Congress to focus on "how these two activities working in tandem can ultimately improve our lives."

Nonetheless, he accused leaders in biology and medicine of creating the false impression "that sequencing the human genome was the key to improving our knowledge of human biology. This statement has led many to believe that obtaining the complete human DNA sequence would mark the end of the project. In fact, the acquisition of the sequence is only the beginning." So, he said that his new company, which he regarded as a recognition by the private sector of the importance of genomics, would help to "get everyone to the starting line a little bit sooner."

He cited events from his own career that had led up to the as yet unnamed new venture. For Venter, as for so many others, finding genes was crucial but had been slow. While a brain researcher at NIH, he had spent ten years finding the gene for the adrenaline receptor. In the late 1980s, he recalled, he had made NIH's first cooperative research agreement with industry, working with ABI on ways to speed up the sequencing process. It was in this collaboration, he said, that he had used the expressed-sequence-tag (EST) method to identify genes more quickly. He expressed disappointment at being rebuffed (by Watson and an NIH review committee) when he proposed that NIH's "intramural" staff scale up sequencing. He laid the blame on researchers, primarily from outside NIH, who make up the majority of such "study sections." According to Venter, "The extramural genome community did not want genome funding being used on intramural programs."

After leaving NIH for TIGR, Venter recalled, his continued leadership of EST work resulted in labeling "more than half of the genes in the human genome" for a special 1995 supplement to *Nature*. In addition, the whole-genome shotgun method had eased the sequencing of a series of microbes that caused diseases or survived in extreme environments. Although his laboratory had been assigned to sequence work on chromosome 16, he continued to find non-NIH support for other projects. A new program to sequence the ends of BACs, sponsored by DOE, would provide "a framework for linking the human genome together."

Venter repeated his vision of a knowledge company, which would make its money packaging data and selling multimillion-dollar data-access subscriptions, rather than developing drugs, and operating with a minimum of proprietary secrecy. Its product would not be just "a highly accurate, ordered sequence that spans more than 99.9% of the human genome," Venter said. "The goal is to develop the definitive resource of genomic and associated medical information." Releasing primary sequence data to the public every three months, making the genome "unpatentable," the company would earn its way by fees for access to the complete database and its related software. But the company would make a focused effort to find "important structures," significant information on what a particular gene did, and patent up to 300 of those.

Maynard Olson went on the attack: "The Human Genome Project has come a long ways since its fragile beginnings a decade ago." Biological and medical researchers could assume that "complete genome sequences will soon be available for all intensively studied organisms." The human sequence, when completed, would fit on a single compact disk. Comparing the genes of the different species already enabled researchers to probe more deeply into the mechanisms of disease. Genome-related industries were thriving. Using both private and public databases, drug companies were redirecting their work to exploit new genomic knowledge.

In Olson's opinion, the academic community not only sought a high-quality "reference sequence" but was also "responsible for the critical task of training a growing cohort of young scientists who can lead genome analysis into its open-ended future."

Early support from Congress, including Rep. Claude Pepper of Florida's patronage of the National Center for Biotechnology Information, was paying off. Indeed, Olson said, "Congress was actually ahead of the majority of scientists in recognizing that it was time to move boldly to create an information-based future for biomedical research." And now, "in this turbulent environment . . . the system is working."

The commercial risks for Perkin-Elmer were large, Olson said. "It remains unclear how the company will recover its substantial investment." He disparaged the "well-orchestrated" publicity campaign for the new venture: "What we have at the moment is neither new technology nor even new scientific activity: what we have is a press release. I believe that I speak for many academic spectators when I say that I look forward to a transition from plans to reality. In short, 'Show me the data.'"

Olson hoped that the commercial project would succeed and make its data "sufficiently accessible to the scientific community" so that biomedical science could proceed more rapidly between 2000 and 2003. But he predicted that this contribution would be "transient." Olson believed that the effort "will fail to produce a sequence of the human genome that will meet the long-term needs of the scientific community." He foresaw "catastrophic problems." There would be 100,000 gaps in the draft. These would leave "uncertainty even as to how to orient and align the islands of assembled sequence between the gaps." And these islands could be "misassembled," that is, they could fail to reflect "the organization of the human genome." Olson asserted that "large amounts of low-quality sequence data are a poor substitute for smaller amounts of high-quality data." To ensure a sequence that "will meet the test of time," Olson said, "it is of the utmost importance that a vigorous public effort be maintained."

Lamenting the idea of Congress mediating a technical dispute, he urged legislators to wait and watch. "Congress is the wrong forum in which to debate the relative merits of capillary-gel electrophoresis versus slab-gel electrophoresis, whole-genome 'shotgun' sampling versus a clone-by-clone approach."

Addressing the committee, Collins also stressed quality and permanence. He was proud that in less than ten years the cost of a single base of sequence had fallen twentyfold, from $10 to fifty cents. He was also proud of the "clone-by-clone" strategy, which catalogued only the stretches of DNA whose location had been "carefully mapped." He said, "You know where the fragment came from." This would make it much easier to reassemble the fragments "back into sentences, paragraphs, chapters, and books." In contrast, Celera's "whole-genome shotgun" procedure would produce fragments that had not been "previously mapped or catalogued prior to sequencing."

Collins continued to find disturbing the prospect of having to wait eighteen months to see if the Celera approach, validated by the fruit-fly test bed, would work. He said, "The industry effort may not deliver the product in the time and manner proposed. The industry approach to sequencing has not been tried on large and complex genomes, such as the human, and depends on newly developed and unproven machines."

David Galas, then working in a small biotechnology company in Seattle, placed himself between supporters and detractors of Venter's enterprise. Earlier he had headed DOE's genome program. He was friendlier to the Celera concept than Olson. He called it "a galvanizing prospect to the entire community." The claims made in the May announcement were "both credible in detail and most welcome and judging from my familiarity within the field, are probably well within reach. . . . Their proposal appears to be well founded and plausible."

Hence, the public project would have to "reform" its strategy and "refocus" its effort, reorienting itself to a new primary goal, a "first draft," as quickly as possible. Then, the project would move on to its "very accurate, high-quality, complete reference sequence of the genome." Government projects should make maximum use of developments in the private sector. Both NIH and DOE realized, Galas said, that neither could "continue unchanged with the strategy currently in effect, [nor could they] reduce the level of their efforts." He endorsed the steps they had already taken to revise their sequencing strategies and their plans to take advantage of the new privately generated data. He believed that "the private sector sequencing effort has served as a beneficial stimulus to refocus the federal effort."

Speed was necessary, Galas said, so that genomics could start sooner to investigate the millions of tiny genetic variations between people, sources of disease risk. Hence, "even a rough 'first draft' would be absolutely invaluable to the broad biomedical community." The drafts would speed the production of "a very accurate, high-quality, complete reference sequence of the genome," which he said would "become the single most important database of human biology." To achieve it, Galas advocated more money for the NIH and DOE projects. It was a brisk and balanced view of an uncomfortable situation. His presentation neatly summed up what would happen over the next two years to bring a full human genome sequence into existence far earlier than expected.[31]

Raising the Stakes, Twice

Soon after the hearing, Rep. Porter, the Illinois Republican chairman of the crucial appropriations subcommittee, gave the decisive response. He said he saw no reason to "change our funding."[32]

Porter's comment underlined the decisive nature of the spring and summer of 1998 in the history of genomics. Craig Venter had contributed mightily to the result in that their perception of insult in the way he presented his plans galvanized the public-project scientists to beat him at his own game, the essence of which was speed, not quality, and then to keep going on the path that had been charted.

The public project, determined to seize back the initiative, set up a committee of highly competitive center directors, which, after much debate, settled on how to proceed. They gulped and agreed to the maximum scope and pace of industrialization. Because Venter's proposal rested on a sequencing strategy that had not been tried on a large animal and a new generation of machines not yet on the market, it soon became clear to Congress that its support for a broad-front genome program would have to continue and—in the climate in which NIH's budget was doubling—increase.

Venter evidently thought that by creating an enterprise centered on the will of one person, he could sweep past the efforts of a committee. But, spurred by anger and refusal to be outdone, the public project spent the summer of 1998 hammering out a plan to finish the sequence two years early (2003) and produce a draft—all that Venter contemplated—two years before that (2001). For the third year of the 1996 pilot program, seven American centers received a further commitment totaling $60 million, along with the directive to study the details of the accelerated scale-up. Besides Lander, Waterston, and Gibbs, grants went to Glen Evans at the University of Texas in Dallas, Richard Myers at Stanford, Bruce Roe at the University of Oklahoma, and Maynard Olson at the University of Washington.[33]

Collins and his colleagues, still running scared despite Congress's reinforced backing, raced ahead to a major speedup, in consultation with DOE. Over the next two months, as directed, the major NIH-supported centers studied the idea of a crash program to produce a draft by 2001. These studies amounted to what the military calls a war game, fueled by a passion to be first and to keep the genome from being patented. Collins was convinced that if the genomic highway had too many tollbooths, scientists would get off it. He believed that patents should only cover uses, not the sequences themselves: "We need to reward people who discover function as opposed to those people who turn the crank."[34]

To prepare for battle, they analyzed possible shortcuts, such as reducing "the depth of coverage," the number of times a stretch of DNA needed to be sequenced to assure accuracy. Another shortcut would be to do some steps simultaneously rather than sequentially. Fast and furiously, initially with great skepticism, the heads of the major genome centers went over the details of a program to catch up and surpass, without junking their ultimate goal of a best-quality sequence that would be "the central organizing principle for human genetics in the next century . . . an essential reference for all biologists."[35]

Lander recalled a meeting at the Baylor College of Medicine just before the decisions were announced. The night before the meeting, he said that Collins was "very much on the fence, because he was getting buffeted between Sulston on the one hand and me on the other hand." A major concern still was whether the will to finish "a high-quality product" was there. "Maynard Olson was concerned we'd never finish. Nobody would have the stamina for finishing." But Sulston, Waterston, and Lander all agreed, "We were going to finish this." A major motive was pride. "We had always signed on to do a finished sequence of the human genome." There was going to be "a genome of record," even if "the paper of record," the *New York Times*, did not understand this. The plan was "to address the [Celera] hype by simply rearranging the order of our work to get a draft out early." He and the others in the public project were not "going to sell our soul."[36]

So the public project would leap forward to a draft that would speed gene finding. Some ten days later in September, their conclusions received official approval from the NHGRI and were announced. In the background, Watson lobbied in Congress for the extra money that would be needed.

The public project would stick to its step-by-step approach of using clones whose positions were known, but it would produce the "finished" sequence in 2003, two years ahead of the previous schedule. About a third of this would be done in 2001, when the public project would also release a "working draft" covering 90 percent of the genome. The Bermuda principles of free access within twenty-four hours would be followed. Collins told Rep. Calvert's subcommittee, which formally approved the new course, "All of this stir in the private sector has opened people's eyes to the possibilities. . . . This is not a reaction. It is action. Every year that goes by without finishing this sequence is an opportunity lost for biomedical science. This is not a time to be cautious and conservative and coast along."[37]

One influence on the public project's planning was a "field trip" to a highly automated factory where Motorola produced cell phones. The trip was arranged by Gerry Rubin, who was impressed with the help Motorola had given him in linking the scattered players in the drive to a sequence for the fruit fly, *Drosophila*. Collins and all the leaders of the sequencing centers were among the visitors. People who doubted the appropriateness of an industrial approach to biology were converted.[38]

The resultant plans swiftly led to proposals that were even bolder than expected. The centers, instructed to hurry up and submit detailed grant requests to a review panel of scientists, saw new ways to speed up. Probably the boldest ideas came from the Whitehead Institute. Lander's group wanted to sequence a third of the human genome. They intended to scrap their existing sequencers and reequip with the new 3700s from ABI. They would commit to automation rather than hire more workers. The trustees of Whitehead agreed to obtain a loan of $40 million to buy the machines at once. Repayment would come from NIH over five years, at the rate of $8 million a year. Other centers ordered large batches of 3700s. By the spring of 1999, orders had reached 700, including Venter's for 230 and Whitehead's for more than 100.

At that point, having raised the stakes on Venter once, the public project raised him again. At a meeting, Collins suggested a new deadline: mid-2000, a year earlier than Venter's announced goal. His audience was the hard-pressed "G-5," the directors of the largest sequencing centers in California, Texas, Missouri, Massachusetts, and England. Collins was still nervous about whether Elbert Branscomb, Gibbs, Waterston, Lander, or Sulston would accept such a leap.

As it happened, they liked the idea of getting a useful tool out to researchers sooner than planned. They also liked the idea of putting the whole sequence in the public domain, to keep gene sequences from being locked up in patents by Celera or such rivals as Incyte Pharmaceuticals in California or Genset in France. But to do it, some of the centers would have to boost their productivity tenfold in just a few months, which was far better than the centers had been doing. And the handling of the DNA clones, the BACs, would have to be highly centralized. It was a picture, Collins recalled, of "acute pressures," particularly on mapping centers that did not expect this pace. But the G-5, who had already begun meeting every Friday by conference call, said yes.

With ambitious proposals from the centers approved, $80 million in new grants were allocated to the centers in Houston, St. Louis, and Cambridge, Massachusetts. The new plan, calling for a price between twenty and thirty cents for each sequenced base, was announced in March 1999, and Vice President Al Gore pronounced himself "thrilled." Not long afterward, the public project enlisted such private firms as Genome Therapeutics of Cambridge, Massachusetts, and Incyte to help out. Venter, himself leading a frantic struggle to get his machines and software working, just laughed, saying that the new nonprofit plan consisted of projections and had "nothing to do with reality." Still, he was nervous. He told one interviewer, "I wish we were talking about this after we accomplish it." A few months later, Venter, too, set a deadline of completing a draft genome during 2000.[39]

The new deadline for a draft was eighteen months ahead of the goal set only six months before. Receiving new machines at the same time as NIH-supported centers, Venter had no choice but to match his rivals. Print and broadcast media eagerly covered this competitive ratcheting. The existence of a race made an arcane competition vivid for a wide audience. It was a good story, and Venter had ignited it.

The Competition Heats Up

In August 1998, Venter announced the Celera name for his enterprise. His scientific advisors were eminent, again demonstrating Venter's ability to ally himself with heavy hitters. The initial list included ethicist Arthur Caplan of the University of Pennsylvania; Arnold J. Levine, who had discovered a key cancer gene, rebuilt molecular biology at Princeton, and was now president of Rockefeller University; Victor McKusick, the dean of gene finders; Richard J. Roberts of New England Biolabs, a 1993 Nobel Prize winner; Melvin Simon of Caltech, leader of the invention of BACs; and Norton Zinder. Roberts, codiscoverer of "split genes," would be chairman.

Venter embarked on a different tack for choosing whose DNA would be sequenced. Four years later, he explained it this way:

> We sequenced five individuals' genomes for our work—three women and two men of self-described ethnicity as African American, Hispanic, Chinese and two Caucasians. This is because race is not a scientific concept and it is not a genetic concept. Many people find this hard to believe because they see differences in skin color and think that means we are different genetically. But we are all part of a continuum. We can all trace our ancestors back to approximately four groups in Africa, all black. There is more variation in the genetic code of the black population than there is between the African American, Caucasian, Hispanic or Chinese populations.[40]

That summer, Incyte Pharmaceuticals declared that it, too, would seek to sequence the whole human genome even faster than Celera—but it would keep its data in a private database open only to subscribers. And in October a team of sixty-four American, British, and French scientists published a human map with the positions of 30,000 genes—twice as many as had been located before. Researchers in Cambridge, England, and St. Louis noted in November that the entire genome could be completed in seven years. The cost would be one ten-thousandth that of health care in the same period.[41]

The "Worm" Sequence Is Completed

In December, there was joy at the St. Louis and Cambridge centers at reaching a major landmark in genomics. Print and broadcast coverage was massive. The two laboratories, having completed their eight-year, $65 million effort to sequence all the DNA of the silvery, transparent nematode *C. elegans* or "the worm," published and analyzed the results. The worm, with 97 million bases (three times as many as yeast and about as many as the smaller human chromosomes), was the first animal to be sequenced, only three years after the first microbes. Excitement was palpable among scientists. Gary Ruvkun of Harvard Medical School exclaimed, "It's the first time we can see all the genes needed for an animal to function." In Dallas, Glen Evans called the worm "a very simple model for how people work." Lander told the *New York Times*, "This is a brilliant innovation of half a billion years ago that we are getting a look at for the first time."

The key to success was the existence of a detailed physical map using the YACs invented in Olson's lab back in 1987. It was a convincing proof of the "clone-by-clone" sequencing strategy that the public project preferred. Technical progress along the way was so rapid, Waterston observed, that the project could now have been done in a single year. His and Sulston's vision had been fulfilled. And so had that of Sydney Brenner, the "father" of worm research.

The worm was already useful for comparisons with other heavily studied organisms, such as yeast. Separated by eons of separate evolution from a common ancestor, yeast (with 6,217 genes) and the worm (with 19,099) had remarkable genomic similarities. David Botstein at Stanford and colleagues reported that "the core biological functions" of each were carried out by "proteins similar in sequence and number." Thus, a large proportion, 40 percent of genes in yeast and 20 percent in the worm, constituted ancestor genes that had been "conserved."

Collins pointed out that "half of the disease genes in humans have identifiable counterparts in this worm." He added, "I don't think that it is an overstatement to say that the hopes of the parents of a child with a birth defect, the hopes of a young man with a family history of cancer and the hopes of a couple caring for aging parents are advanced."

The weekly journal *Science* devoted an entire section to the worm sequence, in part because the work with the worm had accelerated the drive to sequence the human genome. The journal billed the achievement as "a landmark in biology" that produced "a treasure trove of information" for studying "a host of problems at new levels of sophistication." The way was open to get below the surface: "This first phase of sequencing is over, but the real research is only beginning." Sulston said, "We are not done. We never shall be done."[42]

The SNP Consortium

Meanwhile, the rewards and risks of patenting, a major means of protecting "intellectual property," were on many people's minds. Of the highest importance was the measurement of differences between people that made one person more liable to a given disease than another. The main type of variation considered in those days was the change of a single nucleotide for another, the so-called single-nucleotide polymorphism (SNP). Such SNPs were presumed to be numerous in the widespread, common diseases that were the principal target of the genome project. The key issue was pinning down the actual causes of illness. Hence, a highly detailed SNP map would be vital for the future medical usefulness of the full human sequence, even though most of the variations lie in the vast, dark, unexpressed regions of the genome.

The ability to compile a catalogue of variations would flow promptly from the complete human sequence. The perception that this catalogue would be more powerful than earlier methods took hold soon after the February 1996 decision to gear up human DNA sequencing. In September 1996, after a year of review, *Science* published a paper by Neil Risch of Stanford and his colleague at Yale, Kathleen Merikangas, entitled, "The Future of Genetic Studies of Common Diseases." A discouraging news report the year before, also in *Science*, by the journalist Gary Taubes had spurred them to write it. According to Taubes, epidemiologists, who study the frequency of disease in different populations, were having trouble in their long struggle to pin down the influence of the environment on human illness. Risch and Merikangas asked themselves whether the situation was hopeless, or if there was a better approach. They concluded that there was. Once the human sequence was complete, they wrote, genomic technology would allow the compilation of a variation catalogue, which, in turn, would strengthen the quest for causes of disease. Their conclusions were widely read.[43]

This paper, which Collins termed "a kick in the pants," put the measurement of SNPs on the front burner, along with the related question of whether this vital intellectual property would be open to the world or captured by one or more private companies. In launching his drive for the human sequence in May 1998, Venter had declared strong interest in

SNPs. Venter looked forward to identifying millions of them. The same month, Lander and others at the Whitehead Institute, along with Robert Lipshutz of Affymetrix, used new DNA chips from the company to map more than 2,000 SNPs across the genome. This "proof of principle" spurred the public genome project to set itself the goal of identifying 100,000 SNPs by 2003. A mere two years later Lander acknowledged, "In fact, this proved way too modest." By then 300,000 had already been identified, with 2 million expected in another year. Many companies beside Venter's, such as Affymetrix and Incyte in California, Orchid Biocomputer in New Jersey, CuraGen in Connecticut, Millennium Pharmaceuticals in Massachusetts, and Genset in France, were also interested in the SNP business.[44]

Balkanization was an urgent threat, and giant private firms soon concluded that this field must not slip into a thicket of private hands, giving rise to perpetual patent suits. Hence, private and public interests in the genome coincided. And so, in the spring of 1999, major drug companies formed a central organization to prevent patenting in the SNP domain. The industry-academic organization was called the SNP Consortium, headquartered in Chicago. The Wellcome Trust and ten of the world's biggest drug firms put up $45 million for the work. Just as the genome project put sequences on the Web every night, the consortium would immediately publish the SNPs it found. The actual cataloguing would be done by four of the biggest academic sequencing labs. The effort was analogous to a significant earlier industrywide collaboration in electronics, the Texas-based Sematech.[45]

Concern about patenting went beyond the desire of scientists everywhere to avoid duplication of effort and help assess the accuracy of the growing databases. Measuring a person's susceptibility to illness, companies could use the knowledge to accelerate development of drugs and approvals by regulatory agencies. Further, SNPs would help identify classes of people who might react well or poorly to a drug. This new field with the fancy name of pharmacogenetics, "stratifying" patients by their response to treatment, would move medicine decisively toward targeted care—and advice about prevention. There was a third potential benefit. Most drugs, developed at great cost in time and money, failed in testing because they did not help many patients—

or hit them with dangerous side effects. Perhaps some of these could be "rescued" and given only to the patients who would be helped. The economic significance of SNPs was huge.

High Throughput

To Lander, the double speedup in the Human Genome Project in 1998 and 1999 exacerbated the inherent challenges and underlined some of the paradoxical rules of a great technical project. To him, the model was not one big facility committed from the beginning to a particular factory design. He thought this was the defect of Walter Gilbert's 1986 proposal to industrialize the sequencing-by-hybridization process he and George Church were working on. He recalled, "It was simply too detailed for a project of this sort." A better model was the transcontinental railroad of the 1860s. "They knew [where] they had to get, but they didn't quite know how they were going to pull off this project." Counsels of perfection were dangerous. The Pacific railway builders "just built the damn thing with cottonwood ties that they knew were going to not survive. But at least they had the railroad. . . . They could bring in other ties later."

Although the genome project had a direction and goals, Lander held, there was a need for "studied vagueness." The details had to be left to a group of competing scientific and engineering groups with "the most creative people." They would have to adopt a particular approach, modify it, and move on to the next, as they did with the evolving choices of particular chunks and storage places for DNA. It was essential to make "Just In Time technology decisions."

The St. Louis and Cambridge, England, centers had expanded rapidly, using a fair amount of manual labor, to do the worm sequence. But the Whitehead Institute center had concentrated on mapping (for which it developed the Genomatron) and was a much smaller player in sequencing. But by 1995, it was clear that the future belonged to sequencing. Lander's team had spent millions less than it had estimated on the mapping and quietly reprogrammed the money to build up "all these high-throughput kinds of methods." This emboldened the Whitehead group for the intense process of submitting new grant proposals in the winter of 1998. These would leap far beyond previous speeds and scales and cut the cost per base even more drastically.

The Whitehead proposal to scale up the pace fifteen to twenty times found favor with the scientific reviewers. But the Whitehead team almost missed its chance to present the idea. As the group headed for Washington, snow at Boston's Logan Airport canceled one departure after another, and they only got to the review board with a few minutes to spare. This gave Lander an opportunity for a little drama. He showed a transparency with the list of flights. After each, he wrote the word "Canceled" until, after the last flight, he wrote, "Success!" The transparency, Lander thought, showed that his team would "leave no stone unturned" to make the genome project a success. "What excited them," Lander recalled, "was that there was a vision to take on the whole problem. It was audacious. It was ambitious and it was what was needed to get the job done."

A new problem arose. The Whitehead sequencing center, Lander was convinced, would be "the best pipeline in the genome project," but most of the libraries of DNA clones were elsewhere, in the hands of researchers reluctant to share them. So the Whitehead sequencing center, deciding once again to ask for forgiveness later rather than permission now, began filling its sequencing pipeline with random DNA clones the group created without telling the authorities down in Bethesda. When the other centers found out about this, Lander told them that "the project would not succeed unless we could take full advantage of the capacity that we had put in place."

The genome project reached its goal, in Lander's opinion, because nobody was in charge. Indeed, Venter "dissed the public project," saying, "Oh, it's just a bunch of disorganized academics and we [Celera] are the organized company." Lander thought the opposite. "Had any one party been able to dictate how the project was done, I think it would have failed."[46]

A Sequencing Center in Full Swing

The Whitehead initiative took shape in a former warehouse on an industrial side street of Cambridge, Massachusetts. There, in a high-ceilinged, brightly lighted hall dominated by massive air-conditioning pipes, some

125 Prism 3700 laser robot DNA sequencers hummed. By the spring of 2000, the machines were reading 200 DNA letters every second with only an occasional visit from the technicians led by Lauren Linton. The near silence contrasted eerily with the control rooms for the flights to the moon, jammed with spring-loaded engineers, each responsible for a single piece of the mission. The machines were reading DNA that had been cloned, purified, and then colored with fluorescent dyes the lasers were designed to detect. The stream of data ran through cables upstairs to nearby banks of powerful computers. At the computers, bioinformatics specialists swiftly readied sequences for that night's posting on the World Wide Web. The electronic information from the MIT–Whitehead Institute sequencing center joined a mounting flood of readouts from the similar centers in California, Texas, Missouri, and Cambridgeshire. The overwhelming amount of information was growing more overwhelming daily, creating a corresponding headache for scientists struggling with the inevitable errors and the challenge of comparing data from different databases.

At the Whitehead center, thanks to numerous running changes throughout the process, the average speed of reading DNA bases was 12,000 a minute, some eight times faster in February 2000 than at the beginning of 1999. Twenty years earlier, reading so many bases would have taken a whole year. In 1996, the time had dropped to twenty minutes (ten bases per minute). The U.S. and English centers reached the 1-billion-base mark in November 1999 (it was a G), and the 2-billion mark (a T) in April 2000. All the while, similar machines were working at top speed on the premises of Celera Genomics outside Washington, D.C., but their data stayed on site.[47]

An Ovation for Venter

Once the teething troubles of the Prism 3700s diminished to a manageable level, the feverish phase of the genome project began in the latter half of 1999. The total length of decoded DNA bases leapt from millions to hundreds of millions in a matter of months. In a duel of press releases, Celera and the public project announced arrival at a series of milestones: so many bases sequenced, so many sequencing phases completed, so much assembly work begun, and so forth.

In September 1999, the Human Genome Project announced it had put together a quarter of the whole sequence and was looking forward to a "finished" sequence of the relatively small human chromosome 22 before the end of the year. The work, involving some 200 scientists on three continents, was led by Ian Duncan of the Sanger Institute in England. When the results were published in December, the public project grabbed the megaphone and shouted loudly about delivering "the First Chapter of the Book of Life." Collins proclaimed, "To see the entire sequence of a human chromosome for the first time is like seeing an ocean liner emerge out of the fog, when all you've ever seen before were rowboats." NHGRI said that the public project had sequenced a third of the human genome.

The excitement about the chromosome 22 announcement spilled over into the private sector. Celera's stock price jumped 15 percent in a single day. The index of biotechnology stocks had doubled in 1999.

Not to be outdone, Celera and its collaborators at Berkeley and other labs announced in September that they had finished the sequencing for the fruit fly, *Drosophila melanogaster*, the test bed for the private-sector human sequence. So they could go on to the assembly process—Celera said it would finish this stage by New Year's—and devote their Prism 3700s to the human sequence. In October, Celera announced it had sequenced its first billion human DNA bases, although these were in many fragments that remained to be assembled. Meanwhile, the drive to sequence the mouse, including a mapping grant to TIGR, now run by Venter's wife, Claire Fraser, was under way.

In January 2000, Celera announced that it had begun assembly of 10 million sequenced fragments, a process expected to take three months. Late in March, Collins announced that the public project had posted more than 2 billion DNA sequence letters on the Internet, all available to Venter as well as everyone else.

Days earlier, eighteen months ahead of schedule, Rubin of Berkeley and some 200 colleagues across the United States and Europe published the fruit-fly sequence, with its numerous similarities to the human genome, in *Science*. Part of the operation had been an intense eleven-day "jamboree" of forty-five fruit-fly geneticists, protein specialists, and computer experts, who assessed the usefulness of the assembled fly genome for the next step, called annotation, or finding genes and figuring out what the gene-specified proteins do. As a gift from Celera, Rubin's for-profit partner, a CD-ROM containing the fly genome, something only dreamed of for many years, was placed on the chair of every one of the 1,300 scientists attending the annual fly biology conference in Pittsburgh. Venter received a standing ovation.[48]

Throughout 1999, the new machines and their chemicals proved devilishly difficult to handle predictably. Parallel problems with computer software and hardware, all experienced against approaching—and highly public—deadlines, appeared almost intractable. And after the sequencing, the task of "assembling" all the sequences in order was so difficult that it was not accomplished until the last moment in June 2000.

NASDAQ Follies: The Clinton-Blair Memorandum

Amid the crescendo of news, a background pressure on the public project was Celera's announcement in October 1999 that it had provisionally applied for patents on thousands of genes it had already found. Was it a race—or a gold rush?

Unquestionably, the excitement of a race built up journalists', and presumably the public's, interest in the story. But the competition and attention also raised the question of why the public and private projects could not collaborate more fully. The looming achievement was being billed as something for the ages, a royal road to accelerated human health benefits. In England, the *Guardian* thought people were living through "an exciting and vertiginous point" in history. Would the moment be tarnished by what looked like squabbles over scientific priority or money? Too much infighting could generate a public relations disaster that would hurt public and private projects alike. John Sulston, favoring cooperation in principle, said the stumbling block was how to publish the results, together or separately.

Intense negotiations to stop the backbiting between the Venter and Collins camps, some of them mentioned in the press, were started, broken off, and started again. The goal was a dignified close to the race for the complete human DNA sequence. Neither side wanted to look petty at what was clearly a momentous juncture in applying biology to human health.

In the summer of 1999, sensing danger, Lander called Hunkapiller (the supplier of 125 of the new machines to the Whitehead Institute) to see if he could promote the idea of face-to-face contact between Venter and Collins on collaboration, that is, combining the information being developed on each side. The result could have both practical and political benefit. According to Shreeve, Venter arrived at a hotel near MIT and found three people already there, Lander, Hunkapiller, and the investor Noubar Afeyan, representing Tony White at Perkin-Elmer. The discussion soon turned hot when Venter said the Celera data would be freely available, just not right away on the Web, and that the company would seek patents on things of medical importance. Venter suggested making a DVD (like the one that would be distributed some months hence for *Drosophila*).[49]

Around this time, it became public that British Prime Minister Tony Blair had convinced President Bill Clinton to draft a joint statement applauding open access to genomic data. Pursuing the hope of collaboration, Lander joined Venter's advisors on a conference call in October, during which data-release policy was the central issue. As the door remained open, at least a crack, leaders of the rival efforts met on December 29. From the public project were Robert Waterston, Francis Collins, and Harold Varmus. Venter was accompanied by Tony White. But Lander was not present. Wade of the *New York Times* called it "a meeting of hawks without a mediator."[50]

An acrimonious session led to a couple of months of silence, ending in a letter from Varmus to Venter demanding an immediate resolution. The letter became public and set off a media frenzy. Venter fired off an immediate and firm refusal, in effect ending discussions.

The fall of 1999 and winter of 2000 were high-stakes times for Celera. Early in March, just in time as it turned out, Celera sold more than 4 million shares on the stock market. The offering brought in some $944 million, creating a war chest for the drive to find the functions of genes and their proteins as a tool for "personalized health/medicine." Stocks in biotechnology and genomics companies were soaring to record highs, having risen more than twenty-eight-fold at Celera since the previous June, at the very moment that the dot.com boom was carrying the NASDAQ stock index to 5,000, with the average stock selling at more than a hundred times expected earnings for the year. To a longtime observer of the biotech industry, speaking with the *Star Tribune* in Minneapolis, the trend appeared "fairly insane. . . . There have been some real advances in biotechnology. But sequencing the human genome doesn't give you a drug to sell."[51]

The bubble was about to burst. On March 14, President Clinton and Prime Minister Blair issued the statement about open access they had been working on for months—and threw investors into a panic. The statement bore strong traces of input from the public genome project on both sides of the Atlantic. Acknowledging the role of patenting in exploiting discoveries relevant to health, Blair and Clinton called for the maximum of "unencumbered access to raw human sequence data." At a medals ceremony in the White House, Clinton spoke further, urging private companies to make "responsible use of patents."[52]

Amid general ignorance on Wall Street of what the genomic issues were, the statement and inaccurate early news coverage had higher voltage than they might have, coming as they did just after the public collapse of the public-private negotiations. Despite hastily issued disclaimers from the White House and Collins, the statement struck nervous investors on Wall Street as an attack on gene patenting and therefore a threat to the profitability of biotechnology and genomics companies. An irritated Venter commented later, "I think they were very confused about what they were saying." The NASDAQ's biotechnology index fell 13 percent in a single day, and Celera dropped 21 percent to a price far below the rich level of only two weeks before.

It was a biotech panic, wiping out tens of billions of dollars in market value. It helped bring down broader stock indexes like the Dow Jones and Standard & Poor's by 1.4 and 1.8 percent, respectively. It also signaled the beginning of the end for the bull market of the 1990s, slashing the technology-oriented NASDAQ index 80 percent over the succeeding months and contributing to a recession. At a congressional hearing early the next month, Venter cried foul, denouncing the Clinton-Blair statement for causing the "loss of over fifty billion dollars in market capitalization in the biotechnology sector in two days," followed by further dramatic declines. Undaunted, however, Celera agreed on March 20 to merge with Paracel, a company in Pasadena, California, that specialized in genome analysis. Venter was still intent on his goal of building Celera into an information company like Bloomberg.[53]

A Script for Life: "The Real Fun Starts Now"

In June 2000, as the public announcement of success in drafting a human genome drew near, many attempted to describe what the world would be like when it had a preliminary blueprint of human DNA. Opening up new insights into human development, health, and disease, the imminent achievement was viewed as a turning point in biology with potentially vast medical significance. Most commentators described the initial human draft sequence as a starting line. Zinder told *New Yorker* writer Richard Preston, "This is the beginning of the beginning. . . . This is like Vesalius [the sixteenth-century pioneer of human anatomy]. . . . With the human genome, we finally know what's there, but we still have to figure out how it all works. Having the human genome is like having a copy of the Talmud but not knowing how to read Aramaic."[54]

Even if it were a prerequisite for all future biology, the human DNA sequence would have to be explained. All the genes and their controls would have to be specifically catalogued, and the way they interacted to develop and operate a human being would have to be mapped. The variations between one human being and another that led to greater or less susceptibility to disease would have to be spelled out. To make sense of the genome would require more sophisticated tools—new generations of hardware and software. A long campaign of sequencing even more "model organisms" to compare with humans was crucial to learning the functions of elements of the human genome and to deepening understanding of evolution.

Such sweeping goals were not merely hyperbole for the public. The priority tasks were also set forward in professional journals. A salient example was a special "Insight" section of the weekly journal *Nature* on June 15, 2000. The section's unified label for the tasks ahead was "Functional Genomics." Because of what the editors called "astounding technical advances," the drive to catalogue all genes and what they do had "leapt from being a surrealistic, or at least futuristic, concept in the 1980s to an accepted (if not yet everyday) part of science in the year 2000." The next phase of genomics would improve both the diagnosis of disease and the productivity of agricultural plants. Study of the genomes of microbes could open a path to new energy sources. Computational biologists would cope with a suddenly data-rich biology. New DNA chip technology would allow the activation, or "expression," of genes to be recorded. Studying SNPs and other variations between people would identify multiple sources of common diseases. And a new field, pharmacogenomics, would probe why patients react differently to drugs.[55]

It was already understood that the number of human genes could turn out to be far smaller than the more than 100,000 claimed by several genomics companies, notably Incyte, Human Genome Sciences, and DoubleTwist. Several groups busied themselves comparing known, "annotated" protein-coding genes with the ensemble of known proteins. As June 2000 opened, the journal *Nature Genetics* carried the estimate of 30,000 genes by Jean Weissenbach of Paris, using techniques from the sequencing of pufferfish, the dangerous Fugu appetizer of Japan. Brent Ewing and Philip Green of the University of Washington, using a method worked out for the worm sequence, estimated the number at 34,000. A higher estimate, around 120,000 genes, came from John Quackenbush of Venter's TIGR.[56]

Leaders of the program were not shy in characterizing what was up. Eric Lander told CNN, "Biology is getting its Periodic Table," referring to the catalogue of chemical elements from hydrogen to uranium that was the rock on which twentieth-century chemistry was built. He went on to say, "For the first time we know what the 100,000 or 50,000 building blocks of the human are. . . . It's going to change everything. Just like every molecule has to be described in terms of 100 or so atoms, every process in the body—how a cancer cell works, how a fetus is developed—all of this is going to have to be described in terms of this measly list of 50,000 or so genes." Looking back to the late 1980s, when "we decided we had to go for it even though we never had a clue of how to do this," Lander observed, "It was a tremendously exciting time, but in some sense it's just a race up to the starting point. The real fun starts now."[57]

Less than a week later, Craig Venter testified to Congress that the human genome sequence "truly holds the promise of ushering in an era of personalized and preventative medicine. This information will also challenge our existing medical professionals. They will need to be re-educated and retrained." He added, "The biomedical research community will not have crossed a finish line at this point. We will just be getting to the starting line for decades of new discovery and advances."[58]

On June 19, public television interviewer Charlie Rose talked to a succession of key players. Comments flowed:

> Watson: "You can say it's the script for life. It's the information for the play of life. And if you have the script, the actors can—you know what the actors are going to do."
>
> Collins: "This beats splitting the atom. This beats going to the moon. This is an adventure with consequences, and it's an adventure into ourselves."
>
> Venter: "It's providing a whole new resource to the world to begin to understand how cancer works, how heart disease really comes about, and what we can do about it. . . . This is science at a level never before experienced in the history of biology."
>
> Hamilton Smith: "It's the most important thing in medicine and biology . . . for the next 100 years or so, I would say."[59]

Not for the first or last time, public comments about what was coming approached hyperbole. Watson told the press, "This may be the most important step we've taken in science. We've got our instruction book: how to build the 747. We can't read it all yet. But that we've evolved to where we can have the book—that's pretty good."[60] Declaring that the practical consequences of the atomic bomb and moon projects remained modest, Collins said, "This reading out of our own Book of Life—those consequences will not be modest. They will be daily. They will be personal. They will be significant."[61] Harold Varmus was more measured. With the full human sequence, he said, "we have gotten to the point where we have most of the information in hand, and now what we need to do is think about what it's saying to us." Looking forward to the central problem of human variation, Arnold Levine, president of Rockefeller University, said, laughing, "The differences between us are going to be very interesting. . . . We don't know how to quantitate yet how much is genetic and how much is environment."[62] David Baltimore, president of the California Institute of Technology, wrote, "The DNA sequence tells us little about the functions of the genes or their role in the body's overall economy. . . . The next half century will be taken up with understanding the roles of each of the human genes."[63]

There was sniping as well. Haseltine of Human Genome Sciences said his company had already catalogued 95 percent of the medically useful human genes. So the full DNA sequence that so many were striving for would end up as "a rather meaningless footnote to a revolution that is well under way."[64]

Assembly: Putting the Pieces Together

Negotiations resumed for some kind of amity in announcing completion of the public and private rough drafts. The catalyst was Ari Patrinos, the current boss of the DOE genome program, which had often supported Venter and TIGR over the years. There were pizza suppers at Patrinos's town house and talks that evolved toward a joint announcement by Collins and Venter—a carefully scripted "draw."

But there had to be something to announce. The 400,000 fragments of human DNA that had been sequenced in the public project and the millions of pieces sequenced by Celera's different method had to be "assembled." This term described the computationally intensive task of building a single coherent genome—with as few stretches flipped and as few gaps as possible. And unexpressed, noncoding "junk" DNA needed to be distinguished from genes. Because of its methodical use of fragments of known position, the public project thought its assembly task would be easier than Celera's job of linking up millions of pieces blown out of its shotgun. But the computational task turned out to be horrendous for both of the racers. Throughout tense days and long nights, the principal fuel was adrenalin.

The leaders of the bioinformatics team at Celera in Rockville, Maryland, were Granger Sutton and Gene Myers (a pioneer of the whole-genome shotgun idea who luxuriated in the chance to make it work). He, Sutton, and superintense colleagues used programs called Overlay and Grande on their enormous Compaq supercomputer, which had four trillion bytes of computer memory.

For the public project, the programming for the assembled draft sequence was done at the University of California, Santa Cruz (UCSC), where Robert Sinsheimer had held the inaugural genome conference fifteen years before. The job was done alone in exactly one month by Jim Kent, who had just passed his PhD oral exams at UCSC. Forty years old, the bearded Kent had recently changed careers. He liked to say he switched after he handled a program that took up a dozen CD-ROMs and reflected that the human genome would fit on one. After a dozen years in computer-animation programming, he became a graduate student in molecular biology. His supervisor was David Haussler, whom Lander had called at Christmas in 1999 with the message, "Your country needs you." Thinking back to this call, Lander reveled in the public project's ability to recruit wherever the talent was. "We could sign up anyone to join on, which meant we had access to the best minds in the world."

But the task was very hard. Haussler's team had made little headway in the early months of 2000. Kent sent Haussler an e-mail asking how it was going. The reply was, "Jim, it's looking grim." In this dire situation, Kent offered to try a simpler strategy, adapting the software he had been developing for the ever-present Human Genome Project test bed, the nematode *C. elegans*. Haussler wished him Godspeed.

On May 22, working in the garage in back of the bungalow where he lived with his wife and two children, Kent began adapting the *C. elegans* program to the thirtyfold-larger human genome. He was awed by his task: "The assembly process is kind of like solving a giant jigsaw puzzle, but it is a much more complicated jigsaw puzzle than you ever imagined." Spurring him was a fear that the genome could be locked up in a maze of patents. He found it upsetting that the U.S. Patent and Trademark Office, by patenting genes, was protecting "a discovery rather than an invention." He said, "We wanted to get a public set of genes out as soon as possible."

To write 10,000 lines of code, Kent used one hundred personal computers with fast processors from Intel Corporation that cost a total of $250,000. Intended for teaching, the computers were borrowed from the university after Haussler appealed to the UCSC chancellor. Getting about half his usual sleep and programming so fast that he had to ice his wrists after long sessions, Kent began building a program called GigAssembler. The work, which called for a string of "all-nighters," was congenial. Kent recalled, "I like to work intensively for a period; I'm not a nine-to-five guy." But he was not so accustomed to having hundreds or thousands of scientists "waiting for my work before they can do theirs." He achieved his "first cut" in about ten days and kept going. Haussler was ecstatic: Kent had taken four weeks to do what "would have taken a team of five or ten programmers at least six months or a year." Venter, too, was impressed. He had thought the public project's data were "un-assembleable." He said, "It was really quite clever. . . . We were truly amazed."

But researchers needed to have easy access to the assembled sequence. In urgent service of a totally accessible public genome sequence, Kent turned around immediately to build the UCSC Browser that would let visitors to the Santa Cruz site view the raw sequence alongside other types of information relating to a particular stretch of DNA. Strengthening the browser with more detailed analyses of successive GigAssembler assemblies would preoccupy Kent for the next two years. In April 2002, when the human sequence was 60 percent "finished," Kent gratefully turned the task over to the National Center for Biotechnology Information in Bethesda, Maryland. By then he had done thirteen "iterations."[65]

"This Is Our Genome"

As the competition between Kent at UCSC and Myers and Sutton at Celera intensified, the high-level amity plans led to the scheduling of a maximum klieg-light announcement at the White House on Monday, June 26. President Clinton had demanded of the genome project leaders that they "fix" what could become an ugly situation. On both coasts, assemblies were achieved at the last minute, on the evening of Thursday, June 22, for Kent and, harrowingly, on the night of Sunday, June 25, for Myers and Sutton.

The ceremony was held in the White House East Room, linked electronically with Prime Minister Blair and many guests in the dining room at 10 Downing Street in London. It was a moment to concentrate on the historic opening of a new window on all life, leaving personal credit aside in acknowledging a triumph of both reason and compassion. Like astronauts of thirty years earlier in orbit around the moon, with its tawny surface gliding beneath them, the scientists were acknowledging that they came for all humanity, not for themselves. President Clinton compared the "wonderful map" of human DNA with the maps Meriwether Lewis and William Clark had brought back to Thomas Jefferson almost two centuries before. He looked forward to a day when cancer would simply be the name of a constellation. Acknowledging James Watson's presence in the room, the President said, "Thank you, sir." Collins and Venter stood on either side of Clinton, together as they also were on the cover of *Time* published that day, and each spoke cordially.[66]

With the ruffles and flourishes accomplished, the klieg lights winked out. From the journalistic point of view, it was "a wrap." Now, biologists across the world could concentrate on refining and using the genomes of humans and a growing number of other species. On July 7, the UCSC Browser for the public sequence was posted on the Web; there were 20,000 hits on the first day, and half a trillion bytes of information went out from Santa Cruz. Less than ten years later, the site had 5,000 to 6,000 distinct visitors a day. Haussler could watch with pride as the parade of colors—he called it "an ocean of A, G, C, T"—swept seemingly endlessly across a computer screen. "Everyone is seeing it," Haussler recalled thinking. "I can't describe that experience." He recalled his sense of wonder: "This is it. . . . This is our genome."[67]

CHAPTER FIVE

Surprises

After a stormy voyage, the international team of genome researchers had safely reached the shore of a largely unknown continent. Glad to escape years of media lights and somewhat more confident of sustained funding, they were eager to push inland to make sense of the genome. They sensed that surprises lay ahead but did not foresee the explosive growth and drastic shifts in perception of what genes are and how they are expressed or silenced. And in the topsy-turvy world that awaited them, the humble molecule RNA took its place alongside DNA as a major operator in the immensely complicated living cell.

The consortium's researchers had some idea of the tasks they faced and approached the unknown with some optimism. Keeping abreast of Celera, stroke by stroke, they had ensured that the sequence of human DNA remained firmly in the public domain, immediately available to thousands of researchers around the world. They had assembled an array of biological and computational tools. The draft sequence of human DNA was the first version of a piece of genetic infrastructure to be used in all future research. The free access for all was intended to cut down on duplication of effort by scientists operating in isolation and thereby to speed discoveries relevant to both human health and biological evolution. A bright hope existed that the resources created would be useful to drug companies struggling to develop new medicines.

Only a "Parts List"

But it was clear that the genomic library contained a very incomplete map of the territory ahead. Shelf after shelf of electronically stored books contained strings of nucleotides, more than 3 billion of them. But how did the letters form the words—particularly active elements analogous to verbs—and paragraphs and chapters of what the researchers liked to call "the Book of Life"? Some stretches of DNA remained unreadable with existing technology, particularly the highly repetitive segments at the center and tips of chromosomes. Even though genes were thought to be discrete entities strung along the DNA, only a fraction had been identified and placed.

What the genes did—their functions or particular jobs—often remained a mystery. So did the interactions between genes and both the proteins they specified and their environment. In an enthusiastic talk at MIT in February 2001, with 475 people in the room in front of him and 200 more watching video screens in adjacent rooms, Eric Lander said all the researchers had at present was "a parts list." He added, "A Boeing 777 has 100,000 parts in it. Having a parts list doesn't tell you how to put it together or how it flies." Nonetheless, Lander was upbeat about what he called "the greatest lab notebook in history." He added jovially, "Evolution has been taking notes for three and a half billion years, and we are getting a chance to look at its notes. The problem is, it's only been taking notes on the successful experiments."[1]

The biologists knew they needed a complete catalogue of the genes and the rest of the functional elements, including the myriad of still-unknown genomic control switches. From 2003 to 2007, hundreds of researchers in some three dozen institutions collaborated on a four-year, $42 million pilot project called "The Encyclopedia of DNA Elements" (ENCODE). Surveying forty-four stretches of DNA that added up to 1 percent of the genome, the cataloguers entered a forest of unexpected complexities (to be described in greater detail below).[2]

Meanwhile, each of the more than twenty remaining chromosomes of the human genome required in-depth analysis. The task took from 1999 and 2000, when the smallest chromosomes were "annotated," until the largest chromosome was completed in 2006.[3]

Onward to Human Variation

With firm support from the U.S. Congress and the Wellcome Trust in the United Kingdom, the global consortium could now keep its vow to "finish" the human sequence. Researchers needed to fill some 150,000 gaps, greatly increase the length of continuous DNA sequences, and keep checking that each segment ran in the right direction.

To keep pace with this and other tasks, the numbers of bases determined each second had to continue rising steeply, with the help of upgraded equipment for preparing samples, reading off the nucleotides, and analyzing them in computers. This essentially industrial process had to become continually faster and smoother.

Alongside this process of continued refinement of the "reference" sequence, which is a picture of the genome that is more than 99 percent the same in all humans, a new challenge grew in intensity. This task, as we shall see in Chapter 6, was the opposite of the first. The draft sequence, based on the DNA of a handful of anonymous people, focused on an enormous degree of genetic uniformity among all the earth's 6.5 billion inhabitants. But getting at the mechanisms of diseases, so as to prevent and treat them, required spelling out the thousands and thousands of variations between one human being and another. This challenge had lent urgency also to continuing to prevent patenting of the single-base differences at myriad points in the genome. Many regarded SNPs as "precompetitive" and hence not suitable for patenting.

Venter Goes His Own Way

Having stayed in the game up to the finish line, the scientists of the public project were happy enough to acknowledge that Craig Venter's impatience, drive, ability to attract and motivate some of the best talent, and knack for collecting headlines had driven the genomic enterprise forward by two years. They also became increasingly ready to adopt Celera's whole-genome shotgun method for much of the massive sequencing that lay ahead. Venter had gambled willfully, even recklessly, in using a sequencing technique that offered the promise of getting around the ponderous gene-mapping procedure chosen by the public project's scientists. To achieve a unique, accurate order, the nonprofit researchers had relied on sequencing carefully mapped clones of DNA much larger than the small fragments in Venter's system. But the whole-genome shotgun procedure had worked in fruit flies (*Drosophila*) and then in humans. Eric Green of the National Human Genome Research Institute (NHGRI), and its eventual director, said, "*Drosophila* taught us a compelling lesson. The hybrid approach has a lot of validity." By the middle of 2000, it was already clear that, for most purposes, the public project should rely on combinations of their clone-by-clone procedure and Celera's whole-genome shotgun technique. Hence, methodological hybrids were used not only on the crucial parallel sequence of the laboratory mouse but also on other sequences of interest, such as the chimpanzee and the dog.[4]

Scientists in the public project were also able to leave Celera to its own problems of marketing genomic information by subscription. As that information became more and more accessible on free databases, the need for major drug companies to subscribe to Celera's dwindled. Celera's parent, Perkin-Elmer, grew uneasy at the lack of a long-term business model, and Venter and his boss, Tony White, clashed more and more often until Venter was forced out early in 2002.[5]

Celera began its evolution from a database company to one focused on diagnostic tools. Signaling the change, Celera posted all the previously proprietary sequences on the Web in 2004. Venter took with him a considerable fortune in stocks, devoting much of it to philanthropic, nonprofit purposes, including the endowment of The Institute for Genomic Research (TIGR), eventually renamed the J. Craig Venter Institute. But he retained a relentless, potentially commercial interest in harnessing microbes to the generation of renewable energy and the remediation of polluted environments. The quest led him to form yet another company, Synthetic Genomics, which in 2009 signed a $300 million, ten-year agreement with the oil giant ExxonMobil to collaborate on turning algae into fuel.[6]

Fewer Than 30,000 Genes

The drive to discover how the genome worked led immediately to the unexpected. As soon as the international and Celera teams sat down to analyze "the first panoramic view of the landscape of our genome," they began uncovering facts that further shook "the neat and tidy genome we thought we had," as one scientist put it in 2007. As we have seen, this was not the first time that the idea of a string of well-defined units of heredity had been challenged. Notable examples were the 1977 discoveries that (1) genes came in pieces whose messages could be read in different ways to make different proteins and (2) genes could overlap each other in a virus.[7]

Now, in February 2001, came the first analysis of the whole human genome. Not without prior disputes over access to Celera's data and subsequent controversy over whether the Celera sequence depended on the public version, the rival teams' separate papers were published on successive days in *Nature* (Lander et al.) and *Science* (Venter et al.). Both groups reached eye-popping conclusions. First, less than 2 percent of the genome encoded the blueprints of the cell's workhorse proteins, leaving about 98 percent of the genome without apparent function. Further, the human genome appeared to include about 30,000 genes, an estimate that continued to fall in subsequent analyses.[8]

This number, only a third of what many expected, had immediate medical and commercial significance. Companies' claims of far larger numbers of genes were exploded, unexpectedly complicating the relationship between genes, proteins, and drugs. The low gene number, foreshadowed the year before by Phil Green in Seattle and Jean Weissenbach in Paris, was smaller than almost all the joshing bets that genome researchers had been making among themselves. It was not enormously larger than the totals for the worm and the fly. So, now the grand problem of explaining the enormous differences between the human species and others was taking a new direction. The number of protein-coding genes alone could not explain the extra complexity of the human genome, the factors that made humans human.

To look beyond the gene to explain the workings of life, it was vital to learn the location and makeup of the DNA regions that tell genes whether to turn on or off and whether to be expressed swiftly or slowly. Such genomic accelerators and brakes are profoundly important in several ways. For instance, the normal development of a fertilized egg into the 100 trillion cells and hundreds of types of tissues in a full-fledged adult human is a process involving subtle regulation at every stage. Of particular relevance to humans is the development of the brain. Another important process is the proliferation of cells, for which elaborate controls determine both normal and abnormal forms. Normal proliferation can be seen when specialized cells multiply suddenly to produce antibodies against invading foreign substances. The abnormal proliferation involved in cancer results from a succession of small genetic changes that remove controls.

Lander, the lead author of the public project's report, and 2,500 colleagues reveled in the paradoxes. He said with anticipation, "Starting today, the real serious analysis of things can begin." It was exciting to have sequenced the first creature with a bone structure, a vertebrate, and one that was twenty-five times larger than any that had been sequenced before. To be sure, the differences between one person and another were minor, but they had already discovered 1.4 million single nucleotide polymorphisms (SNPs) that pointed to different levels of risk for particular diseases. Humans shared many genes not only with the mouse, a fellow mammal of very different size, shape, and abilities, but also with yeast, the worm, and the fly. However, the unexpectedly small number of human genes seemed to make more proteins than exist in other creatures and to do so "more creatively and with more variety." Indeed, the totality of proteins seemed a better predictor than genes of the greater complexity of humans.

The genes were not distributed evenly across the chromosomes like lights along a street. Instead, they clustered in some places like genetic "mountains" and avoided others that were "gene deserts." In this "lumpy" distribution, some regions' content was enriched with guanine-cytosine pairs, which correspond to the dark bands across the chromosomes that can be seen through a microscope. Other regions constituted cytosine-guanine pairs, called CpG "islands." Repeated sequences accounted for at least 30 percent of the total DNA. These were among the genomic hitchhikers acquired over eons, jumping from one part of the genome to another and surviving as largely inactive "transposons." Lander called them a "spectacular fossil record of the last billion years of evolution."[9] Remarking on the small proportion of the DNA that carried blueprints for proteins, Phil Green said, "It almost looks like we are not in control of our own genome." In 2004, he remarked, "As science has proceeded, our view of ourselves in the universe and our importance in the universe has gradually diminished. From Copernicus we learned we were not the center of the universe; we were not even at the center of our own solar system. Then Darwin showed that we're not unique among organisms either: we're just one descendant branch of a fairly complex evolutionary tree." A human genome filled with parasite DNA, Green said, "diminishes our view of our importance. In a sense, we are only minority players in our own genome."[10]

Commenting in *Nature*, David Baltimore, who had opposed the project at the start, said that chills "ran down my spine" as he contemplated the surprisingly low number of protein-coding genes, which he placed between 26,000 and 31,000, only four times the total in yeast and only double that of the fruit fly. He said that accounting for the much greater complexity of humans "remains a challenge for the future." Expressing a wide consensus, Baltimore foresaw "hard work" in the years ahead: producing as precise a human genome as possible; comparing it to a growing number of other sequences, such as the mouse and the chimpanzee; analyzing the activity of each gene; and helping the pharmaceutical industry develop new drugs by probing the vast number of small differences in DNA between people.[11]

The newly examined human genome signaled that the new world of genomics would be one of surprises. The cascade of the unexpected has never stopped. In the last ten years, biologists have many times been forced to alter their picture of what a gene or a genome really is.

The View from the Mouse

From the beginning of the project, its planners gave the laboratory mouse top priority among the model organisms for comparison with the human. Bred and inbred by the millions since the 1920s at places like the Jackson Laboratory in Bar Harbor, Maine, the mouse had been enlisted in countless medical and genetic experiments. An exceptionally useful medical model for humans, the mouse was usually the creature chosen for the first round of testing the probable effects of a chemical on humans; it was central to genetic experiments focusing on cancer and other diseases. Thus, the mouse was fated to be the second mammal sequenced. Its genome was almost as large as the human's. The two sequences had already been shown to be very similar, even though mice and humans had branched off from each other at least 75 million years ago—when dinosaurs still dominated the earth.

A central question was, What did the human genes actually do? How did they work to foster or prevent a disease—singly or in cooperation? In this search for function, there was obviously little opportunity to experiment on people, other than by studying individuals or families afflicted by a particular mutation. People cannot be bred experimentally as microbes and animals can. So, to understand function, to begin to decipher the genomic library, scientists knew from the beginning that they needed to go beyond rapidly multiplying species with life cycles measured in hours, days, or weeks to increasingly larger mammals of significance to medicine, energy, and agriculture. The nonprofit centers had learned so much from the crash buildup of sequencing that they and many smaller labs could plunge immediately after 2000 into deciphering numerous parallel sequences of the model organisms. They also knew that their plunge into what is called comparative genomics was essential to finding all the human genes.

The way had been cleared in the early to mid-1990s by the campaign to map the location of genes across the mouse genome in increasing detail, using both genetic and physical techniques to determine the relative and absolute distances between genes. The effort had spurred the inevitable large-scale automation of the drive to sequence the genome. Thus, it was not too surprising that in the fall of 2000, as papers analyzing the human drafts neared completion, sequencing capacity was found for the mouse. The strain chosen was the classic inbred strain C57BL/6J, ultimately descended from "fancy mice" bred a century earlier on a farm in Granby, Massachusetts.

Although leaders of the public project like Francis Collins denied it, the move was driven in part by the evident interest in the mouse at Craig Venter's Celera. There was plenty of justification for hurry. A prime notion behind producing the drafts was the hunger of biologists and medical researchers to start using human data on a large scale. Automatically, this raised the pressure for data from the mouse. Why? Because the human sequence evidently would be written "in a language we cannot read," as Collins put it.[12] To help decipher that language, biologists needed to develop the ability to compare the human genome with that of a close mammalian relative, particularly the mouse. Not only had the mouse been studied intensely for decades, but the existence of a detailed map made it closest to being ready for a big push.

As with the human sequence, the genomics leaders were determined to keep the mouse sequence totally public. And, as had happened before, some industrial firms shared this determination. Six NIH institutes joined with the Wellcome Trust in England, two major drug companies (GlaxoSmithKline and Merck), and DNA chip maker Affymetrix in a consortium. They scrambled together $58 million to do a rough draft in six months, and Arthur Holden, veteran of the SNP Consortium, led the effort along with Collins. The bulk of the sequencing would be done at the big centers in St. Louis, Missouri; Cambridge, Massachusetts; and Cambridge, England. The urgency lay behind the setting aside of the "hierarchical shotgun," or clone-by-clone approach, in favor of the whole-genome shotgun technique that Celera used on the fruit fly and human sequences. But the sequence would not be completely assembled.[13]

The effort began none too soon. In February 2001, at a press conference held as the rival scientific papers were published in rival journals, Mark D. Adams of Celera announced that the company had assembled a mouse sequence covering 99 percent of the genome—and would reserve it for its database customers. He added that Celera could find only 300 human genes with no obvious counterparts in mice.[14] Late in April, the company confirmed that it had begun working with DNA from three laboratory mouse strains a year earlier and had achieved sixfold coverage, with each nucleotide sequenced six times. It had used none of the mouse data the public consortium had begun posting on the Web. The whole-genome analysis would not be published in a peer-reviewed scientific journal as this would probably require release of proprietary data. Venter commented tartly, "Those are other people's publication rules. We're going to set our own publication rules." But late in May 2002, Venter's team did publish its comparison of mouse chromosome 16 with human chromosome 21, which showed that the highly parallel sequences had "been remarkably conserved during the more than one hundred million years that have elapsed since the lineages leading to humans and mice diverged." Few genes were unique in mouse or human: 14 of 731 in the mouse and 21 of 725 in the human.

Less than two weeks later in 2001, NHGRI announced that a mouse draft with "threefold coverage," with each nucleotide sequenced three times, now covered about 95 percent of the genome. The scope of mouse-human comparisons had widened greatly, speeding the identification of unknown genes and the determination of how genes operate or contribute to disease. Notable genetic differences between mice and humans were underlined: The mouse has more genes for the sense of smell, the immune system, and sexual reproduction. Collins said ebulliently, "This is a great day for finding genes in the human." He added, "Now we need to finish the work so the mouse sequence is as accurate and complete as the human sequence." Others were less polite about the public project's need to catch up. The data on the Web were not user-friendly, causing some exasperated users to pay Celera's academic-rate subscription fee. Researchers told Eliot Marshall of *Science* that the public mouse DNA collection "is so fragmented, unordered, and full of gaps that it doesn't enable [us] to study the structure or function of genes."[15] In July, Venter derisively said, "N.I.H. could provide a subscription to every scientist in the U.S. to see the complete [Celera] annotated mouse genome for less than it costs them to sequence it."[16]

Not deterred, NHGRI pushed on with a new phase, one that would again feature the clone-by-clone approach much more prominently, and invoked labs like David Haussler's at UC Santa Cruz, and Ewen Birney's European Bioinformatics Institute in England to develop more user-friendly formats.

A year later, just in time for the fifteenth annual genome conference at Cold Spring Harbor, and just before the Venter team's comparison of mouse and human chromosomes was published, the public project posted an assembled, user-friendly, sixfold mouse draft on the Web, and the scientific surprises began multiplying. The work went much more smoothly, as the sequencing centers used upgraded machines and sample preparation. Advances in computerized information-handling accelerated the process, although it was still largely based on whole-genome shotgun sequencing. Two laboratories, the Whitehead Institute in Cambridge, Massachusetts, and the Wellcome Trust Sanger Institute in England, had both developed software. The Whitehead's was called Arachne, and the Sanger's was called Phusion. The end result, Collins said, was far better than the human draft in 2001. Venter, now back at his nonprofit scientific work, said he felt "wonderfully vindicated."[17]

Speaking at the Cold Spring Harbor meeting, Robert Waterston plunged into the surprises that foreshadowed a sea change in how genes and genomes were perceived. The human sequence reported the year before, with its small number of genes occupying only 1.5 percent of the whole genome, had been arresting. But now, a particular finding about the mouse crossed an intellectual threshold. An important question in comparing the sequences of the mouse and the human concerned how much of the two genomes had been "conserved" over the last 75 million years. By a long-established consensus among geneticists, this information would point to indispensably important stretches of DNA. The notion was that conservation arose from an evolutionary version of the maxim: "If it ain't broke, don't fix it." Features that are adaptive in a wide range of environments are retained.

A computer-science undergraduate student of Haussler's, Krishna M. Roskin, found that an astounding 5 percent of the two genomes, under heavy evolutionary pressure, remained unchanged. A portion of the conserved DNA sequences represented coding regions, amounting to 1.5 percent of the genome, roughly the same as in the human draft of 2001. But what was the function of the other 3.5 percent of noncoding, "silent," essential DNA? Presumably, it contained the myriad of controls that determined when genes were turned on, up, down, or off, implying that the genome contained more than twice as much DNA for regulation as it needed for spelling out proteins. Haussler observed, "To have an undergraduate contribute so substantially to a very hot topic in science is pretty stunning." Roskin said, "I guess you could say that my senior thesis research was published in *Nature*, which is pretty cool."[18]

Surprisingly, the mouse had 178 gene-poor regions in common with the human. And evidence was mounting that many control segments lay at a considerable distance from their genes. Researchers like Edward

Rubin and Inna Dubcak in California wondered if the essential noncoding DNA helped to keep the structures of chromosomes together or to manage the pairing of chromosomes just before cells divide to make two daughter cells.

The view from the mouse made the complexity of genomes strikingly evident. Lander commented, "The mouse genome turns a spotlight on all sorts of features that were not apparent from the human genome."[19] This was not wholly disconcerting because it was proving difficult to define what pieces of human genetics determined how humans differed from their fellow mammals, including close relatives like chimpanzees and rhesus macaques. The noncoding DNA emerged as the probable seat of human uniqueness. But there were many puzzles. The mouse genome showed that noncoding DNA, supposed since the late 1970s to be unessential, did not undergo mutation faster than the coding segments, which resided in slowly mutating genetic "canyons." Instead, the noncoding DNA, dwelling in so-called "plateaus," mutated at different rates in different regions. Waterston bemoaned "a complication we didn't anticipate." The ground was shifting under geneticists' assumption that DNA mutations occurred at a constant rate, as if according to a "ticking" evolutionary clock. Pui-Yan Kwok, a geneticist at UC San Francisco, exclaimed, "Genomes are evolving in a completely nonuniform way."

Haussler recalled, "We were very interested in finding out how much of the human genome is still very similar to the mouse genome, suggesting that it is being conserved through natural selection." Interviewed by Elizabeth Pennisi of *Science*, he said, "We're seeing there's more to the story."[20]

The discoveries about the mouse dealt an emphatic blow to a simplistic "gene-centered" view of the genome, bringing both exasperation and delight to the genomic community. The field was being turned upside down. To be sure, most of the discoveries pointed up new complexities that could place barriers in the way of understanding living systems and using that knowledge to alleviate suffering. But with so many assumptions being challenged and overthrown, as Lander put it, the young people in the field knew that important work lay before them. The frontier was not closed.

James Watson of the DNA double helix (Nobel Prize 1962), left, with Frederick Sanger, right, first to decipher a protein (Nobel Prize 1958) and inventor of the sequencing method automated for the Human Genome Project (Nobel Prize 1980). *Courtesy Wellcome Images*

Hamilton O. Smith, Nobel Prize winner (1978) for restriction enzymes, pictured where he works on artificial life. *Courtesy J. Craig Venter Institute*

Walter Gilbert, physicist, experimental biologist, sequencer (Nobel Prize 1980), entrepreneur, venture capitalist, collector, and photographer. *Courtesy Walter Gilbert*

Montana-born Leroy Hood, leader in automating Sanger's sequencing technique, advocate of personalized medicine, founder of leading biotech companies, and, in 2000, the nonprofit Institute for Systems Biology, Seattle. *Courtesy Institute for Systems Biology, Seattle*

David Botstein, yeast researcher, whose 1978 brainstorm was a landmark in mapping the location of genes on DNA, early skeptic about the Human Genome Project. *Courtesy Lewis-Sigler Institute, Princeton University*

Alec Jeffreys of Leicester University, 1984 inventor of DNA "fingerprinting," which is used worldwide to establish parentage, identify fragmentary remains, pinpoint criminals, and absolve the innocent. *Courtesy Sir Alec Jeffreys*

Maynard Olson, "skeptical chemist," on occasion consensus philosopher of the Human Genome Project he helped plan. *Courtesy University of Washington*

Robert Waterston, student of "the worm," *C. elegans*, leader of Washington University's sequencing center, and a frequent broker among fierce competitors. *Courtesy University of Washington*

Pieter J. de Jong, builder of "libraries" containing DNA clones, stored in scores of freezers and used by researchers around the world. *Courtesy Children's Hospital of Oakland Research Institute*

Board of Management, 2000, of the sequencing center at Hinxton near Cambridge, England, where one-third of the human genome was sequenced. Front (left to right): Jane Rogers, John Sulston, David Bentley, and Richard Durbin. Back (left to right): Bart Barrell, Murray Cairns, Alan Coulson, and Mike Stratton. *Courtesy Wellcome Trust Sanger Institute*

Sequencing factory at Hinxton, early 2000s, showing rows of machines made by Applied Biosystems, Inc. of California; in other rooms, semiautomated preparation and "amplification" of DNA samples and computers to crunch the data posted for free every night on the Internet. *Courtesy Wellcome Trust Sanger Institute*

Eugene Myers, advocate of the "whole-genome shotgun" method he applied at Celera Genomics to assemble one of the first two human sequences, completing it in the early morning of June 26, 2000. *Courtesy Eugene Myers*

Jim Kent, programmer at University of California, Santa Cruz. In a month, cooling his wrists in ice after long sessions on the keyboard, he completed the assembly of the global, nonprofit, public project sequence on June 22, 2000. He then developed the first browser for researchers to exploit the database posted online on July 7, 2000. *Courtesy David Haussler, UCSC*

Strolling the red carpet outside the East Room, June 26, 2000, moments after President Bill Clinton (center) praised the "most wondrous map ever produced by humankind." Left, J. Craig Venter of Celera Genomics; right, Francis Collins, director of the National Human Genome Research Institute. *Courtesy William J. Clinton Presidential Library*

Eric Lander, leader of Whitehead Institute sequencing center, December 2002, expounding the completion of the surprise-laden sequence of the mouse (in celebration of scientific publication in *Nature*). *Courtesy National Human Genome Research Institute*

At the Smithsonian Institution in April 2003, near the fiftieth anniversary of Crick and Watson's double helix, Francis Collins celebrating the "finishing" of the public project's human sequence. *Courtesy National Human Genome Research Institute*

Richard Gibbs, leader of the Baylor College of Medicine sequencing center, left, and David Cox, right, coinventor of the Radiation Hybrid method used in gene mapping. *Courtesy Richard Myers, HudsonAlpha Institute for Biotechnology*

Victor McKusick of Johns Hopkins University, a founder of human genetic research and an early advocate of the human genome project. *Courtesy Richard Myers, HudsonAlpha Institute for Biotechnology*

Richard Myers, coinventor of the Radiation Hybrid method, a leader in genomics at University of California, San Francisco and at Stanford, and inaugural director of a new research center in Huntsville, Alabama. *Courtesy Richard Myers, HudsonAlpha Institute for Biotechnology*

HUMAN GENOMIC SEQUENCE GENERATED BY LARGE SCALE CENTRES:

RELEASE

- Automatic release of sequence assemblies >1kb (preferably daily)
- Immediate submission of finished annotated sequence
- Aim to have all sequence freely available and in the public domain for both research and development, in order to maximise its benefit to society.

POLICY

- The funding agencies are urged to foster these policies

March 1996 note outlining the policy of "automatic release" of sequences longer than 1,000 nucleotides, determined at the Hamilton Princess hotel, Bermuda. *Courtesy Richard Myers, HudsonAlpha Institute for Biotechnology*

Bruce Alberts of University of California, San Francisco, chair of the 1986–1988 committee that designed the Human Genome Project, and later president of the National Academy of Sciences and editor in chief of the weekly journal *Science. Courtesy University of California, San Francisco*

Ronald Davis, a player in biotechnology since the 1970s, leader of development of techniques used in advanced sequencers announced in February 2010. *Courtesy Stanford Genome Technology Center*

Thomas Gingeras, a leader in discovering "pervasive transcription" of apparently functionless DNA, one of the major surprises of the 2000s. *Courtesy Michael Englert Photography/Cold Spring Harbor Laboratory*

"Jim is the first of the rest of us." May 31, 2007, Baylor College of Medicine. Richard Gibbs, left, and Jonathan Rothberg, founder of 454 Life Sciences, center, handing over a twelve-gigabyte hard drive containing the DNA sequence of Jim Watson, right, who promptly posted almost all of it on the Web. *Courtesy Human Genome Sequencing Center, Baylor College of Medicine*

Yale team that sequenced the "exome" of a five-month-old Turkish boy in 2009, finding that he suffered from a treatable disease. Front (left to right): Irina Tikhonova, Murim Choi, and Richard Lifton. Back (left to right): Ute Scholl and Shrikant Mane. *Courtesy Jerry Domian, Yale University*

Cancer geneticists (left to right) Bert Vogelstein, Kenneth Kinzler, and Victor Velculescu of Johns Hopkins Kimmel Cancer Center, Baltimore, explorers of pathways linked to cancer. *Courtesy Johns Hopkins University School of Medicine*

Standing amid the inevitable banks of computers, the Washington University team that in 2008 compared the complete sequence of both normal and cancerous cells from a victim of acute myelogenous leukemia (AML). Left to right: Richard Wilson, Timothy Ley, and Elaine Mardis. *Courtesy Washington University School of Medicine*

Honors and Celebrations

People often asked whether the genome project would result in Nobel Prizes, a problematic notion. The rules for Nobel Prizes in physics, chemistry, and medicine limit a particular award in a particular year to three individuals. The genome project was a vast international collaboration, with both commercial and philanthropic support, in which a battery of technologies was gradually perfected to create increasingly refined catalogues and directories of genetic information. Nobel Prizes tend to be awarded for a specific discovery or invention, not for leadership or lifetime achievement. To be sure, the decisions of the awards committees in Sweden, even with the help of solicited recommendations from around the world, can be quirky. For example, Albert Einstein received his expected Nobel Prize in physics for the photoelectric effect, not for his special and general theories of relativity. Otto Hahn was honored for discovering atomic fission, but not his coworker of twenty years, Lise Meitner, who had coauthored the first published interpretation of the discovery.

To a considerable extent, the Nobel Prizes related to the genome project had already been awarded for discoveries that opened the way. The discovery of restriction enzymes led to the 1978 Nobel Prizes in medicine for Werner Arber, Hamilton Smith, and Daniel Nathans. In 1980, the Nobel Prize in chemistry was shared by Paul Berg, a pioneer in recombinant DNA research, and the discoverers of two different DNA sequencing methods, Frederick Sanger and Walter Gilbert. Kary Mullis's discovery in the early 1980s of the polymerase chain reaction brought him a 1993 Nobel Prize in chemistry. The 2002 Nobel Prize in medicine was awarded to Sydney Brenner, John Sulston, and Robert Horvitz for their work on *C. elegans*, the nematode that played a vital role in the genome project.

Of course, there were other genome-related prizes. Collins and Venter won the 2002 Warren Prize from Massachusetts General Hospital, and in 2009 they received the National Medal of Science. Also winning the National Medal of Science were Paul Berg (1983), Herbert Boyer (1990), Daniel Nathans (1993), James Watson and Robert Weinberg (1997), Victor McKusick (2001), and Marvin Caruthers (2006).[21]

Leading figures in genomics also received the genetics award from the Gruber Foundation, the Allan Prize lectureship of the American Society of Human Genetics, MIT's Lemelson Prize, the Novartis Drew biomedical research award, and the Dan David Prize in Israel.[22]

The most encompassing honors for genomics came in 2002 from the Gairdner Foundation in Canada. International Awards of Merit went to Watson and Collins. Eight others received the annual Gairdner Award. Two were pioneers in "bioinformatics": Philip Green of the University of Washington and Michael Waterman of the University of Southern California (who attended Robert Sinsheimer's Santa Cruz meeting in 1985). Representatives of the public project on the list included Sulston from England; Lander, Waterston, and Olson from the United States; and Jean Weissenbach from France. Venter represented the private sector. Each of the ten received a check for $30,000.[23]

At a lunch in Toronto to announce the Gairdner prizes, Venter was the keynote speaker. Interviewed by a local reporter, he bemoaned the public's lack of understanding of genetics: "People have been getting too much of their science from TV." The genome, he cautioned, "is not the blueprint for life." In the drive for genetic knowledge to help cure disease, he added, "genomics is just the starting point." He denounced bans on stem-cell research by saying, "Public policy is being directed by confused religious beliefs."

Reporters asked him about the rivalry with the nonprofit genome project. Before the lunch, he said, "Jealousy is one of the strongest of human emotions. . . . Some on the public project say I've stolen their fame. But the fact is that they had too many centers and labs, and some of them wanted to make a career out of sequencing the genome." With another Toronto reporter, Venter was milder, "There are some of my colleagues there that are sort of upset with my success; they feel it was at their expense, although I think this award helps prove that it wasn't." Lander, asked for his opinion, said, "The public and private rivalry focused attention on the project . . . but the rivalry seems to be over."

A portion of Venter's lecture was printed a few days later in Toronto's *Globe and Mail.* He recalled being told as a graduate student in the 1970s that there remained little to discover in biology. But now, with so few genes in human DNA and so much uncertainty about how they are controlled, "our biology is even more complex than we imagined. Probably 99 percent of the discoveries in biology remain to be made." He spoke of the paradox that people are highly similar—"we are all virtually identical twins"—and yet the tiny differences between people are "significant." He mentioned that a recently discovered variation, conferring resistance to AIDS, had been traced back about 700 years, when the variation provided resistance to the plague.[24]

"From a First-Attempt Demo Music Tape to a Classic CD"

Meanwhile, the drive to "finish" the human sequence was rushing to its culmination, two years earlier than the original deadline of 2005. Olson told a reporter, "We wanted to get the sequence accurate enough so a scientist working with it wouldn't be limited by its quality." He added, "It was a messy process but we finally got there."[25]

On April 14, 2003, amid celebrations in Washington, D.C., of the fiftieth anniversary of the Watson-Crick double-helix model of DNA, the Internet posting of the new version was announced. This version, which would be refined continually over the years ahead, had been pulled together by a "complex, iterative process" six days earlier, only two hours after the project's self-imposed deadline. Just over half the sequencing was done in the United States, and another 40 percent took place in England. The sequence was now 99 percent complete, compared with only 28 percent less than three years earlier. Instead of 147,826 gaps, the finished version had just 341.

Errors had been reduced to about 1 in 100,000, a rate ten times better than the goal the program set originally. Continuous sequences had lengthened 334 times, from 82,000 bases to an average of 27.3 million. The overall cost was similar to that of the 1998–2000 race for a "rough draft," but the price per base had sunk to five cents.[26]

And the refined sequence was uncovering many mysteries, including a great deal of unexpected RNA. Lander commented, "There is probably no 'junk' in the genome." The consortium's analysis was published in a special issue of *Nature* in November 2004. The researchers wrote, "The human genome sequence is intended to serve as a permanent foundation for biomedical research."[27]

Despite a host of new ambiguities, which could be viewed as delaying the application of genomics to medicine, the genome project leaders spoke in upbeat tones. Collins called the achievement "a remarkable gift to all of humankind—all of the letters to our human construction book." Saying that "people will wonder how we got by without it," he added, "We'll be using this for decades, possibly even for centuries." Jane Rogers, director of sequencing at the Sanger Institute in England, observed, "It's a bit like moving on from a first-attempt demo music tape to a classic CD." Watson said, "I knew it was something that could be done and that must be done. . . . We've seen the best of human beings in the way the consortium came up with the sequence." Collins said the new, more accurate human sequence was "an astonishing adventure into ourselves," adding, "We can figure out my glitches and your glitches." Ari Patrinos, leader of the Department of Energy's genome effort, said the human sequence "has created a revolution, transforming biological science far beyond what we could imagine." Lander was delighted that "biology is being transformed into an information science."[28]

Bruce Alberts, chair of the committee that recommended going ahead with the genome project back in 1988, recalled his skepticism that the error rate would improve beyond 1 in 1,000 bases or that uninterrupted sequences would run more than 1 million bases. Now that accuracy and the length of contiguous sequences were hundreds of times better than his expectations, Alberts celebrated a "fabulous outcome."[29]

Gerald Rubin, leader of the fruit-fly genome project, praised the multitude that had finished the human sequence: "The people who stayed in the trenches deserve a lot of credit."[30] Richard Gibbs of Baylor College of Medicine, leader of one of the "Big Five" sequencing centers, was delighted that the comparisons now possible between human and nonhuman sequences would yield "valuable new insights into human evolution, as well as human health and disease."[31] Waterston ebulliently defined the genome as "the instruction set that carries each one of us from a one-celled egg through adulthood to the grave."[32]

But many arduous tasks lay ahead. Arno Motulsky of the University of Washington said in 2003, "It gets very complicated very fast. All we have now is the basic building blocks."[33] Ahead lay the tasks of identifying every gene and its controls, pinning down the interpersonal variations linked to disease, linking proteins to specific genes, and tracing the intricate interactions of proteins. *The Economist* wrote, "The exciting stuff now is not the sequence but what the sequence has to say about biology and medicine."[34]

RNA World

The estimated number of human genes kept sinking in subsequent years to around 20,000 in the 2007 calculations of Michele Clamp of the Broad Institute, narrowing the gap in gene number between humans and supposedly less sophisticated creatures. It was clearer than ever that most genes had more than one job—in fact, an average of at least five. Most genes, via dimly seen signals and controls yet to be deciphered, specified the structure of several proteins. Evidently, there were many alternate ways of reading not only the stretches of DNA that were expressed in messenger RNA but also the noncoding introns between them, where controls presumably lurked. In the process referred to as "alternative splicing," at least several protein-specifying messenger RNAs could be edited and spliced from one gene's sequence.

Inexorably, biologists were being drawn further into the world of DNA's close, possibly ancestral relative, RNA. Besides having one more oxygen atom than DNA, RNA also substitutes a base called uracil for thymine. Many scientists speculate that an RNA world lies closer to the origin of life billions of years ago than DNA. In the RNA world of today lies a tangle of different operations in which RNA appears often to edit itself rather than use enzymes for the job. Of course, RNA was not a new character in molecular genetics. The discoveries that messenger RNA carried copies of the protein blueprints to the factories where proteins are assembled, that transfer RNA escorted each of twenty amino acids to the site of assembly, and that ribosomal RNA was part of the platform where the assembly took place dated back to the 1950s and 1960s. In later years, additional RNAs showed up in specialized roles.

But few expected the blizzard of tiny RNAs that burst onto the scene in around 2000 and continue to fascinate biologists today. In a few years, whole new types of RNAs joined those that banded together in the making of proteins. Among them were the scores of small interfering RNAs (siRNAs). Double-stranded and little more than twenty bases long, they interfered at several stages along the cell's informational route from gene to protein, slowing down gene expression or stopping it altogether. For example, the tiny RNAs could block the first step in gene expression by attacking messenger RNAs soon after they were created. The blueprints for these tiny RNAs lurked in stretches of genomic "dark matter," not only in introns within the gene but also between genes. A boon to basic researchers who wanted to see what happened when a single gene was "knocked out," RNA interference also appealed to applied researchers who hoped to harness the siRNAs to cure diseases. Companies such as Alnylam of Cambridge, Massachusetts, sprang up to develop these technologies, which soon became almost ubiquitous in labs.

The discoveries were considered so important that they were honored with Nobel Prizes in 2006 to Andrew Z. Fire of Stanford University and Craig C. Mello of the University of Massachusetts Medical School, and with Lasker Awards in 2008 to Victor Ambros, also of U Mass Medical, David Baulcombe of the University of Cambridge, and Gary Ruvkun of Harvard University and Massachusetts General Hospital.[35]

New micro RNAs kept appearing. An online paper in *Nature* in February 2009 reported the finding of more than 1,000 of a whole new class of long noncoding RNAs, nicknamed lincRNAs, which seemed to play a variety of roles in cells, particularly in the copying of DNA into blueprint-carrying messenger RNAs. Among the authors were Mitchell Guttman, John Rinn, and Lander of the Broad Institute of Harvard and MIT. Guttman commented, "This new class of large RNAs really opens up a new avenue to understanding the genome." Rinn's take was, "We've opened up the wardrobe and entered into the mysterious world of noncoding RNA. Now we need to find the mechanisms by which these creatures act." Most ebulliently of all, Lander said, "We've known that the human genome still has many tricks up its sleeve. But, it is astounding to realize that there is a huge class of RNA-based genes that we have almost entirely missed until now."[36]

Complexity Increases

In just a few years, as technical advances expanded the kinds of questions that could be asked, consensus beliefs kept taking blows. The picture of how living cells operate became subtler and more complex. Genes were no longer seen in isolation from numerous mechanisms for controlling their activity. In this picture, "receptors" at the cell surface sense changes in the environment outside the cell, and chains of slightly modified proteins send chemical messages down to the DNA vault in the cell's nucleus, where the DNA is regulated in many ways. As had been discovered in the early 1960s, proteins called repressors lying across genetic start buttons could be lifted to allow genes to be read.

The biologists also began to find many "epigenetic" influences—factors outside and around the genes—on the operations of genes themselves. These epigenetic changes, helping to keep a gene silent or allowing it to turn on, include the addition or subtraction of methyl, acetyl, or phosphoryl groups, not only in the DNA itself but also in the family of histone proteins that wrap tightly around it. Methylation and other changes influence how tightly the DNA is wrapped and the DNA's readiness to be copied into messenger RNAs. The wrapping must loosen for a gene to be expressed and bind more closely to silence it.

Rearrangements

The little RNAs were not the only new denizens of the genomic jungle. More emerged from the industry of comparing animal and human genomes, which took off after 2001. The sequence of the dog, prey to many human diseases, was rapidly joined by the sequences of even more closely related primates, such as the chimpanzee, the gorilla, and the rhesus macaque. Posted constantly on the Internet, the nascent sequences from those animals could help in figuring out the functions of particular human genes. Comparisons to the gradually more detailed human sequence extended to puffer fish, a vertebrate species more than 400 million years old; zebra fish, of particular use in studying development from egg to adult; and a solitary and ancient marsupial, the duck-billed platypus. To enrich the picture from earlier model organisms and look more deeply into the evolution of life, a dozen species of fruit flies and a swarm of different yeasts were sequenced.[37]

There was particular interest in the effort of Svante Pääbo in Leipzig to sequence the rival humanoid species, the Neanderthal, which went extinct about 30,000 years ago. The job looked hopeless, given the heavy contamination of any Neanderthal bones with modern human DNA. But in recent years, relatively uncontaminated samples were found in European caves. Taking elaborate precautions to exclude all but Neanderthal DNA, Pääbo and his colleagues put together a rough version and pushed on to greater detail. The Neanderthal sequences were expected to throw more light on what was unique about today's humans as they arose in Africa around 150,000 years ago. The work, frequently reported by journalists, provided a good deal of fun alongside all the new genomic puzzles, helping sustain popular interest.[38]

The orgy of comparisons led to the discovery of unsuspected new elements in the human genome. The story began as soon as a human chromosome, such as number 21, was "finished" even before the first genome-wide sequence. The main tools were increasingly discriminating DNA chips, microarrays that could look at targets too small to see in the microscope and too large to be captured by existing sequencing methods. Very soon, the SNPs had company as badges of genetic variation. Little-suspected, large new classes of genetic differences, within and between species, had to be considered alongside those that previously preoccupied genomic researchers. The catchall names for the numerous mutations included "rearrangements," "structural variations," or "copy-number changes." As early as 2003, it was evident that the evolution from primates to humans had brought changes not only within known genes but also in the noncoding spaces between them.

Chromosomes fused. Segments of DNA, sometimes rather long ones, were duplicated, but not always the same number of times in individual humans or model organisms. In other segments the normal order was reversed. Still others involved DNA insertions or deletions of considerable length. The newly discovered rearrangements created an unexpectedly richer picture of the evolution of humans and opened a new window onto the sources of disease, helping explain why one person was in greater danger than another.

In 2004, more sensitive methods uncovered many additional copy-number variations of sizes ranging from 100,000 to 2 million bases. Among the discoverers were Jonathan Sebat and Michael Wigler at Cold Spring Harbor Laboratory (CSHL) and John Iafrate, Charles Lee, and others at Brigham and Women's Hospital in Boston. Some of the variations were linked to inherited neurological disorders or cancer, and some appeared to have arisen in the lifetime of parents of children suffering some form of autism. By the annual genome meeting of 2005, researchers like Evan Eichler of the University of Washington were reporting that all the rearrangements varied between people. Birney, a leading analyst of genome data, said, "We knew these variations existed, but this year we're asking, do they matter? The answer seems to be yes."

It soon appeared that the mere number of copies could turn out, as *Nature* noted, to be "too much of a good thing" and to foster illness. It was only a short step to a concerted effort by researchers in a dozen labs stretching from Tokyo to Toronto to Barcelona to make a genome-wide map of copy-number variation. Richard Redon of the Sanger Institute in England and his colleagues reported that the 1,447 variations they found in 12 percent of the genome amounted to more nucleotides than all the SNPs in the same stretch. The researchers said their map "demonstrates the ubiquity and complexity of this form of genetic variation." They called for further refinement of the reference human sequence, a process that continued after 2003, "to enable further unraveling of the complexity of human genomic variation." Commenting on the work, two scientists at Duke said the copy-number variations discovered so far were "probably the tip of the iceberg."

A year later, the monthly journal *Nature Genetics* ran a special section on the new forms of variation. Stephen Scherer of the Hospital for Sick Children in Toronto and his coauthors termed the discoveries "an explosion" and called for standard ways to analyze the information. The maps would have to grow far more detailed, and the work would have to go beyond the DNA chips into full-scale sequencing.[39]

"Extra" RNA—Lots of It

The rapid evolution of the DNA chips, increasingly generated by commercial suppliers rather than "home-brewed" in academic labs, brought even more sweeping surprises. One of the microarray companies, Affymetrix in Santa Clara, California, played a major role. A team led by Thomas Gingeras, a compact, quiet man, was also in the business of studying what happened in small human chromosomes like number 21. They began finding a far more complicated architecture in the genome than anyone expected. Most of the supposedly "silent" DNA was transcribed into RNA, not just the DNA of the genes and their presumably close-by controls. This discovery was so startling, so contrary to the neat, linear view of the genome, that colleagues first wondered if the finding was real or just an experimental "artifact." Later, when the biology community began accepting the reality of "pervasive transcription," people wondered at such apparently purposeless prodigality in the cell. What was the extra RNA for?

Gingeras and others in the international consortium cataloguing all the functional elements of the genome scratched their heads along with the doubters but said they would have to face up to the findings. They accepted the designation of "heretics." By 2007, it was clear that many controls, such as signals to start making a particular DNA copy, lay far from the genes they controlled, even on other chromosomes. Such "transcription start sites" could reside within other genes. Genes could overlap, sharing expressed sequences, the exons. The reading of genes did not just go in one direction from a starting point; the process could go the other way too. A particular protein blueprint could draw not just from one DNA strand but the other as well.

In addition, there was trouble for neat ideas about evolutionary conservation. A great many conserved segments of DNA did not have any obvious function. And many other segments that did have a function were not conserved. There seemed to be a great reservoir of DNA in which changes neither helped nor hurt but which might be drawn upon for beneficial mutations in the face of prolonged changes in the environment.

Graduating from Catholic University in Washington, D.C., in 1970, Gingeras went on to New York University, where his doctoral thesis in 1976 concerned the chemistry of the reproductive system in the fruit fly. Richard Roberts of CSHL, whose lab was discovering one restriction enzyme after another, recruited Gingeras because of his interest in fruit flies. Soon, however, after an obligatory stint finding a restriction enzyme, Gingeras found himself immersed in genetically mapping adenovirus, for years a major research tool in the lab's cancer research. Not long after, other studies of adenovirus led to the epochal discovery of split genes. Roberts saw a chance to sequence the entire 32,000 nucleotides of adenovirus. James Watson, director of the lab, seized on the idea and, with typical precipitation, told the *New York Times* the job would be done in six months.

With Roberts off for a sabbatical, the task fell to Gingeras, who endured numerous prodding visits from Watson. The sequencing job was not finished in six months—the last 20 percent was always hard—and Watson was "beside himself," Gingeras recalled, and "never forgave" him when another lab, in Uppsala, Sweden, also worked out the sequence in the early 1980s. From Gingeras's point of view, the project had positive results. He learned how to develop programs to "assemble" genomes, searching for where genes started and stopped and working out the amino acid sequence of proteins. Such programming used some of the first computers at Cold Spring Harbor, which were landing on more and more desktops despite Watson's intense suspicion. A result was a program for identifying the places where particular restriction enzymes did their cutting. Gingeras recalled, "He just thought we were wasting our time playing with this TV screen."

Moving on to the Salk Institute in La Jolla, California, Gingeras simultaneously worked for a biotechnology company named Sibia and at the University of California, San Diego. He became involved in isolating RNA from viruses like HIV in individuals. The work led to a method for rapidly multiplying RNA samples for analysis, a method that made money for Sibia. In the early 1990s, after some two years at a drug company interested not in research but only in making "a lot of money fast," Gingeras was recruited by Stephen Fodor, founder of Affymetrix and pioneer of commercial DNA chips. Fodor was interested in whether the microarrays could be used to sequence the DNA of the AIDS virus, HIV. Although Gingeras thought not, he came over to try the experiment and stayed some fifteen years, rising to vice president for biological science before returning in 2008 to Cold Spring Harbor as head of "functional genomics."[40]

Around 2000, he recalled, he and his colleagues tried "a very simple experiment" that drew him away from commercial research and deep into the RNA world, which he came to call "RNA space." The experiment was only possible now that a "reference" human sequence was available to all. In an instance of the frequent private-public partnerships in genomics, the National Cancer Institute (NCI), interested in probing differences between normal and cancer cells, became a sponsor. To work on this, the Affymetrix team's idea was to "interrogate" the "finished," that is, highly "annotated," human chromosomes 21 and 22. They would do a census of the RNA transcribed from DNA in these chromosomes and found in the outer domains of cells (the cytoplasm or cytosole, where proteins are made). Their straightforward expectation was that the RNA they detected would match up with the genes that had been identified in the most accurate human sequences so far.

The two chromosomes contained at least 770 "well-characterized and predicted genes." About half of these, 127 on chromosome 21 and 247 on chromosome 22, were regarded as "known" genes, containing a total of 4,564 protein-coding exons. The exons were expected to be transcribed into RNA messages for making particular proteins. In their survey, the researchers used new DNA chip probes (which they helped develop) to study eleven different lines of human tumor cells. Their target was RNA with a string of adenine nucleotides attached in what was called a "poly-A tail." To their astonishment, they found ten times more RNA than they anticipated.

As they tested their results for flaws, they found that the previously unsuspected RNAs "are present at low copy number per cell," perhaps explaining why nobody had picked them up before. They concluded that "a significantly larger proportion of the genome is transcribed and specifically transported as mature cytoplasmic poly $(A)^{+}$ RNA than previously considered."

The first report of Gingeras's team, which included Philipp Kapranov and Affymetrix founder Steve Fodor, came in May 2002, just as researchers were reading of the remarkable similarities between the human and mouse genomes. The response was not positive. Their findings, Gingeras recalled, "triggered a very large push-back from the scientific community," which divided into two camps. One said, "Yes, we *knew* that this was happening [and] absolutely *had* to be true—that the rest of the genome was being used for something." But the majority said, "This is all garbage and just simple technological noise."[41]

Clearly, this extra RNA's purpose was a mystery that demanded more research. Within months, NCI committed an additional $4 million to Affymetrix to probe more deeply into the puzzle. With such collaborators as Kevin Struhl of Harvard Medical School and Daniela Gerhard at NCI, Gingeras, Kapranov, Jill Cheng, and others at Affymetrix began finding more mysteries. These included (1) unusual locations for swarms of signals to start transcription, (2) apparent overlapping of coding and noncoding sequences, (3) apparent transcription from the opposite DNA strand, and (4) the tendency of many of the RNAs to stay within the cell's nucleus. Labs in Chicago, Baltimore, and Cambridge, Massachusetts, reported related findings. By the fall of 2003, Gingeras's team was writing of "the possible reconsideration of current concepts of the structure and functioning" of genomes in cells with a nucleus (eukaryotes). A year later, the message was, "The observations derived from these studies provide some pause as to the state of our understanding concerning where and how the information from the human genome is organized."[42]

These observations caught the attention of others in genomics, and Gingeras found himself giving a keynote address to the 2005 annual "Biology of Genomes" meeting at Cold Spring Harbor. He advised his listeners in Grace Auditorium to "sit back and put away earlier notions" of the functioning of genomes. He told them that transcripts of unknown functions, or TUFs, vastly outnumbered transcripts from known genes. Attitudes must change, he said, from "gene-centric to transcript-centric." After a recital of the surprises, he jocularly advised, "Don't worry! Be happy! It will all work out!!!"[43]

"Now We Are Reading the Spaces in Between"

By this time Gingeras was one of the leaders of the international ENCODE project, along with Birney. Their work recalled Lander's comment that the new sciences of genomics were building a periodic table of biology, analogous to the chemical version put together in the nineteenth century. Even the pilot ENCODE study of selected pieces of the genome, published in 2007, reinforced the sense of ground constantly shifting under the biologists' feet.

The members of ENCODE had focused on about fifteen regions where knowledge was already substantial; fifteen others were chosen at random. They pulled together 200 different databases. The ENCODE report confirmed the finding of "pervasive transcription" into RNA. Many of the newly discovered noncoding sequences overlapped with coding sequences. New details emerged about DNA's interaction with the surrounding protein packaging in the chromosome. The editors of *Nature* defined the ENCODE task as finding "how the cell actually uses [the genome] as an instruction manual," and *Nature* reporter Erica Check phrased the challenge as working out "how our cells make sense of the DNA sequence in the human genome." Commenting on ENCODE's results, John Greally of the Albert Einstein College of Medicine in New York wrote of "some surprises that challenge the current dogma of biological mechanisms. . . . The simple view of the genome as having a defined set of isolated loci transcribed independently does not seem to be accurate."

With massive effort, the ENCODE consortium had not only compiled a list of functional elements but also compared human DNA transcription with that of twenty-eight other animals. They found that some 40 percent of the sequences constrained by evolution seemed to lack functions. And some 50 percent of the functional segments showed little evidence of constraint. Greally wrote, "Surprisingly, many functional elements are seemingly unconstrained across mammalian evolution." Biologists, reporter Check observed, should look at "the idea that important DNA might be unstable." But Greally was not discouraged. On National Public Radio, he observed that ENCODE's finding of many important sites in "the wilderness" between genes "broadens the opportunities available to us" for genetic testing for risks of disease. Collins was not quite so ebullient. "It will take us a bit longer to sort out what is the mechanism of disease risk," he said, adding, "We were intended to be complex and we obviously are." He was "rather awed by what we are seeing."

Birney, who was delighted by the demonstration "that we can gain this information genome-wide," spoke of a "quite astonishing shift" in thinking about the way the genome operates. He also said, "People involved in genomics knew this stuff was not hanging around for the hell of it. The junk is not junk. It is very active. It does a lot of things." In the manner of the gene-number lottery a few years earlier, Birney made a "beverage of your choice" bet with John Mattick of the University of Queensland in Australia about the proportion of these RNA transcripts that actually did a "biologically important" job. Mattick said 20 percent, Birney said less. Gingeras was convinced that most of the RNAs were functional.

Reporter Elizabeth Pennisi of *Science* observed that the definition of a gene was no longer clear. There had been a major change from the old view of the genome as "compact, information-laden genes hidden among billions of bases of junk DNA." Now genes appeared to be "neither compact nor uniquely important." She added, "On the whole, many genetic changes don't affect overall biology." These unconserved changes—"gate crashers at a party"—served, Birney said, as a "warehouse" that "evolution can tap into when it needs to." Collins said, "It's like the clutter in the attic. It's not the kind of clutter that you would get rid of without consequences, because you might need it."

A survey of some 400 genes by Alexandre Reymond of the University of Lausanne in Switzerland showed that 80 percent of them had previously undetected coding exons and that some of the RNAs derived from exons in two genes, not just one. The promoters of genes involved in starting transcription could be found at either end of the genes, not just one. Reymond commented, "What we have found is that everything is interconnected like a web."

ENCODE's results received wide attention in the popular press, many members of which attended or listened in to the announcement press conference. Rick Weiss of the *Washington Post* wrote of "a dauntingly complex operating system" in which genes were not "discrete packets of information arranged like beads on a thread of DNA." Mark Henderson of the *Times* of London wrote, "An in-depth examination of the human DNA map has turned basic biology concepts upside-down." Steve Jones of University College London told the *Daily Telegraph*, "Now biologists are beginning to face up to the uncomfortable truth that they have only been looking at the nouns in life's lexicon—the crudest and most basic elements of any tongue. Now we are reading the spaces in between—verbs, adverbs, adjectives, pronouns and all the rest, and they are complicated indeed. Worse, the genome babbles, stutters and mangles its pronunciation and now and again seems to speak utter nonsense." *Le Temps* in Geneva said the task of making sense of the "ribbon of our genetic patrimony" was proving "harder than expected."

Amid the questions and ambiguities arising from the first round of ENCODE, there also was confidence that the consortium should go ahead with a second four-year effort to study the whole genome. With almost startling advances in the efficiency of genomic technologies, the costs were estimated at $23 million a year.[44]

Two years later, sitting on a terrace outside Grace Auditorium at Cold Spring Harbor, Gingeras said that his ideas were still developing. He was seeing evidence of "protein factories" in multiple locations in the nucleus of the cell as well as in its outer cytosole region. They would draw, from a variety of locations, the information for starting the transcription of DNA into RNA, for making proteins that do the transcription, for the boosters or repressors for the process, and for such processes as editing RNA. Because the new knowledge was still in the hands of specialists, Gingeras said, it was difficult to convince other scientists that many or most of these RNAs have work to do and to get the new insights into the biology textbooks. He added, "We are still in this world of un-integrated knowledge."[45]

CHAPTER SIX

Human Variety

To harness genomics to the control of common diseases like cancer or diabetes, biologists and their computer-savvy allies had begun changing focus even as they rushed to complete a draft human sequence in 2000. They pointed away from the universal to the particular, from what makes everybody nearly identical to the myriad of tiny changes making everyone different. The new focus would be on human variations—the changes that predispose one person rather than another to an illness or better-than-average health. Searching for such polymorphisms would transform genomics from an arcane and specialized enterprise into one requiring the cooperation of many thousands of volunteer human subjects willing to donate their DNA to studies of human variation.

This was not a new phenomenon. The classical subject called epidemiology, the measurement of the incidence and spread of disease, has long required the cooperation of many people—not only those suffering from a disease but also those who are not. Epidemiology serves not only as a warning system, as in an epidemic, but also as a tool to get at causes of disease. A famous early example was the tracking of an outbreak of cholera to a particular water well in London in the 1850s—and the shutting of that well against fierce opposition. Since the 1970s, epidemiology has been at the heart of intense debates about whether chemicals in the environment affected cancer rates. The wide variation in disease rates from one country to another—such as low rates of heart disease in Japan and high rates among Japanese Americans—strongly underlined the environment, not just genetics, as a factor in illness. But it proved very difficult, particularly in a highly mobile country like the United States, to get precise measurements of a person's exposure to a toxin—with the notable exception of cigarette smoking.[1] For the time being, at least, the search for causes of disease would have to rely on genetics.

After 2000, with a complete human sequence as a reference, and with increasingly powerful laboratory equipment and software to catalogue human variation, it was time to combine genomics with epidemiology. Genomics would become participatory.

The Paradox of Human Variety

Variation is a major paradox, not only for the field of genomics but also for members of the public struggling to understand what genomics means. Despite remarkable similarities in their DNA endowments, people also differ from each other in millions of small ways and, as became clearer with ongoing research, some big ones. During the first decade of the twenty-first century, as mentioned in the previous chapter, researchers found that these differences entailed not just mutations in single nucleotides but also the menagerie of "structural variations." These included larger deletions and insertions, reversals of sequence, and copy-number variations—all of them increasing or decreasing a person's chances of suffering a particular illness.

The paradox of similarity and variation was underlined in a talk late in 1999 by Eric Lander at a White House "Millennium Evening" for President and Mrs. Bill Clinton and their guests.

> Quite apart from its medical significance, the texture variation in the human genome holds great fascination. Any two human beings on this earth are 99.9 percent identical at the DNA level—only one difference in 1,000 letters. So as you look to your neighbor to the left and to the right, you should appreciate how nearly identical you are. [Laughter.]
>
> On the other hand, one difference in 1,000 letters in a genome of 3 billion letters still translates to 3 million differences between any two

> individuals. So if you look to your left and your right again, you can also revel in your absolute uniqueness. [Laughter.]

The stubbornly held concept of "race" fades in the face of such genomic knowledge. Any two people, Lander said, have one tenth of one percent difference from each other, but two ethnic groups, although they may often be at war, have much less.

> In fact, since we all left Africa seven thousand generations ago, there just hasn't been a lot of time to build up large amounts of genetic variation. We do see differences. In fact, we're cued into seeing differences between people. That's very misleading about what is really going on at the genetic level. You may think two humans look very different from each other, but the truth is they're much more genetically similar than two chimpanzees are. Chimps have much more genetic difference within their species than we do, because we are such a new, young, small species.[2]

These remarks impressed President Clinton so much that he immediately began to mention them in speeches and kept doing so, as in a "Class Day" speech at Harvard in 2007.[3]

"Get Outside of Hypothesis-Limited Science"

Even before the draft human sequences of 2000, one of the first to plunge into human variation was David Altshuler, a physician specializing in endocrinology and internal medicine at Massachusetts General Hospital. The son of Alan Altshuler, who became transportation secretary of the Commonwealth of Massachusetts and dean of the Harvard Graduate School of Design, he received a bachelor of science degree from MIT in 1986, then went on to take a PhD at Harvard in 1993 and an MD at Harvard Medical School the next year. Joining the Whitehead Institute as a postdoc in the late 1990s, he became a founding member of the interdisciplinary Broad Institute of MIT and Harvard in 2003. Like others who followed the central tenets of genomics, he was dissatisfied with a medicine of "one size fits all." To be truly useful, he thought, medical research must identify the actual mechanisms of disease and, according to a person's particular genetic makeup, find and deliver specific treatments for those mechanisms. The inadequacy of a blunderbuss medicine aimed at symptoms, not causes, came up every day in the clinic. "Each time I write a prescription for a new patient," he told a reporter in 2000, "I know it might not help at all and could do harm."[4]

Although environment, including lifestyle, was, and would remain, crucial in treatment, Altshuler was profoundly pessimistic about exhorting patients to change their behavior. What fraction of people could be persuaded to eat less or walk more to avoid such rising scourges as obesity and diabetes? It was not the same problem as smoking, Altshuler recalled in 2006. "You don't need to smoke to live, but you have to eat to live, so it's quantitative as opposed to qualitative." He continued, "If those things could be done, I think they could actually have huge impacts, so I'm very much in favor. However, as a clinician, having patients come to you and simply doing what is done today—just say, 'You should lose some weight. Let me send you to a dietician, and, you know, and try and walk more,' is completely ineffective. The studies have shown over and over again that it doesn't work. . . . Diabetics in America aren't controlled [with] available medicines and lifestyle modifications." In such a situation, Altshuler was convinced, it was necessary to move beyond "wishful thinking" to admit that "we should invest in knowing more about the disease" so that it could be detected, prevented, and treated.

He had not come into genomics as "a genomics person." Instead, "I'm interested in disease. I came to the conclusion that the best hope for disease was to find the genes. . . . The only way we were going to find the genes was to understand genetic variation." Altshuler's path to this involved the same attitude that brought about the genome project, a desire to "get outside of hypothesis-limited science," in which a huge number of people pursued the same leads, the same "very small number of reductionist models." Instead, there must be a complete catalogue of genetic variation, unbiased by educated guesses. To Altshuler, "What we really care about, as medically oriented people, is the disease in the population." He went into genomics, he said, because "I saw a total disconnect between the science I saw and the patients I took care of in the hospital."[5]

The Haplotype Simplification

At the Whitehead Institute, just one stop on the Boston Red Line subway from Mass General, Altshuler joined the growing multilaboratory group that was not only building a catalogue of millions of human single-nucleotide polymorphisms (SNPs) but also seeking ways to simplify the analysis so that insights about disease would emerge more quickly. They were working toward imperatives that were clear even before the great scale-up in human sequencing of the late 1990s. In *Science* in October 1996—a month after the paper by Neil Risch and Merikangas mentioned in Chapter 4—Lander wrote of assembling a "comprehensive catalogue of common variants" in the human genome. This would be achieved by sequencing the coding regions of one hundred people and comparing these with the first human sequence. With the sequence, he argued, one could go beyond the modest-sized studies of affected families to larger studies, comparing the mutations of particular genes in a well-studied group of patients.[6]

At first, it was feared that the vast domain of human variation would be too complicated to map. The SNPs, points at which a single base in one person's DNA differs from that of another, numbered in the millions. These small mutations touch tens, even hundreds of millions, of lives because they are involved in cancer, heart disease, diabetes, and other common maladies. Was it realistic to look for new precision in the decades-long task of finding and isolating many more genes related to disease and the proteins that many of these genes specify?

The immense number of SNPs—10 million affecting at least 1 percent of the human population—quickly drove Altshuler and such colleagues as Mark Daly to a major simplification. Exploiting a bit of luck, they discovered that clusters of SNPs tended to be inherited together and thus to reside close to each other along the chromosomes. These groups of DNA segments, which tended to pass as a group from one generation to the next, formed blocks called haplotypes. This insight reduced the nightmarish 10 million variables by a factor of twenty—to only 500,000 locations along the DNA.[7]

This number fell into the range just being reached by DNA microarrays. These "DNA chips," manufactured by companies like Affymetrix,

Illumina, Agilent, and NimbleGen using the methods of Silicon Valley in making semiconductor circuits, would soon be able to sample a person's genome at half a million points. So a "HapMap" project could carry the hunt for SNPs to a new level. Costing more than $130 million in its first stage alone, the project began in 2002 with 270 volunteer human subjects, evenly divided among Asians, Africans, and Americans of European ancestry. The first round, analyzing about 1 million SNPs, was completed in the fall of 2005, and the second study, published two years later, analyzed more than 3 million SNPs. Altshuler was delighted at new knowledge about both human disease and human evolution.[8]

The effort had been foreshadowed in February 2001, when the SNP Consortium reported, in the same issue of *Nature* that analyzed the draft human sequence, that it had already catalogued 1.42 million SNPs. With DNA chips of increasing power, the consortium noted, the shortcut way was open to haplotype mapping of variation across the human genome. Not far behind lay a genomic version of the classical "case/control" studies of epidemiology, including tests of the safety and efficacy of drugs. With the reference human sequence in every researcher's hands, they could use microarrays—followed by sequencing of short segments of DNA—to test the genomes of many people who suffered from a disease against the DNA of control groups of many people who did not. Echoing earlier researchers, they wrote, "We anticipate immediate application to studies of human population genetics, candidate-gene studies for disease association, and eventually unbiased, genome-wide association scans."[9]

The course from HapMap to genome-wide association studies, which happened over about five years, had a heavy freight of advocacy behind it. Director Francis Collins of the National Human Genome Research Institute (NHGRI) declared, "This is the single most important genomic resource for understanding human disease, after the sequence. We needed it yesterday, as far as I am concerned." He didn't agree with critics who dismissed the HapMap as a boondoggle to keep the machines humming at the big sequencing centers now that the human genome sequence had been assembled. The new project was launched into a crosswind of old resentments against the genome project, with its focus on data rather than hypotheses and its concentration of effort so as to carry out the project.[10]

Academia and Industry Compete—and Cooperate—on Chips

The power of DNA microarrays to probe human variety kept growing dramatically. These tiny rectangles of glass coated with a grid of thousands of DNA "probes" had been invented around 1995 by such pioneers as Stephen Fodor of Affymetrix and Patrick Brown of Stanford University. Soon, in intense competition as well as cooperation with university laboratories, the DNA chip could measure thousands of genes at once. The implications for such diseases as cancer were profound. Medical researchers could now view a mosaic of genes that were "on" in a cancer cell and others that were not. The activity of healthy cells could be compared to that of cancerous ones, in what were called expression studies.

The chips powerfully extended what the clinical pathologist could see through the eyepiece. Cells that looked like the same type of cancer now divided into several classes—distinguishable varieties of genetically linked disease. This was an important clue to a huge problem in treating cancer patients. Drugs helped, say, a third of patients, did nothing for another third, and harmed the rest. Now the researchers and doctors had a hint as to why. They could also ask a related question: Did differences in patients' genetic makeup affect how they responded to a particular drug? Could the treatment be tuned to the patient?

The competition and collaboration between academia and companies like the California-based Affymetrix and Illumina arose both from varying research interests and money. The university researchers could not afford to pay for large numbers of commercial chips and often wanted them custom-built for particular research questions. So they embarked on a considerable movement for "home-brew" chips. But the companies were not idle. They boosted the capacity of the microarrays again and again and sometimes formed collaborations with such large laboratories as the Broad Institute. One aim was to extend the chips' analytical reach beyond SNPs into the longer structural changes in DNA.[11]

Before the final rush toward the human DNA sequence in 1999 and 2000, the DNA microarrays had soared from a novelty to become a $1 billion industry that was expected to explode in sales as the chips' price came down and they became routine tools of diagnosis. Besides helping doctors identify patients who would be helped or harmed by a drug, the microarrays had also begun assisting drug companies to search for new drugs, seed companies to develop new varieties of food plants, and scientists to identify genes that predispose a person to a disease.

Establishing Trust

Cooperation from disease victims and healthy people over many years had always been central to human geneticists seeking the cause of diseases that ravage families or "defined" populations living in small areas. Such geneticists as Victor McKusick, who died in 2008, a founder and leader of the field for decades, have had to inspire and maintain confidence and trust in afflicted populations amid emotional whipsawing. On the one hand, an individual may be highly motivated to help find preventives, palliatives, or cures for a disease that has struck him or one or more beloved relatives. But the family may feel fear or shame, or siblings and other relatives may not want to know of a curse that could strike them too.

Given the anguish of families suffering from a rare but catastrophic illness, their collaboration with geneticists can be considered astonishing. An explanation may be a form of the altruism that motivates millions of people to donate their blood to people who need it now, so as to maintain a system they or those dear to them may need one day. Many people taking part in these studies express the noble hope that understanding the roots of their suffering will aid fellow human beings.

McKusick for many years studied genetic illnesses in the Old Order Amish in the United States. Mary-Claire King won the collaboration of breast cancer sufferers in her search for genes implicated in the disease. Nancy Wexler, in her battle to find the gene that caused Huntington's, won the collaboration of the 1,000 related people living around Lake Maracaibo in Venezuela. Wexler has taken and occasionally shows films of the sufferers, who, with twisted limbs and halting movements, show in their faces an unquenchable human dignity.

A DNA World

Soon after 2000, hundreds, then thousands and tens of thousands of people around the world began to feel the effects of the new world created by ordering the complete sequence of human DNA subunits. Even as more and more human research subjects donated their DNA, genetic science and its applications became increasingly salient in popular consciousness. Drugs involving manipulations of DNA, byproducts of a new biotechnology industry, continually entered the market. They included treatments for breast and lung cancer. Each day, an estimated 20 million people take cholesterol-regulating drugs called statins, which sprang from genomic research conducted since the early 1970s by Michael Brown and Joseph Goldstein of the University of Texas.[12]

On July 25, 2008, the first test-tube baby, Louise Brown, who had married in 2004 and had a two-year-old son, celebrated her thirtieth birthday. The clinic where she was conceived staged a party with thirty IVF babies, one for each year since 1978, to meet her. They represented at least 3 million others.[13] Meanwhile, the practice of prenatal genetic screening for a widening array of catastrophic inherited disabilities became more and more general. A growing number of newborns was screened for several dozen hereditary problems that called for emergency action or prolonged routines to stave off disease.[14]

To connect people to crimes or exonerate them and to resolve disputes over parentage, samples from millions of people—4.5 million in Britain and nearly 7 million in the United States by 2009—were stored for "DNA fingerprinting." Based on recording a dozen or so patterns in the DNA that distinguish one person from another, this invention of the English scientist Alec Jeffreys of the University of Leicester has been used since 1985 in establishing guilt or innocence of crimes. The tests have identified many who were found guilty but also led to many dramatic exonerations of prisoners on Death Row. Since 1989, the U.S.-based Innocence Project has calculated, DNA evidence has cleared more than 220 prisoners in thirty-three American states.

The first use of such DNA forensics was in proving that a boy from Ghana really was the son of a woman in England who claimed she was his mother. Not only did DNA fingerprinting settle paternity cases, but, as in the destruction of the World Trade Towers, it identified people whose remains could be determined no other way.[15]

Because DNA identification reaches so far into the criminal justice system and popular consciousness, it continues to arouse controversy. Huge backlogs of DNA samples have built up in crime laboratories, delaying thousands of criminal cases. Many prosecutors have resisted taking DNA samples or repeating tests with more sophisticated versions to check for miscarriages of justice. In Britain, human rights groups have campaigned vigorously to destroy DNA samples taken from people who are never charged with an offense. Two men even took their case all the way to Europe's Court of Human Rights in Strasbourg. "Struck by the blanket and indiscriminate nature of the power of retention in England and Wales," the court ruled in December 2008 that Britain did not have the right to retain records on the two, both arrested in 2001. It said that people "were entitled to the presumption of innocence." Charges against one of the English men had been dropped, and the other had been acquitted.

The British government stalled on changing the rules, arguing that many of the 800,000 "innocents" in its DNA database might commit crimes later; it eventually said that all samples could be kept for twelve years. Jeffreys was indignant: "My view is very simple. . . . Innocent people do not belong on the database. Branding them as future criminals is not a proportionate response in the fight against crime." In the fall of 2009, further public outrage forced the government into a "U-turn"—an agreement to drop the requirement to keep DNA records on the uncharged. Meanwhile, new protests erupted over the plan of the British Borders Agency to use DNA tests on people seeking asylum to determine if they are telling the truth about their nationality. Many human rights advocates objected that the methodology was flawed. Jeffreys was one of many who protested. "My first reaction is this is wildly premature, even ignoring the moral and ethical aspects."[16]

A Yoruba Community Enlists

For the HapMap project, the need to build trust extended to the highly literate Yoruba people, who number about 30 million and live mostly in Nigeria but also in several countries immediately to the west. Centuries of slave trade brought many Yoruba to the New World, along with aspects of their culture. Volunteers were sought in Abu Alamu, a town not far from the university city of Ibadan. People in the town had taken part in a study of breast cancer in families. Now, in July 2003, they were being offered the chance to participate in the global effort to build a haplotype map of the entire genome, an effort that would help make sense of the now-completed total sequence. Seeking genes not captured by family studies, geneticists from Nigeria and the United States, including Professor Charles Rotimi of Howard University, requested DNA samples from ninety residents in "trios"—a father, mother, and child—all tested to be sure they were free of HIV, which is killing millions throughout Africa. To assure diversity, their samples would be collected alongside those from Han Chinese in Beijing, Japanese in Tokyo, and Utah residents of northern European ancestry. The Yoruba were essential to the project because humanity originated in Africa, where, as Lander told the Clintons, variations have built up far more than in the groups that migrated relatively recently to different latitudes and continents.

The people in Abu Alamu were accustomed to the leadership of a council of elders and big town meetings every month or so, where 200 to 500 people gathered to discuss both community security and family quarrels. Thus, there was nothing unusual in holding focus groups to explain the HapMap project, interviewing a dozen "key informants," surveying some 300 people, and convening a "working group" of religious leaders, educators, practitioners of traditional medicine, and a local community advisory board. Town meetings were held at the start and end of the consultations.

During individual counseling, while they studied the consent forms translated into Yoruba, many people sensed a chance to make a contribution for the future of humanity. This was true even though those sampled would be neither paid nor given any personal medical advice. Indeed, to preserve privacy and avoid stigma, no information that identified a particular donor was being kept. The investigators mentioned these issues in part to dampen expectations of personal or community benefit and to prevent social pressure to take part.

As the study started, the community advisory board issued a very friendly letter of welcome to the geneticists, but added a formal request. Its members hoped for better medical care in their town. The local clinic was open only two days a week. The nearest health center, ill equipped, was some kilometers away, and rushing seriously ill patients there at night was always a bad experience. Could the already established breast cancer center be upgraded to a health center? Could the project build the town what was called a cottage hospital? These were highly practical requests, but the investigators, while sympathetic, did not have the money in their grants for such projects. They did facilitate extending the local clinic's hours, but otherwise they could only offer a promise to send the request back to their funding agencies. Collins of NHGRI arranged a full description of the consultations in Abu Alamu at the seventeenth annual meeting of genome scientists at Cold Spring Harbor in 2004. Noting that fulfilling this request would set a precedent, Collins said the issue had not yet been resolved.[17]

The HapMap Catalogue

The success of the worldwide HapMap effort to map the common variations in human beings was a major step in human genetics, and the scientific journal *Nature* featured the HapMap Consortium's report of its first comprehensive results on the cover of its October 27, 2005, issue. Simultaneously, a HapMap event was staged at the annual meeting of the American Society of Human Genetics in Salt Lake City, complete with a press teleconference. Michael O. Leavitt, a former governor of Utah and then U.S. Secretary of Health and Human Services, was on hand. He said that he had become interested in genetics and medicine as governor, when he received "medically significant" families at the governor's mansion. One guest told Leavitt that his grandfather, father, and now he had been diagnosed with age-related macular degeneration. Leavitt praised recent genetic work on the disease.

Collins then introduced nearly a score of the project's leaders, from Beijing to Tokyo to Ibadan, Houston, Montreal, Baltimore, Boston, and Oxford, and said that the project had uncovered "virtually every heritable cause of human disease." As a physician he said he had dreamed of this moment for twenty years. Now that research had escaped the trap of "secondary" symptoms and gone "straight to the illness," Collins hoped the HapMap catalogue of variation would inspire "rational drug design."

Altshuler of the Broad Institute, Thomas Hudson of McGill University, and Peter Donnelly of Oxford all said that planning had begun to exploit the HapMap in genome-wide association (GWA) studies. Microarrays had become cheap enough to test samples of thousands of people for their risk of particular common diseases. Noting a cut of more than a hundredfold in the cost of a genome-wide analysis of a person's DNA, Kelly Frazer of the California company Perlegen said that the data for the second phase of HapMap were already in hand. Well before that analysis was published two years later, many genome-wide studies were under way. And the next level of exploring variety, with an ultrafast new generation of sequencing machines, was rising rapidly on the horizon.[18]

Protection Against Genetic Discrimination

Volunteering for the studies was encouraged, at least somewhat, by laws forbidding genetic discrimination in employment or insurance, such as the one enacted in the United States in 2008. Collins, who had lobbied ceaselessly for the U.S. legislation along with such legislators as Rep. Louise Slaughter of New York, called it "a moment of celebration." He was among those photographed around President George W. Bush as he signed the bill into law. Delighted, Collins told the press, "Your skin color, your gender—all of those are part of your DNA. Shouldn't the rest of your DNA also fall under that protective umbrella?" For him, a "real cloud" had cleared for families with a history of particular genetic diseases. They could now step forward and be tested in the interest of science and their own medical care. "Hooray for that."[19]

With or without such protection, thousands of volunteers participated in the search for genetic causes of disease, using increasingly robust technology. The research no longer sought the single genes implicated in the 1980s in rare diseases like muscular dystrophy and cystic fibrosis. Indeed, finding common diseases linked to multiple genetic factors had been the principal motivation for the great gamble of sequencing the entire human genome. It was now possible to ask what caused diabetes, heart disease, cancer, irritable bowel disease, and mental disorders like schizophrenia and autism.

To get at these multigene illnesses, the researchers had to move beyond the relatively small numbers of people in families with a history of a particular disease to much larger numbers of unrelated patients and unaffected "controls." In genome-wide association studies, the genomes of each of these large human cohorts would be sampled to find genetic factors with a significant statistical relationship to a disease. The guesswork of probing for "candidate" genes of interest would be laid aside in what the researchers called unbiased studies.

Guilt by Association

In March 2005, after years of waiting for practical means of carrying them out, genome-wide association studies started up on a big scale. The opening salvo was fired by three independent groups, reporting simultaneously and online in *Science*, announcing that they had found a risk factor for a widespread disease of the eyes: age-related macular degeneration. As much as half the risk of the disease appeared to come from a mutation in a gene for the immune-response protein Complement Factor H, at a location called 1q31 on chromosome 1, the biggest human chromosome. The finding raised hopes of earlier detection and more effective treatment. This was the work that Secretary Leavitt praised that fall when the first stage of the HapMap was announced.

The first group, led by Josephine Hoh of the Yale School of Medicine, also included researchers from Rockefeller University, the EMMES Corporation, and the National Eye Institute. A second group, led by Margaret Pericak-Vance of Duke University, included colleagues from Vanderbilt University. In the third, leader A. O. Edwards of the University of Texas Southwestern Medical Center was joined by scientists from Boston University School of Medicine and the company called Sequenom.

The disease involves the so-called macula at the center of the retina, where light is focused so that we can perceive things sharply, straight ahead, as we need to do in reading, driving, watching movies and television, or playing a sport. In *Science*, Stephen Daiger of the University of Texas in Houston described the macula as containing cone cells, that is, "photoreceptor cells that are specialized for distinguishing colors, resolving closely spaced objects, and detecting motion." The macula, he wrote, "is critical for the full vision that enriches our lives." As people age beyond fifty, yellowish waste deposits called drusen begin to accumulate in the center of the retina. If the fragments are unusually large and numerous, they promote cell death in and around the macula, gradually suppressing central vision. An uncommon, or "wet," form of the illness leads to blindness, while the more common "dry" form of the illness progresses more slowly. The previous year, the U.S. Food and Drug Administration approved a drug called macugen to slow the progression of the wet form. Age-related macular degeneration afflicts some 10 million Americans.

The three groups of scientists knew that macular degeneration ran in families where the risk of the disease was several times that of the general population. But the cause remained a mystery. They moved beyond family studies of twins, siblings, and close relatives that had been zeroing in on chromosome 1. For more detail, they used the "case/control" approach of comparing groups of scores to hundreds of unrelated people, some with the disease and some without it. With the help of the HapMap catalogue of SNPs and the sequencing of particular stretches of DNA, they did not just pinpoint the protein but identified the particular amino acid altered by mutations in the SNP designated as rs1061170. When the amino acid tyrosine at position 402 in the protein chain changed to histidine, a person's risk rose almost threefold.

On National Public Radio, Hoh said that the research would have been impossible without the Human Genome Project. "I think this is the first step towards really understanding the pathogenesis of this disease." Daiger chimed in, "It's just one of many, many examples of how incredibly effective and useful the Genome Project has been." Responding to disappointment that cures appeared to lie far in the future, Daiger admitted, "We were expecting too much. . . . Humans are much more complicated than we originally thought. . . . To be honest, there's a large gap from the basic science understanding to actually treating people." Nonetheless, the wave of genome-wide association studies was off to a dramatic start.[20]

"Breakthrough of the Year"

In the ensuing cascade of scientific publications, hundreds of risk factors for the most common human diseases were found and reported. By late 2009, a catalogue kept at NHGRI listed nearly 400 published studies of GWA from around the world. The results from one group of volunteers usually were cross-checked against those in other locales. As a further check and in search of more detail, the researchers also sequenced specific stretches of DNA. In 2007, *Science* magazine called the flood of studies finding new disease-related factors (and potential targets for drugs) the "Breakthrough of the Year."[21]

By that time, several major studies of diabetes had been followed by the Wellcome Trust Case Control Consortium study of seven diseases that enlisted a total of 17,000 people—2,000 patients with each disease, and 3,000 controls who were free of all seven disorders.

The series began with a report from Iceland. There, since the late 1990s, 140,000 out of the nation's population of 300,000 have donated samples of their DNA for genetic research. Analyzing the DNA repeatedly was a specially chartered private company called deCODE Genetics, headed by the Icelandic medical researcher Kari Stefansson, who returned from the United States to start it. Tall and imposing, with a shock of white hair and a white beard, fond of black T-shirts and black jackets, Stefansson was called once by Altshuler the Johnny Cash of variation studies. "Like his ancestors," Altshuler said, "he sailed on uncharted waters." Stefansson convinced Icelanders and their center-right parliament, Althingi, that their country could be a laboratory whose work would benefit all humanity. Although a court decision limited access to medical records, deCODE could still analyze donors' samples in depth. As part of its search to develop drugs and diagnostic tests, deCODE has become a world center of association studies, uncovering dozens of factors that increase the risk of common maladies, such as heart attacks, stroke, and pancreatic cancer.

Alas, deCODE's scientific achievement of identifying scores of genetic factors associated with disease has not been matched by commercial success. The company's efforts to find marketable drugs fell short and its diagnostic test sales could not overcome growing losses. Some observers said the underlying problem was that, so far, genetic tests were uncovering only a small fraction of the risks of inherited disease. In November 2009, deCODE sought bankruptcy protection. Stefansson reflected ruefully that perhaps the company had been founded ten years too early. He hastened to add, however, that the Icelandic human genetics work would proceed under the established privacy controls. He said that the samples and data continue to be the property of the donors and will remain in Iceland.[22]

In January 2006, *Nature Genetics* published online deCODE's report of finding a mutation on chromosome 10 associated with adult-onset diabetes, or type II diabetes. Although the disease presently afflicts only a small percentage of the world's population, including some 20 million Americans, that proportion is expected to rise as humankind ages and eats more high-calorie and fatty foods. The unwieldy name of the protein made by the gene was transcription factor 7–like 2, abbreviated as "TCF7L2." Building on previous work, the researchers found their target in the third intervening sequence, or intron, of the gene. They compared the DNA of 1,185 Icelanders with 930 unrelated controls and found a noticeable increase in risk.

About one-third of the volunteers had one copy of the mutation, which boosted their type II diabetes risk by 45 percent. One out of fourteen had two copies, carrying an extra risk of 141 percent. To verify the results, deCODE drew on two other populations. One was a cohort of Danish women who had volunteered for Denmark's Prospective Epidemiological Risk Factors study, where the extra risk was 41 percent; the other was a group in Pennsylvania undergoing cardiac catheterization, where the extra risk was higher, at 85 percent—smaller than Iceland, but still substantial.

This result triggered a front-page account in the *New York Times* that quoted a favorable reaction from Altshuler. After hearing a lecture by deCODE's chief statistician, Augustine Kong, Altshuler called the deCODE study "a beautiful piece of work and as convincing as any initial publication could be. . . . In terms of the epidemiological risk of diabetes, this is by far the biggest finding to date."[23]

But Altshuler wanted to be sure. He and colleagues in the Diabetes Prevention Program Research Group found the findings in the deCODE study "rather provocative" and were nervous because frequently an intriguing finding did not pan out when other groups tried to reproduce it. They reflected, "Inconsistent reproducibility has been a vexing problem" and conducted a new study of the diabetes project population, published six months later in the *New England Journal of Medicine* (*NEJM*). They used samples from more than 3,500 people, ethnically more diverse than in Iceland, who were overweight but not diabetic and had been followed for three years between 1996 and 2001.

The researchers ran genotyping tests to see whether the two most salient mutations, called rs12255372 and rs7903146, speeded up full-fledged diabetes. The impact was sharply reduced in patients who complied with recommended changes in lifestyle and also went down in patients taking the drug metformin. They also found indications that the main problem was impaired production of insulin to metabolize sugar, rather than cells' resistance to insulin. Altshuler and his colleagues said their results were "a strong confirmation of the original genetic association."[24]

Diabetes: "It's Gotta Get Out There"

Diabetes was again the focus of a quartet of studies published in April 2007. The first came from an English-led group that participated in the seven-disease Wellcome Trust Case Control Consortium. Writing in *Science*, they reported that a variant of a chromosome 16 gene discovered in type II diabetes patients, called FTO, was associated with obesity and a propensity to develop type II diabetes. The SNP involved was numbered rs9939609. The team studied 1,924 British type II diabetes patients and 2,938 controls. Both adults and children with two copies of the mutation weighed nearly seven pounds more than people without it. The effect could not be blamed on birth weights. The children had a 70 percent higher risk of obesity. The actual function of FTO was not yet known, but Mark McCarthy of the University of Oxford said, "Our findings are a source of great excitement" because they could lead to treatments for obesity.

The team compared their findings with no fewer than twelve other cohorts of patients and controls totaling almost 39,000 people. Illustrating the complexity of assembling enough volunteers for statistical strength, the researchers came from Exeter, Bristol, Oxford, London, Cambridge, and Newcastle-on-Tyne in England; Dundee in Scotland; Oulu in Finland; and the U.S. National Institute on Aging.[25]

Two weeks later in *Nature Genetics*, the deCODE team weighed in again with the discovery of a second "susceptibility gene" associated with type II diabetes. A mutation in the gene CDKAL1 cut the body's output of insulin. The group had tested five European-ancestry groups of volunteers (one in Iceland, two in Denmark, one of Dutch blood donors, and one in Philadelphia), along with a sixth group in Hong Kong and a seventh in Nigeria and Ghana, a total of 20,000 people.[26]

The same day, three papers from collaborating groups appeared online in *Science*. One came from the Wellcome Trust study, and two derived from a collaboration among the Broad Institute, Lund University in Sweden, and Novartis. Eleftheria Zeggini of Oxford, Altshuler of the Broad Institute, Laura J. Scott of the University of Michigan, and a host of coauthors had found three new susceptibility variants in and around CDKAL1 on chromosome 6, which the Icelanders also found; CDKN2A/CDKN2B on chromosome 9, and IGF2BP2 on chromosome 3. The studies' research also confirmed the previously identified FTO and TCF7L2, as well as four others. The number of "guilt-by-association" factors linked to diabetes had risen to ten.

A sign of the intense competitive pressure to find diabetes-related factors occurred during Laura Scott's ski vacation the previous February. She spent most of the time in the lodge in front of her computer, racing to finish her paper and "very occasionally trying to go out and ski." Caught up in the gold rush atmosphere of her field, she thought anxiously, "It's gotta get out there." One author, Mark McCarthy of the University of Oxford, crowed that the three reports had found three new genes, whereas deCODE had found only one. Stefansson retorted that the extra genes had such a small effect that he didn't have time to go after them. Still, he admitted that the competition was heating up. "I would be a fool if I thought these guys would never pull their act together."[27]

Seven Diseases

The culmination of this phase was the report from the Wellcome Trust Case Control Consortium, led by McCarthy and Peter Donnelly of Oxford. The participating scientists came from more than fifty institutions in a score of places in England, Scotland, and Wales, as well as Australia, France, and Gambia in western Africa. Aside from the shared 3,000 controls, drawn from a study of people born in Britain in 1958, there were cohorts of about 2,000 people for each of seven diseases. These included manic depression or bipolar disorder, coronary artery disease, the irritable bowel syndrome called Crohn's disease, hypertension, rheumatoid arthritis, juvenile-onset type I diabetes, and adult-onset type II diabetes.

And the study found mutations significantly linked to disease in six of the seven. There was one each for bipolar disorder and coronary artery disease, nine for Crohn's, three for rheumatoid arthritis, seven for type I diabetes, and three for type II. Some of these mutations along the DNA were risk factors for more than one disease. Of the twenty-four associations found, twelve confirmed previous findings, but twelve were novel. And "replication" studies like those published in April had confirmed all but one association. The consortium's report was very upbeat. It said the results were a "thorough validation of the GWA approach" and that the participants and the new knowledge created "a powerful resource for human genetics research." Among the projects that would benefit were three in the United States: the Genetic Association Information Network, the Framingham Genetic Research Study, and the Women's Health Study.[28]

The big challenge now was to tighten the statistical associations further and to go beyond finding SNP markers close to the site of mutations to studying the specific variations so as to determine their "exact nature." The consortium was not through with its work. A year later, Donnelly was telling colleagues at a Nature Publishing Group conference on common diseases in Cambridge, Massachusetts, about plans for Phase Two. At least seven more diseases would be studied in Britain, and tuberculosis in Gambia. The expanded research would march under the banner of "fine mapping" and reach beyond SNPs to copy-number variations. Many key regions of DNA would be sequenced. The number of human subjects would be 120,000, and the cost would exceed $50 million.[29]

Prospective Studies

Frustratingly, the factors that were unearthed each explained only a small part of a person's total risk of a disease. They hardly constituted a genetic smoking gun. So, where were the mutations that really mattered?

Scientists bemoaned the level of this "missing variation" and immediately sought ways to find it. They intensified their hunt for rare variations in people's genes that could heighten risk more than common variations. This led them beyond the world of SNPs, the variations at a single point along the genome. They began exploring the dizzying array of other DNA changes: deletions or insertions, sequence reversals, and the pervasive genetic stutter called copy-number variations. All of them showed statistical links to a person's genetic health risks.

Going even further, they again began to put more faith in studies of diverse groups of people living in a particular area—such as northern Finland, Rotterdam, Copenhagen, Dallas, or Framingham, Massachusetts, where environmental factors were common. Participants could be followed over many years in what is called a "prospective" study. Examples of communities lending themselves to long-term study of their physical health and genetic makeup are legion. Three generations of inhabitants of Framingham have been followed for sixty years to discover genetic and environmental causes of heart disease.

The prospective studies, focusing on an entire population in one area, have begun to tease apart the influences of heredity and environment. For example, in two "genetically isolated" northern provinces in Finland, the health of more than 12,000 people born in 1966 has been followed ever since. A recent study of 5,000 of the cohort measured blood pressure, sugars, lipoproteins, insulin, and body-mass index—factors important for both heart disease and adult-onset diabetes. A conclusion was that environmental factors, such as diet, contributed less than 30 percent to the variation in risks.

Before the first decade of the twenty-first century ended, the scale grew larger. In collaboration with U.S. scientists, who included Leroy Hood of Seattle, the grand duchy of Luxembourg committed $200 million to a program focusing on the ultimate goal of administering "the right drug to the right patient at the right time and in the right dose."[30] In Britain, after years of debate, Biobank UK began recruiting National Health Service members aged forty to sixty-nine in 2006. Three years later, 250,000 of a planned total of 500,000 had enlisted.[31]

In 2007 in northern California, the health-maintenance organization Kaiser-Permanente announced plans to recruit half a million participants in a prospective study. Codirectors would be Neil Risch of the University of California, San Francisco, the early advocate of genetic variation studies, and Cathy Schaefer of Kaiser. By late 2009, more than $30 million had been granted by the Robert Wood Johnson Foundation and the National Institutes of Health (NIH) to begin analyzing DNA samples from 100,000 of the 400,000 who had filled out questionnaires. All were Kaiser-Permanente patients whose medical records—every visit, every test, every prescription, every outcome—had been recorded electronically for a dozen years.[32]

Rare Mutations in Heart Disease

Alongside the blizzard of microarray-based GWA data, a drumfire of price reductions in sequencing sounded constantly. Successive new generations of sequencing machines were on their way to cutting the cost of an entire human sequence by some 10,000 times between 2000 and 2009—with dramatic further cuts expected within a few years. So, genomic researchers could look beyond the idea of sequencing as a way to check their microarray results or delve deeper into particular DNA regions. The window was opening on the rare variants that might dominate a person's risk of disease. Leaving HapMap far behind, researchers began sequencing thousands of volunteers to get closer to the causes of common diseases.

A notable harbinger of this trend came from Dallas, Texas, where the Dallas Heart Disease Prevention Project began sequencing particular regions in more than 3,500 randomly selected and racially variant volunteers, more than half of them African Americans, looking for rare variants that might add large risks of disease.

The study was led by Helen Hobbs, a Boston-area native with both an MD and a PhD, who started out in Dallas in 1980 as an internal medicine resident at Parkland Memorial Hospital. There, according to the renowned heart researcher Michael Brown, "she became famous . . . for her ability to function in a crisis and treat very sick people, and she also had the ability to inspire and lead the others around her." Nonetheless, she took the advice of her mentor, Donald Seldin, head of medicine at the University of Texas Southwestern Medical Center, to abandon plans to specialize in endocrinology and go into research. Seldin asked Brown and his coworker Joseph Goldstein to take her in.

Despite the skepticism of Brown and Goldstein, whose work would lead to a Nobel Prize a few years later, she shifted to their lab. She later recalled, "I'd always been attracted to people who did science—I loved to hear their stories. . . . But I really had no experience in the lab at all." Gregarious and fond of the outdoors, she feared she would feel confined in the laboratory environment: "It seemed like such a singular, lonely pursuit."

Hobbs concentrated on the genetics of a problem that afflicts tens of millions of people around the world: fat metabolism. After years of stumbling, she "started to have results" and went on to be named an investigator of the Howard Hughes Medical Institute and director of the McDermott Center for Human Growth and Development at UT Southwestern. She could now say, "I get a lot of pleasure out of running a laboratory and seeing people grow and develop as scientists. . . . It's much like the pleasure I used to get in the wards."

In February 2007, the Dallas study reported online in *Nature Genetics* that it had sequenced a gene involved in the body's metabolism of fat in its multiracial sample, which includes about 29 percent European Americans and 17 percent Hispanics. At the level of As, Ts, Cs, and Gs, Hobbs, Jonathan Cohen, and their colleagues could compare one person to another and see what effect a particular DNA pattern had on an individual's blood sugar levels. The role of the gene they probed, ANGPTL4, in fat metabolism had shown up in mice.

About 3 percent of people of European ancestry carried a variant called E40K that lowered their blood sugar and raised the proportion of high-density lipoprotein, or HDL, the "good" cholesterol in their blood. In the same issue of the journal, Kelly Frazer and Eric Topol of the Scripps Institute in San Diego commented that this finding "suggests that rare nonsynonymous variants may be important and are readily identified by resequencing approaches." Was this result real in a group of people who had been followed for only seven years? Fortunately, older and larger cohorts had been followed for longer. One of these, the twenty-year-old Atherosclerosis Risk in Communities Study, had enrolled a total of more than 15,000 subjects in four U.S. locales. The other, the Copenhagen City Heart Study, with more than 10,000 participants, was started in Denmark in 1976. Among randomly chosen cohorts of 4,000 people between the ages of forty-five and sixty-four in each of the comparison groups, the Dallas results held up.[33]

The Genome Goes Retail

As early as 2007, the era of complete individual genomes began. James Watson and Craig Venter were the first to consent to post their DNA sequences on the Internet. Both said they had stepped forward to combat fears about personal genomics and pave the way for the thousands of sequences that a truly personalized medicine would require.

The posting of Jim Watson's diploid sequence, containing all the 6 billion bases he had inherited from both his mother and father, was turned into a celebration on May 31, 2007, at the Baylor College of Medicine in Houston. The main site was the Cullen Auditorium in the middle of Houston's vast concentration of medical schools and hospitals. On opposite walls were portraits of the late Texas oil magnate Hugh Roy Cullen and his wife Lillie. On opposite sides of the speakers' platform hung the flags of Texas and the United States. There, before an audience of more than 400, Watson, wearing a light yellow jacket, a white shirt, and a borrowed blue tie, was handed a 12 GB portable hard disk containing his DNA sequence. It had been produced, with several duplications to ensure accuracy, over two months at a cost of $1 million.

This was not done with the classical Sanger sequencing of the past, which handled ninety-six samples at a time, but with a "massively parallel" second-generation machine from 454 Life Sciences in Connecticut that deciphered 800,000 microscopic samples simultaneously. To analyze the data, 454 had collaborated with David Wheeler and others at the Human Genome Sequencing Center at Baylor, one of the so-called Big Five sequencing factories. Wheeler said that he'd had little sleep for a month.

Jonathan Rothberg, 454's founder, told an emotional story of how he began putting together the ideas for the new machine. In 2001, Rothberg's newborn son, Noah, was unable to breathe and was rushed to neonatal intensive care. Rothberg waited in agony for what proved to be a good recovery. He showed a snapshot of a now-thriving Noah between his older and younger sisters. At the hospital, he found himself asking, "Why can't I have Noah's genome?" This would help figure out what was wrong. Rothberg started jotting notes on how to achieve this quickly and cheaply, by doing sequencing on special chips that would "get rid of all that up-front work." He held up one of the chips for the audience. By 2005, 454's machine could sequence the simplest of bacteria, *Mycoplasma genitalium*, in four hours. To do the job with Sanger sequencing in 1995, it had taken Claire Fraser and her colleagues at The Institute for Genomic Research 24,000 operations spread over four to six months. But now, Rothberg said, "The cat's out of the bag" on sequencing costs, which were sinking fivefold every year. "Jim is the first of the rest of us."

Rothberg had already been thinking how to extend the method to bigger organisms, including humans. He contacted Richard Gibbs, head of the Baylor sequencing center. Who would be 454's first human subject? Gibbs suggested Watson. But would the iconic figure of DNA science consent? Rothberg told his Houston audience that he had prepared frenziedly to convince Watson and wrote anxiously to seek an interview. Later, as he strolled in a Boston mall, Rothberg's cell phone rang. A secretary commanded, "Hold for Jim Watson!" Watson would see him in Cold Spring Harbor. Arriving with a bundle of forms, "shaking," Rothberg entered Watson's office. Watson began the conversation with one word, "yes," and "Project Jim" was born.

In Houston, Watson declared himself "thrilled and grateful," saying he had kept only one section of his genome private, the region where susceptibility to Alzheimer's disease lay. His mother's mother had died of Alzheimer's. He knew, Watson said, that he "risked possible anxiety" over his genome. If he did have a predilection for Alzheimer's, he would dread every forgetful moment. But he didn't want to hold anything else back. The war in Iraq was a much more likely reason to lose sleep. If he held back ten genes, he would be more anxious: "Lack of knowledge is worse."

Watson's sequence did confirm that he had a variation in the gene called BRCA1 related to breast cancer, of which his sister Elizabeth had died. He said, "I never thought until today that it was good that I didn't have daughters." Project Jim also revealed a tendency toward basal cell carcinoma, which had plagued him since he was twenty-five and forced him to wear an endless series of floppy sun hats. His intolerance of lactose was confirmed. But overall, his DNA would be hard to interpret medically until thousands of people had also been sequenced to decipher the mechanisms of disease in detail.

When the human genome was "finished" in 2003, Watson told his hearers, he had been happy, "but how could it change medicine? It couldn't until personalized medicine became cheap." To make further progress, Watson counseled setting "well-defined goals" and taking them to Congress with a price tag of $1 or $2 billion. He said Congress was bored with just adding 10 percent to the NIH budget. Instead Congress should be excited with a program to "make people in other parts of the world think the U.S. is well run." One goal would be a five- to ten-year drive to pin down the causes of mental illness. "The techniques are sufficient to understand autism," he said. A second goal would be a similar push on cancer. "We need diagnosis of cancers before treating them. We're going too slow. There is no sense of urgency to get these tests out. . . . We should get on with it. We have to have a DNA diagnosis to determine if a drug will work." Gripping the lectern, the seventy-nine-year-old Watson said, "People my age are impatient."

After his talk, he and the other speakers went up to the fourteenth floor of the neighboring Alkek Building, where the Baylor genome center was located, to consume celebratory punch and cake. Clusters of workers in the lab swarmed around him to get their pictures taken with the man whose DNA they had helped sequence. Watson was presented with a jaunty, flat-brimmed Texas stockman's hat, which he donned without cracking a smile (he was pleased enough with it to wear it again a year later at a Cold Spring Harbor lawn party to toast his eightieth birthday). Asked for a comment, Watson replied, "Work harder."[34]

The complete scientific analysis of Venter's genome was published in September 2007, before that of Watson's. His sequence, part of which had contributed to the Celera draft of 2000, was analyzed over several years by the latest Sanger sequencers from Applied Biosystems, called 3730xl's, at a cost of tens of millions of dollars. He regarded this exercise as a swan song for Sanger sequencing in his laboratories. Almost simultaneously, in his widely publicized 2007 memoir, *A Life Decoded*, Venter published a score of short sidebar essays describing mutations of interest. He claimed to have written the first genomic autobiography.

Almost immediately, researchers began comparing the Venter and Watson sequences with the standard reference sequence of anonymous volunteers. One of these comparisons solved a diagnostic mystery for Watson. Late in 2009, Watson told television interviewer Charlie Rose that his doctor had prescribed a beta blocker to control his blood pressure, which was rising as he grew older. He recalled, "Within a week I was falling asleep in the middle of the afternoon, you know, middle of a play or something. You know, it was just grim. And then they gave me another beta blocker. I fell asleep again. And then the doctor, you know, put me on to an ACE inhibitor, which didn't seem to work."

Researchers at the J. Craig Venter Institute, comparing Venter's DNA with Watson's, found the answer. They discovered, Watson said, "that I have two copies of an Asian variant which metabolizes beta blockers at only 10 percent the rate [of] most Caucasians." So Watson had retained far more than the usual amount of beta blocker. For Rose, Watson drew the lesson: "Certainly within 10 years, no one will . . . get a beta blocker without their looking at the DNA first."[35]

Meanwhile, at Harvard Medical School, George Church began enlisting volunteers for his Personal Genome Project, which focused on the protein-coding regions that constitute less than 2 percent of the genome (the so-called exome), and introduced the results for ten of them to the public in the fall of 2008. One of these volunteers, psychologist Steven Pinker of Harvard, published his reflections about what he was learning in the *New York Times Magazine* early in 2009.[36]

In 2007, yet another class of volunteer grew larger. For some years, people had been joining programs designed to use their genetic information to trace their ancestry and places of origin—some of which Michael Thomas of University College London denounced in 2009 as "little better than genetic astrology." Much more respected was the Genographic Project, a partnership between the computer company IBM and the National Geographic Society. The project uses DNA from the mitochondria, small structures in the cell that are inherited in the female line. According to its director, Spencer Wells, Genographic's work with its 350,000 members focuses not on "medically relevant genetic markers" but instead on "regions of DNA that tell us about historical connections and migration routes" across the planet from Africa, starting about 60,000 years ago. To get a broad sample of humanity, the project conducts expeditions to enlist participants from remote populations. Wells said in February 2009, "I would hope that our results, showing clearly that we are all part of an extended African family separated by only two thousand generations or so, would lead people to reevaluate their concepts of 'race.'"[37]

The participatory model spread into the domain of information that might have medical significance. In 2007, three companies, 23andMe and Navigenics in California and deCODEme in Iceland, began offering personal genetic tests to customers who could view the results in privacy and discuss them with their doctors or not, as they chose. The customers were given reports on their risks of a battery of diseases with genetic markers that were being detected in the wave of GWA studies performed in 2006 and after. The companies used advanced versions of DNA chips like those of the research studies, which were capable of scanning half a million SNPs or more.[38]

The new direct-to-consumer genetic-testing companies publicized their services by such means as celebrity "spit parties" to collect enough DNA. Hundreds of business leaders, journalists, and scientists were among those paying from $400 to $2,500 for their risk profile, which was posted on a personal, password-protected website. These profiles grew more detailed as GWA studies linked more factors to particular diseases. 23andMe began by pinpointing a score of individual health risks, but the total soon grew to a hundred.

This retailing of the genome impressed the editors of *Time* so greatly that they named it number one on their list of "the fifty best inventions" of 2008. They said that 23andMe's system "does the best job of making [DNA tests for the public] accessible and affordable." Of such personal genomics, *Time* advised, "Say hello to your DNA—if you dare. . . . The curtain has been pulled back, and it can never be closed again."[39]

"My DNA Is My Data"

The founders of the direct-to-consumer testing took a populist stance, arguing that people owned their own DNA and had a right to know what could be discerned in it, whether they discussed it with their doctors or not. Facing challenges from state regulators in California and New York, they denied that their service constituted medical advice that demanded the equivalent of a drug prescription. They emphasized the value for people of knowing about health risks they could avoid and the fun of learning about their ancestry and sharing their genetic information in social networks. They offered such follow-ups as genetic counseling and updates as the technologies discerned more risks more precisely. They pledged to use certified laboratories. One indignant Internet user commented, "My DNA is my data." The tone reflected the individualistic culture of the Internet, one of whose key forces was Google, the original investor in 23andMe. Anne Wojcicki, cofounder of 23andMe, was married to Sergey Brin, Google's billionaire cofounder.[40]

Public attention already was high when founders of all three consumer genetic-testing companies met on the same platform for the first time at the 2008 annual genomics meeting at Cold Spring Harbor. In a ninety-minute session, they explained their programs and plans, heard comments, and answered questions. Alongside Linda Avey, who headed 23andMe along with Wojcicki, were Dietrich Stephan of Navigenics and Kari Stefansson of deCODEme. Some months earlier, Stefansson had told *US News*, "I'm convinced that in the next five to 10 years every educated person in the western world is going to have a profile like this."

Before taking seats at a table, the three speakers stood at a blond wood podium decorated with a stylized DNA double helix, a few feet in front of the blackboard on which, just shy of twenty-two years earlier, Harvard Nobel Prize winner Walter Gilbert had written, "$3 billion"—his rather accurate guess of the cost of the Human Genome Project. Two speakers responded to the DNA-testing entrepreneurs. One was Kathy Hudson, then head of the Genetics and Public Policy Center in Washington, D.C., a prominent critic of insufficient regulation of genetic testing. The other was Joseph McInerney, head of the National Coalition for Health Professional Education in Genetics, representing physicians, nurses, and others struggling to educate themselves about the implications of genetics for medical care.

Mindful of an audience that included researchers recruiting thousands for GWA studies, Avey asked, Why not sequence everyone's DNA? She thought many of her customers might volunteer for the research. Then Stephan of Navigenics carried the argument further. Identifying people's risks for disease, he said, was not done just for the fun of finding your ancestors or the reassurance of contact with other people with your genetic problem. It was essential, he said, to the urgent cause of making medicine both predictive and preventive. Because of the soaring medical-care costs of a rapidly aging global population, humanity was gripped by an unprecedented health-care crisis. Hence, diseases must be caught and treated early, while people cut their exposure to harm from their environment. "In this generation, we need a solution," Stephan said. "We cannot wait another generation."[41]

"What Should You Do About It?"

Convening the session was Collins, NHGRI director since 1993. He would leave the directorship just weeks later to write a book on personalized medicine before becoming NIH director in 2009. Ebulliently citing the pace of genomic technology and knowledge, he exclaimed, "What genomics hath wrought!" The impact on the public, he said, "has gone far beyond the expectations of two years ago." He observed, "The marketplace has also taken notice of this sudden explosion in genetic information." Reflecting the considerable skepticism of medical geneticists, Collins asked if consumers of DNA testing would "be educated, helped, or alternatively misled"; he further wondered, "How will health care providers grapple with this information when patients come to them seeking assistance in interpreting it?" He told *US News*, "Many of the claims that are being made are quite fanciful."

Other observers had begun raising doubts. Could individuals and their doctors gain, at this stage of genetic knowledge, personally useful, "actionable" information? A nagging question, Collins told an interviewer, was, "What should you do about it?"[42]

Beneath that question lay a more fundamental one: What could the genome tell a scientist, let alone a medical professional or a patient? A decade after the first draft human sequences, and after a flood of sequence comparisons with other mammals, such as mice, rats, dogs, and even platypuses, knowledge of what the genes actually do remained sketchy. How do the substances specified by the genes and their numerous controls cause a disease? Do they act by themselves or, far more likely, in collaboration with other genes along various biochemical pathways and under the influence of the environment? Since the new disease-related factors each added only a small amount to a person's total risk of an illness, a big portion of the genetic influence on disease remained undiscovered. It was still unclear when the often-expressed hope of "personalized medicine," with prevention and care tailored to an individual's genome, would touch most people's lives.

Some of the negative comments were scathing. Three physicians wrote in the *New England Journal of Medicine* that a doctor to whom a patient brought "a genome map and printouts of risk estimates" should start with "a general statement about the poor sensitivity and positive predictive value of such results." The authors continued, "For the patient asking whether these services provide information that is useful for disease avoidance, the prudent answer is, 'Not now—ask again in a few years.'" The authors were David J. Hunter, an epidemiologist at Brigham and Women's Hospital in Boston, Muin Khoury of the Centers for Disease Control and Prevention in Atlanta, and Jeffrey Drazen, *NEJM* editor in chief.

Not every geneticist bowed before this authoritative blast. Susanne Haga and Huntington Willard of Duke University's Institute for Genome

Sciences and Policy wrote to *NEJM*, "Consumers appear to be embracing this opportunity to uncover the secrets of their genomes. . . . Everyone has a genome. . . . If physicians are not able to provide information on genomics, patients will simply turn elsewhere."[43]

The editors of *Nature Genetics* were even more upbeat. They wrote, "It would be wrong to underestimate the motivational potential inherent in handing people their genomes and asking them to participate in finding out more about their variation and phenotypes."[44] Thousands would be "participating in one of the most exciting areas of biomedical research," and many might join long-term prospective studies—as Avey hoped.

CHAPTER SEVEN

The Struggle for Medical Relevance

As the second decade of the twenty-first century opened, the flow of cheaper DNA chips and faster sequencing machines toward the clinic and the patient's bedside speeded up. Cancer specialists at Massachusetts General Hospital (MGH) in Boston and the Memorial Sloan-Kettering Cancer Center in New York had begun using chips to obtain the detailed genotypes of their cancer patients, starting with those suffering from lung cancer, and other centers were preparing to join them. The aim was to get beyond the shot-in-the-dark overtones of cancer therapy, developing a clearer picture of a patient's prognosis and picking the initial treatments more rationally.[1]

In the same quest, using ever-faster machines, several large genome-sequencing centers continued their programs of deciphering the DNA of cancer tissue from scores of patients, living and dead.[2] The search for rare mutations that might cause more genetic disease than SNPs also intensified as sequencing became easier. A major effort was the so-called 1,000 Genomes Project, which had begun the shift from a 2008–2009 pilot phase to determining the sequences of 400 people.[3] Sequencing also was invoked to obtain detailed information on the expected time of onset, severity, and pace of a patient's illness. To do this, the Institute for Systems Biology in Seattle had announced plans for California-based Complete Genomics to sequence all the DNA of one hundred members of families afflicted with Huntington's disease, whose gene had been found in the 1990s.[4]

Sequencing had begun entering the realm of diagnosis. In just ten days, researchers at Yale sequenced and analyzed the entire protein-coding exome of a desperately ill, five-month-old Turkish baby boy. Born nine weeks prematurely to closely related parents, the infant was thought to be suffering from a rare kidney disorder. The sequencing located the mutation for the disease, identifying it as a completely different disorder, congenital chloride diarrhea, which could be treated. The researchers estimated that this method of hunting for rare but devastating mutations would cost ten or even twenty times less than sequencing an entire genome.[5] In a race with heavy commercial overtones, given the size of the potential market, rival researchers at the Chinese University of Hong Kong (CUHK) and Stanford University had continued to push development of a test based on sequencing to detect the DNA of a Down syndrome fetus in the mother's blood, in order to avoid invasive prenatal tests with a risk of miscarriage. If the sequence-based test proved practical, it would make routine prenatal screening for Down's possible.[6] Researchers across the world now routinely sequenced viruses like SARS and H1N1 to monitor their migrations and evolution from mild to virulent strains.[7]

One of the world's biggest sequencing centers, the Sanger Institute outside Cambridge, England, had begun shifting its emphasis from large-scale projects to a more clinical orientation. From now on, the institute decided, "the major challenge will be to understand the significance of the variation identified and what it means to an individual."[8] Continued study of genes had concentrated on finding weak points in the biochemical pathways of cancer cells. In lung cancer cells in mice, researchers at MIT's Koch Institute and the Broad Institute of MIT and Harvard had found new vulnerabilities along a biochemical pathway linked to a cancer gene called Kras. This oncogene, discovered back in the 1980s, is apparently involved in one-third of all cancer cases. But the early hopes of fighting cancer through direct assaults on Kras had been dashed. Now, the search for drugs to attack the related pathway could reopen.[9]

Gene therapy, which had languished for many years after the highly publicized deaths of several patients, showed new promise with several inherited diseases. For example, a French team used a disabled form of the HIV virus to transfer genes into bone-marrow stem cells of two boys afflicted with a progressive, fatal brain disorder called adrenoleukodystrophy, or ALD. Therapy halted the progression, highlighting the importance of the earliest possible diagnosis. Testing continued with the two boys and other patients to watch for dangerous side effects and to see if the positive effect persisted. The team leader cautioned, "This is only the beginning."[10]

These developments, as the tenth anniversary of the first human sequences approached, showed intensified determination to apply genomic insights and techniques to human medical care. This was clear even though the impact of genomics remained modest after decades of effort. The vast, multifront genomic research effort continued to expand across the world, and tens of thousands of volunteers were participating in the search for genetic causes of disease. Genomic work had led to a handful of expensive drugs targeted at specific types of cancer. Diagnostic tests for a range of diseases multiplied. Work like that at MGH and Sloan-Kettering signalized the determination of cancer specialists to group patients according to their genetic makeup in deciding the best way to begin therapy.

But several of the disease genes discovered in the 1980s, such as that for cystic fibrosis, had not yet led to cures. New genetic tests used directly by consumers, experts complained, did not yet give them information they could act on. Could drug dosage be based on tests that showed when patients were hypersensitive to a particular drug? Regulators and physicians alike remained skeptical. More generally, pharmaceutical companies continued to be dismayed by the notion of a personalized, preventive medicine. Markets for "blockbuster" drugs could shrink.[11] Regulating how all the genes and their controls work together in sickness and in health remained a serious, possibly even more complex challenge. Yet, public and governmental support for genomics did not falter.

"Like No Other Area of Human Endeavor"

Late in the morning of Wednesday, September 30, 2009, U.S. President Barack Obama paid a visit of affirmation to the National Institutes of Health (NIH) campus in Bethesda, Maryland. Addressing the new NIH director, Francis Collins, and many of his Bethesda coworkers, Obama said, "The work you do is not easy. It takes a great deal of patience and persistence. But it holds incredible promise for the health of our people and the future of our nation and our world." He added, "We know that the work you do would not get done if left solely to the private sector. Some research does not lend itself to quick profit."

Obama took the occasion to celebrate NIH's granting of $5 billion in special economic stimulus money to nearly 13,000 U.S. biological and medical research projects, including many in genetics and genomics. The President called this "the single largest boost to biomedical research in history."

The grants represented half of the $10 billion the federal government had set aside for NIH in 2009 from its $800 billion emergency program to reduce the damage from a near-collapse of the world's financial system a few months before. To be spent over two years, the money represented a whopping one-third increase in NIH's normal annual budget of about $30 billion. The prospect of this money had roused hundreds of laboratories to send in a torrent of new grant applications, representing a backlog of research ideas that had built up over half a decade of shrinking NIH purchasing power. There were so many proposals that 15,000 scientists had to be recruited hurriedly to evaluate them.

This new money was one of the most emphatic signs ever of continued political support for health-related research, driven by both curiosity and a desire to minimize the effects of disease. Over the past two centuries, such work has traditionally gone forward both rapidly and slowly, raising immense hopes and painful frustrations. For example, age-specific death rates from heart disease have fallen sharply in the United States and other wealthy countries over the last half century, while the cancer death rate, except for children, has changed little. In such a situation, one might expect sharp fluctuations in support for health research—a budgetary rollercoaster.

But medical research touches on each person's fitness and survival. And so, the support continues, with fervent backing from the political right and left alike, although at a modest overall level. The annual budget of the National Human Genome Research Institute is about $500 million, a bit under 2 percent of the total annual budget of NIH, which in turn represents little more than 1 percent of the total yearly American health-care spending of $2.5 trillion. U.S. pharmaceutical companies reported spending $65 billion in 2008 on research and development; funding from other government agencies, foundations, and individuals adds significantly to the total. The Howard Hughes Medical Institute, for example, reported annual spending of $730 million in the year that ended August 31, 2009.[12]

Human genomics, particularly the genomics of cancer, was a major focus of the grants President Obama announced in Bethesda. "One of the most exciting areas of research to move forward as a result of this investment," he said, "will be in applying what scientists have learned through the Human Genome Project to help us understand, prevent, and treat various forms of cancer, heart disease, and autism." Although cancer treatments based on genomics are emerging, he said, "we've only scratched the surface of these kinds of treatments, because we've only begun to understand the relationship between our environment and genetics in causing and promoting cancer."

Although $750 million would go to diseases of the heart, lungs, and blood, and $100 million to studying autism, a White House fact sheet announced that "more than one billion dollars of the grant funding was dedicated to research applying the technology produced by the Human Genome Project." The biggest item was a total of $175 million over two years for The Cancer Genome Atlas (TCGA). This grant is for building on a three-year pilot project, costing $100 million, that examined the genomes of cancers of the ovary, lung, and brain in great detail. Now TCGA would go at a faster pace, studying the genomes of twenty types of cancer. One hope was that the tiny RNA molecules called microRNAs could help diagnose which cancer patients face the greatest danger of their cancers spreading, and thus need aggressive treatment that others could be spared.[13]

Obama issued a ringing endorsement of biological and medical research:

> We can only imagine the new discoveries that will flow from the investments we make today. Breakthroughs in medical research take far more than the occasional flash of brilliance, as important as that can

> be. Progress takes time; it takes hard work; it can be unpredictable; it can require a willingness to take risks and going down some blind alleys occasionally—figuring out what doesn't work is sometimes as important as figuring out what does—all of this needs the support of government. It holds promise like no other area of human endeavor, but we've got to make a commitment to it.

President Obama took note of the long, slow tide of genomics entering medicine. Helping to speed the process was an extraordinarily rapid rise in technical capacities. These included both "next-generation" sequencing machines for looking at every letter in DNA and the DNA chips, or microarrays, for sampling a person's genome at half a million or more different points. With the faster new machines, the entire DNA sequence of cancer cells could now be deciphered—more and more cheaply—and compared to the corresponding DNA of normal cells from the same person. In a step that would have been impossible only a year or two earlier, the genetic changes leading to cancer could now be identified much more precisely.[14]

Alongside these advances, the chips were speeding a new discipline that had acquired the name of pharmacogenomics. It aimed to tailor treatments, such as drugs and their dosages, to an individual's genetic makeup and metabolism. By sampling the genomes of many patients with chips, researchers could test whether particular patients would respond well or poorly to a particular treatment. Impatience was building to carry this technology into the clinic and the hospital.

Pervasive Collaboration

The massive additional support for TCGA underlined how much genomics depends on large-scale and intricate collaboration among many different specialties. No medically relevant field showed this trend more dramatically than cancer. The drive created an intense competition and cooperation between two main technologies, DNA microarray chips and sequencing machines.

Soon after 2000, with a complete draft human sequence in hand, a discovery spearheaded by Todd Golub of the Dana-Farber Cancer Institute in Boston had already intensified interest in microarray-based pharmacogenomics. Golub also was a founder of the Broad Institute of Harvard and MIT. Harnessing DNA chips for "expression studies" to detect which genes in cancer cells were turned on or turned off, Golub and his colleagues found that expression was not the same in all tumors that looked exactly the same under the pathologist's microscope. Instead the tumors fell into subgroups. This finding raised hopes that expression studies could go beyond classifying tumor cell types. By probing differences between patients, one could determine who would benefit from—or be hurt by—a particular drug in chemotherapy.

The topic seemed so important that Eric Lander of the Broad Institute pushed for a massive program of sequencing all types of tumor cells. After many scientists recoiled at the projected cost of $1.5 billion, the TCGA program was pared down in 2005 to the $100 million pilot study including many laboratories, which focused on three types of cancer: glioblastoma, lung cancer, and ovarian cancer. The three cancers led to nearly 200,000 deaths in the United States in 2006.[15]

"It's Just This Total Black Box"

One of the many scientists involved in the cancer atlas pilot program was a leader in finding genetic links to cancer, Bert Vogelstein of the Ludwig Cancer Center at Johns Hopkins University. His laboratory is not too far from Pikesville, Maryland, where he grew up. In a 1997 interview for an organization called the Academy of Achievement, Vogelstein said that he did not definitely settle on a life of searching for causes of cancer until he was twenty-seven. He described himself as a "pretty serious" boy, whose family expected him "to do extremely well in school."

He had a strong drive to educate himself. Two or three days a week, instead of attending classes at a private school, he went to the public library next door for the day and read the books that interested him. After some years, it was suggested that he return to public school. When he finished high school and went off to the University of Pennsylvania, he intended to go into medicine. The standard premed courses, however, were a lot less interesting than mathematics. He majored in math, finished college in three years summa cum laude, and went on to a graduate year. Again, he found himself discontented. "I began to feel, even though math was incredibly intellectually stimulating, it didn't have the practical edge that I wanted. I wanted to be able to do something for people." After starting medical school at Johns Hopkins in the mid-1970s, he realized that research, not clinical work, interested him. A summer in a research lab convinced him. "I immediately knew this was fun."

He went on to a medical internship, nonetheless. One of his first patients was a four-year-old girl, whose cancer he had diagnosed. "The look on her parents' face is something that has [been] indelibly etched in my mind," he recalled. They wanted to know why their little girl was going to die—but he could tell them nothing. He told her mathematician father, "I don't know. Nobody knows. It's just this total black box." Vogelstein knew that this was the puzzle he wished to work on.

Joining the Johns Hopkins faculty in 1978, Vogelstein immediately plunged into work on genes associated with human cancer. He wanted to identify and clone them. He was thrice refused grants by review committees and did not get a grant until, using other funding, he had already found a way to study human cancer genes. He recalled ruefully, "You almost have to prove that what you're trying to do is feasible before you can get funding to try it."

In his work, he found that scientists have "the world's best toys" and that he reveled in hunches that hit him just before he got out of bed in the morning. But most of the time the experiments did not work. "In science, you're always failing." He and his colleagues formed a band, with Vogelstein at the keyboard and Kenneth Kinzler on drums. Students played guitars and sang. It became "a great diversion" because "cancer is a very serious disease." If all the experiments failed that day, "at least your music worked."[16]

Eventually, Vogelstein became one of the best-known cancer genetics researchers, focusing on colon cancer, and big prizes began to rain down. In 2006, he and his colleagues published a major study of genes implicated in breast and colon cancer. Drugs based on the genetics of tumors had already gone on the market, including Gleevec for chronic myelogenous leukemia (CML), Herceptin for breast cancer, Iressa and Tarceva for lung cancer, and Avastin for colon, lung, and other cancers. These drugs went after cancers triggered by a few mutations. They were costly, and usually the tumors developed resistance, sometimes quickly. But most cancers involved several or even many mutations.

One key study demonstrating these myriad factors came from Vogelstein's lab. Using tissue from eleven patients with breast cancer and eleven more with colon or rectal cancer, the researchers studied more than half of the known human genes, 13,000. They came up with nearly 200 possible cancer genes, although which ones were significant remained unclear. Still, the finding fit in with the insight that there are many genetic pathways to cancer and supported the assumptions of TCGA.[17]

In September 2008, Vogelstein and such Johns Hopkins colleagues as the drum-playing Kinzler and Victor Velculescu, collaborating with others at Duke University, reported their detailed study of patients with pancreatic cancer and glioblastoma, the particularly intractable brain cancer. The latter disease, from which U.S. Senator Edward Kennedy died in 2009, was also the subject of the first phase of TCGA, led by Linda Chin and Matthew Myerson of the Dana-Farber Cancer Institute in Boston.

In cells from twenty-four patients with pancreatic cancer and twenty-two patients with glioblastoma, Vogelstein's group sequenced a total of 20,000 genes, that is, almost all the known protein-coding genes. As did TCGA, the Hopkins team found a swarm of factors linked to the two cancer types, not just single-nucleotide polymorphisms (SNPs) but other changes, like copy-number variations (CNVs) in particular DNA sequences. Some patients had fewer than the average number of CNVs, and some had more. Yet more significant changes were implicated—among them was the pattern of rearrangements in chromosomes, long known in many hereditary diseases.[18]

Other changes involved epigenetics, a rapidly expanding field of biology. It was christened in 1942 to denote events that genetics could not explain. Epigenetics research probes events not only on or around DNA but in the family of histone proteins that wrap the DNA, tightly or loosely, and affect gene expression. DNA and the histones make up the material of the chromosome, the chromatin. A prominent type of epigenetic event is the addition or subtraction of a methyl group (CH_3) from cytosines in

DNA, which influences whether a gene is turned on or off and the rate at which it is copied into instructions to make particular proteins.

Epigenetics was considered important enough to be the theme of the entire 2004 annual Symposium on Quantitative Biology at Cold Spring Harbor. Later, no less than $190 million was committed to an international epigenome project. An early report from this effort in October 2009, issued by the Salk Institute and colleagues in Wisconsin, reported differences in the pattern of methylation in two different types of cells, embryonic stem cells and fibroblasts. The researchers speculated that this could help explain why the stem cells keep a sort of eternal youth, ready to develop into a wide variety of specialized cells.[19]

The studies of glioblastoma and pancreatic cancer by Vogelstein and his colleagues blasted hopes that one or two genes would be the major factors in these cancers. There were some high-frequency mutations, which Vogelstein and his colleagues called "mountains," but there was a distressingly large number of low-frequency mutations that they called "hills." Velculescu said, "Now we know there are lots of little enemies."

Debate broke out immediately about whether the glass was half empty or half full. Such complexity could push genomic help for cancer patients into an indefinite future. Experienced reporter Sharon Begley wrote a despondent column in *Newsweek* titled "We Fought Cancer . . . and Cancer Won." Bernadine Healy, the former NIH director, drew the opposite conclusion. Not only had the researchers catalogued the major DNA mutations involved, she wrote, but they had also found factors explaining why the glioblastoma resisted treatment. Even more importantly, in the forest of genetic changes in cancer, they had found common mutations and pathways.

The "cascade of genetic changes," as one reporter called it, caused more than a few observers to despair of ever untangling the complexity of cancer. But Vogelstein noted that many of the genes he had found arrayed themselves along certain pathways in the cell that had to do with adhering to other cells or to preparing the cell to move. Also, many of these pathways were common to several types of cancer.

Thus, Vogelstein saw the good news rather than the bad. To be sure, he and the others had found a large number of rare changes and an incredible variety between tumors from different patients. He lamented,

"If you have 100 patients, you have 100 different diseases." It was the same complexity he and others had found a year earlier with breast and colon cancer.

His study identified twelve common pathways in pancreatic cancers. "Maybe we shouldn't even be focusing so much on the individual genes that are mutated. Instead, we should be thinking about the functional pathways in which these genes operate." Taking heart from "genetic alterations of a large number of genes that function through a relatively small number of pathways and processes," he called for more work on "discovery of agents that target the physiologic effects of the altered pathways and processes rather than their individual gene components." Developing many more narrowly targeted drugs like Gleevec was unlikely. But pathway-blocking drugs, affecting many types of tumor, had a chance of helping many more patients. It was an example of the terms of genomics shifting from individual genes to systems, pathways, and networks. Vogelstein told a reporter, "It points to a new way of looking at cancer."[20]

Sequencing Tumor Cells

The hope of finding cures for cancer remained a principal justification for genomic studies. Cancer results not only from the environment a person inhabits but also from his or her individual genetic endowment. Inherited susceptibility to cancer is involved in perhaps 10 percent of all cases. But genetics also appears in a second process, as modest numbers of mutations accumulate, one after another, during a victim's lifetime until one cell loses its normal tight control over its replication. In a deadly process of adaptive learning, the proliferating cells gradually beat back the body's immune system, invade neighboring tissue, begin circulating in the blood, and metastatically settle in distant tissues.

But which genes' mutations unleash the process? Many cancer researchers began asking which are the "drivers" and which are the "passengers." Since the 1960s, when studies of viruses causing cancer began, it has been apparent that the number of genes involved could be quite small. Since the early 1980s, when the first genes linked to cancer were discovered, researchers were long limited to looking at "genes of interest," suspected malefactors in a genetic police lineup. When the microarrays of ever-increasing power were mobilized for the gene search, genetic factors such as SNPs turned up by the hundreds in DNA samples taken from thousands of volunteers. As with other common diseases, each implicated factor raised a person's risk of cancer by a small fraction. Geneticists suspected that rare—or "penetrant"—mutations might be more important. But existing methods seemingly provided small hope of finding them. As the cost of sequencing a genome sank by several orders of magnitude, however, people began hoping that sequencing human DNA—and comparing it to the "reference" human sequence that became available in 2003—would provide answers.

Molecular geneticists concluded that it was time to try sequencing, starting with genes of interest. For example, in 2002, the National Cancer Institute (NCI) granted funds to the genome center of the Washington University (Wash U) School of Medicine in St. Louis to sequence genes implicated in cancer. The targeted disease was acute myeloid leukemia (AML), which one scientist called "one of the nastiest forms of cancer." Each year in the United States, 13,000 new cases are reported, and nearly 9,000 patients die of it. Frustratingly, treatment has changed little in twenty years, and the five-year survival rate is about one patient in five.

Physician Timothy Ley of Barnes-Jewish Hospital, affiliated with Washington University, Elaine Mardis of the Wash U genome center, and their colleagues began searching through hundreds of genes of interest in a group of ninety-four patients. They took DNA from both normal and cancerous tissue. But no arch-criminal stood out.

In 2006, the team's thoughts turned to something bolder, sequencing all the DNA of a patient's tumor and comparing it to the complete DNA of normal tissue from the same patient. No longer would the inquiry be "biased" by limiting it to a guesswork list of suspects. With NCI hesitant, the money for starting the project came from a St. Louis businessman, A. J. Siteman.

For their "proof-of-principle" study, the Ley-Mardis team chose the tissues of a woman in her fifties, hitherto healthy, who sought medical attention because of a combination of easy bruising and fatigue. A biopsy showed that each millionth of a liter of her blood contained

105,000 white blood cells, of which 85 percent were lymphoblasts—the immature cells found in bone marrow that proliferate out of control in AML and other leukemias. The patient would need to begin chemotherapy at once. Hence, to avoid contaminating the study, the researchers needed samples taken before the chemotherapy began. The woman had only a short period of remission. Her cancer recurred after only eleven months, and she died just two years after her initial diagnosis.

Her tissue was chosen in part because she suffered from M1, one of the more common of eight types of AML. About 20 percent of AML patients have the M1 type. Her cells were largely free of the disorganization that often appears in cancer cells.

To simplify this first study, the St. Louis researchers studied only the 1 or 2 percent of DNA that codes for proteins, the exome. They made multiple comparisons, not only between the woman's healthy and diseased tissue but also with a huge catalogue of single-nucleotide variations—the SNPs—and two previously obtained human sequences, those of James Watson and J. Craig Venter (both disclosed in 2007). Wanting to exclude all but the most significant variations, they set high statistical barriers. In 2008, they found that the patient's tumor cell contained ten candidate mutations compared to the normal tissue. Two were known previously, but eight had not appeared on anyone's list.

Encouragingly, the eight novel SNPs lay along known pathways to the induction of cancer. But the picture was not entirely hopeful. The vast noncoding regions of the woman's DNA might contain hundreds more changes that would have to be investigated. And when the Ley-Mardis team compared the woman's sequence with DNA from 187 other AML patients, none of the mutations showed up. The team reported these results in a special section of *Nature*, which also carried accounts of the first sequences of men from China and the Yoruba tribe in Nigeria.

Despite the evidently vast genetic variability of cancers, the St. Louis group was not discouraged. Their results led them to assume that they would have to sequence hundreds of cancer patients to pin down the causative factors. In August 2009, they reported sequencing the complete genome of normal and cancerous tissue in AML 2, a man in his thirties who, after chemotherapy, had survived. In the *New England Journal of Medicine*, the researchers noted that with numerous improvements in their equipment and procedures, the job had been far faster, easier, and more complete than the same process only months earlier with AML 1. Her sequence had cost about $700,000 and AML 2's $200,000. With the price continuing to fall, Mardis announced at Cold Spring Harbor that the St. Louis genome center would do 150 sequences between April 2009 and April 2010. In an interview afterward, she observed, "Complexity should be appreciated.... [There's] huge complexity we're missing."[21]

Making Technology As "Bulletproof" As Possible

Mardis's career was emblematic of the incessant technological demands for automation that have dominated genomics for the past twenty-five years. Her enthusiasm for science began early. "I'm one of those really weird people who knew since I was very young that I wanted to be a doctor or a scientist," she recalled. Fostering her interest was her father, who taught chemistry at a junior college in North Platte, Nebraska, for more than thirty years. DNA and RNA captivated her during a biochemistry course at the University of Oklahoma. The teacher was Bruce Roe, one of the pioneers of DNA sequencing. Mardis recalled her reaction to the macromolecules: "It was like, 'Oh, this is it!' You always hope for a eureka moment, and that was mine." Supervised by Roe, she went on to a doctorate in chemistry from Oklahoma. She worked on automating machines, not only those that prepared samples for sequencing but also the sequencer itself, which distinguished the four bases of DNA by detecting the different dyes attached to each. She recalled that it was an Applied Biosystems 370, the company's early Model T sequencer. It was the first machine of this type that Applied Biosystems had placed in an academic laboratory.

Her enthusiasm for sequencing put her on the spot when she received her PhD in 1989. Although the international human genome project was just starting, it focused not on sequencing but on mapping the locations of genes in model organisms and humans. Mardis was neither trained for nor interested in mapping. So she worked in industry for four years at Bio-Rad Laboratories of Hercules, California, a company interested in equipment to streamline DNA sequencing. Her job was to bring in "commercially available products, or products that we came up with, and test them to make sure they were as 'bulletproof' as possible." Her background was thus the ideal preparation when, in 1993, Robert Waterston and Richard Wilson of the rapidly expanding genome center at Washington University approached her about coming to St. Louis. Wilson was already very impressed with Mardis. He told an interviewer later, "We really needed someone who had been thinking about the technology 100 percent of the time. I had followed Elaine's work in grad school and at Bio-Rad, and I thought she was clearly the only person in the genome community who could get it right."

Offered the directorship of technology development, Mardis decided that "it was a great opportunity that was too good to pass up." She moved into the Wash U genome center as a "research track investigator," switching to the "tenure track" in 2000. Her technology group's "very customer-oriented" mission was "to experiment, optimize, test, and do the training, and pass on the methods [to the production group] that actually work out to be the most robust." Her instinct for collaboration kept her work expanding in many directions, such as the genetics of a lung disorder in infants, the genomes of fungi, and advancing the technology of DNA chips. She also led in preparing the tools to exploit the complete human sequence when it was "finished" in 2003. She took part in TCGA and the 1,000 Genomes Project. Rising to be codirector of the sequencing center, she soon began speaking at major conferences and writing review articles on the future of sequencing.

Her mentor, Bruce Roe, had instilled a fanatic interest in pure samples of DNA. He once remarked, "You wouldn't want to waste good thoughts on dirty DNA." The key to DNA stripped of impurities would be tireless robots that would speed things up and achieve a reliability that humans in the laboratory could not manage. Hence, in the 1990s, wrestling with problems of combining an industrial facility with a research laboratory, she slaved away on a "plaque picker," a robot designed to gather DNA to be mapped or sequenced. This cloned DNA lay within bacteria in dot-sized colonies, or plaques, on shallow laboratory Petri dishes. Each colony lay on a "lawn" amid vast numbers of others. A target plaque was normally extracted using a toothpick. It was exhausting for the human operator to eyeball the plaque of interest, aim the toothpick correctly, jab the dot of bacteria accurately, and then move it to instruments for mapping and sequencing—hundreds of times a day. The process, performed by the lowest-ranking lab worker to the highest, was slow and error-prone, and thus clearly incompatible with large-scale genomics.

Mardis had to defend what she was attempting. Exhausted after a few days—and late nights—of discussion at Cold Spring Harbor, she faced a visiting committee reviewing the work of the Wash U sequencing center. When it came time for the committee to question her, one member asked sternly if she knew how many people had tried and failed in the quest for an automated plaque picker.

Determined to prove the skeptics wrong, she latched onto a machine invented in England that spun out thin tubing for surgeons, with its own tiny knife for cutting off short sections. The tubing picked up plaques well. But how could the machine recognize the right plaques? A computer-enthusiast undergraduate modified the machine's software so that it would pick up only the wanted plaques, without human aid. In six months, the $40,000 machine was operating twenty-four hours a day, freeing humans for other tasks. "Getting the robots, setting them up and training researchers to use them does take some time," Mardis observed, "but once you have the thing running, that's where you get going." She recalled joyfully, "We went from 1.5 million reactions a month to three million."

The achievements of Applied Biosystems in moving from its first sequencing robots of the late 1980s to the much faster "capillary" machines of the late 1990s put added pressure on Mardis and such other leaders of genomic automation as Lauren Linton of the Whitehead Institute in Cambridge, Massachusetts, Donna Muzny of the Baylor College of Medicine in Houston, and Jane Rogers of the Sanger Institute in England. The robots to prepare samples and move liquids had to become faster and faster. And genome centers like Wash U found themselves expanding into new buildings. By 2005, the center was acquiring some of the earliest "next-generation" sequencers and putting them to work in cancer genetics and the sequencing of viruses and microbes to understand their virulence. Mardis observed, "Data on small genetic differences between isolates of bacteria, viruses and other pathogens can be key to understanding changes in virulence, infectivity, antibiotic resistance and other factors important to the control and treatment of disease." The types of new sequencers kept proliferating, and their capacities evolved rapidly, keeping Mardis and her colleagues hopping as they strove with the suppliers to break them in and then adopt the new wrinkles.

Meanwhile, she began teaching at several levels, including giving an advanced short course on sequencing technique at Cold Spring Harbor. One of the alumnae of that course was biochemist Sarah C. R. Elgin, a Wash U colleague. Mardis joined Elgin in teaching a one-semester undergraduate genomics course for a dozen or so "bright biology" juniors and seniors. To cut costs, the course drew its material from computerized databases. During the first half of the semester, the students worked at the Wash U genome center, sequencing and then assembling one or two genes from the X chromosome of fruit flies. The students used classic quality-control software. In the second half, they went to the main campus to annotate their sequences, distinguishing genes from noncoding DNA. In 2008, Elgin told the *St. Louis Post-Dispatch*, "It's a great way to get kids involved in science because it's all web-based. That makes it easy to share information and to find information." The course was designed, Mardis said, to help the students "understand what it takes [including] the science behind it. It's training in analysis."[22]

Crisis in Pharmacogenomics

The nascent field of pharmacogenomics continued to generate puzzling issues. Just before Christmas 2008, researchers in Paris and Boston reported in two medical journals that a substantial fraction, perhaps 30 percent, of patients taking the anticlotting drug Plavix were not benefiting from it because of genetic mutations. The suspect in the case was a reduced-function variation of a gene called CYP2C19. The finding came from a genomic study of 2,200 French and 1,500 American patients who had heart attacks—which the drug was intended to prevent. The *New England Journal of Medicine* commented, "This observation is not trivial." The news about the drug, with the generic name of clopidogrel, set off a red alert across the genomics industry—and at the U.S. Food and Drug Administration (FDA).

For some years, in light of the genomic revolution, the FDA had confronted the fact that genetic changes could make people, notably those suffering from AIDS, cancer, and heart disease, supersensitive to some drugs and undersensitive to others. As a result, a blood thinner could lead to hemorrhages in some patients, or cancer patients could suffer needless side effects from a drug unlikely to help them. Amid protests from some who saw difficulties, the agency had begun to require warning labels to encourage doctors to order genetic tests before administering certain drugs, as it did in August 2007 with the blood thinner coumadin (also known as warfarin). The FDA also had begun to approve specific genetic tests that doctors could order for up to a few hundred dollars.

But the news about Plavix was bigger. World sales totaled more than $7 billion—$5 billion in the United States alone—second only to the cholesterol-control drug Lipitor. Plavix, with more than 90 percent of the U.S. anticlotting market, was routinely given after heart attacks, particularly those followed by the installation of blood-vessel-opening stents in angioplasty—a procedure performed more than half a million times each year in the United States alone. The number of Plavix prescriptions in the United States was nearly 25 million in 2007. Because Plavix was an everyday fact in millions of lives, physicians and regulators hastened to warn against anyone's stopping treatment.

Questions arose immediately. Could patients with the mutation benefit from higher doses, or would they suffer from internal bleeding? Was the rival drug nearest to commercialization, prasugrel (Effient) from Daiichi Sankyo and Eli Lilly, a reasonably safe substitute for people taken off Plavix? And would a newer drug with a longer-running patent drive up the costs of care? Should Plavix labels be changed to warn patients against taking certain heartburn medications (as the FDA did in November 2009)? Should candidates for Plavix routinely take a genetic test before starting on the drug—or would this endanger patients by delaying care? How much money would be saved by not giving Plavix to those for whom the drug offers no help?

It was a crisis in pharmacogenomics, one of the many consequences of achieving a human DNA sequence. The goal is to use genetic information to help decide whether particular patients will do well or poorly on a drug. The Plavix study showed that clear-cut answers may be elusive. Ordering fresh studies of the issues surrounding Plavix, Larry Lesko, director of clinical pharmacology at the FDA, was as baffled as everyone else. He told the *Wall Street Journal*, "What I think we're struggling with is what is the label going to say in light of all the ambiguous data out there." Douglas Weaver, president of the American College of Cardiology, noted that no substitute for Plavix had yet been approved. He asked, "Once you know the answer, what do you do?" Paul Gurbel, director of cardiovascular research at Baltimore's Sinai Hospital, said that the era of "trial-and-error medicine" was over. "The blind administration of these drugs is rapidly coming to an end." James Calvin, director of cardiology at Rush University Medical Center in Chicago, said, "We have to start to become very, very aware of how big an issue this is."

One consequence was that the FDA accelerated its consideration of whether to approve prasugrel. Agency staff said yes in January 2009. The agency's nine-member Cardiovascular and Renal Drugs Advisory Committee met shortly afterward and unanimously recommended approval—but with strong warnings on the label about risks of bleeding and cancer. Rival drug companies strove to introduce yet other blood-thinning drugs to the American market. A key issue was whether they were enough better than Plavix to justify higher costs. The patent on Plavix is slated to expire in 2012.[23]

Next-Generation Sequencing

Early in 2009, two of the major American centers for decoding genomes, the Broad Institute in Cambridge, Massachusetts, and the J. Craig Venter Institute in Rockville, Maryland, announced layoffs of more than a score of sequencing workers because the much greater capacity of new sequencers made them redundant.[24] The breakneck advance of next-generation machines opened the way to rapid spelling out of genes of interest and whole genomes, including those of a growing number of humans.

As the decade ended, a new generation of ultrafast sequencers was being harnessed to the task of completely sequencing the DNA of thousands of people, not just one or two at a time. By "resequencing" stretches of human DNA, the new machines also routinely checked the accuracy of results from the other major tool of genomic research: the DNA chips (microarrays) that help uncover genetic causes of common diseases. As discussed in Chapter 6, the chips, from companies like Affymetrix and Illumina, were the key to the flood of genome-wide association studies (GWA) in many countries in 2006 and 2007.

The need for such resequencing had become clear almost as soon as the human genome drafts were published in 2001. People like Leroy Hood, now in charge of the Institute for Systems Biology he founded in Seattle, constantly repeated that it was a vital step toward a truly personalized medicine able to predict diseases and find ways to prevent them. This approach would encourage patients to participate more actively in their own care. Genomic researchers were sure that it was time to realize the hopes of the late 1980s and early 1990s for a completely different, faster, and cheaper approach to sequencing.[25]

The classic machines based on the methods Frederick Sanger worked out in the 1970s, especially the 3700s manufactured since 1998 by Applied Biosystems of Foster City, California, had been used to produce the "working drafts" of 2000 and then a "finished" sequence of the human genome in 2003. They also turned out a blizzard of other genomes from viruses on up. The engineers of the 3700 had not been laggard. Their machine advanced rapidly to new models with higher speed and accuracy. Over fifteen years their capacity, or "throughput," doubled every twenty-two months, on average.[26]

But in the last years of the first decade of the twenty-first century, the Sanger-style machines were overtaken by the new generation, operating in a "massively parallel" fashion, crunching through sequences more and more quickly every few months. The throughput of the second-generation machines multiplied tenfold a year for several years running. Soaring in speed, with one new model after another, the different sequencer designs from 454 Life Sciences in Connecticut, Illumina in England and California, and the Applied Biosystems Center in Beverly, Massachusetts, needed fewer workers to tend them.

First out of the gate was 454's machine. The company to make it, founded in 2000, took its name from its project number at the parent company, CuraGen. Not much bigger than a toaster oven, the new sequencer grabbed attention in August 2003 when it used just a single sample to sequence adenovirus, a workhorse of cancer research, in less than a day. 454 used a process called "pyrosequencing," invented in Sweden, which it was licensing for millions of dollars a year. Richard Gibbs, head of the genome center at Baylor College of Medicine and an adviser to 454, was enthusiastic. He said, "It's a real threshold moment." More skeptical was Eddy Rubin, head of the Department of Energy's (DOE) Joint Genome Institute in California. He thought going to a much bigger organism, such as a bacterium, would be "a challenge." But the 454 shareholders were enthusiastic and put up an additional $20 million in capital.

Two years later, the 454 system sequenced the genome of the tiny bacterium *Mycoplasma genitalium* in four hours. A Yale study used 454 sequencing to detect drug-resistant strains of the AIDS virus. 454's process received top honors in the *Wall Street Journal*'s 2005 Technology Innovation Awards. In 2007, as mentioned in Chapter 6, 454 Life Sciences collaborated with Gibbs's group at Baylor College of Medicine to sequence the genome of James Watson.

454 machines, like their rivals, went into testing at major sequencing centers around the world. At the Broad Institute, Lander commented, "We're going to be testing all the others, too. . . . I'm in favor of all clever ideas." Nonetheless, "it takes quite a long time before clarity emerges around a new technology platform. We are going to have to see how they perform in many, many respects." Smaller centers, such as the one at the University of Washington where Maynard Olson worked, began sending samples to 454 headquarters in Connecticut for sequencing.[27]

Impatient to take genomics much deeper into medicine, biologists continued with the approach of sequencing a gene of interest in a large number of diseased people and comparing it to the same gene in people free of a particular disease. Olson's favorite hope in 2005 was sequencing the human leukocyte antigen region of human chromosome 6. In that region lie the genes associated with autoimmune diseases like type I (juvenile) diabetes, multiple sclerosis, rheumatoid arthritis, and lupus. He found it frustrating that the "precise molecular explanation" for these diseases "has remained elusive for decades." While he worried that industry would not find the money to develop and market the new machines rapidly, Olson foresaw a "lurch," a leap forward after years of "a great grinding of gears."[28]

By then, rival "next-generation" machines had entered the market. One of these came from Solexa, based near Cambridge, England, and in the San Francisco Bay area. A leader in its work was David Bentley, formerly of the Wellcome Trust Sanger Institute. In a stock-swap deal valued at $600 million, Illumina, a leading microarray maker based in San Diego, bought Solexa in 2007. Illumina swiftly went through the severe tests of making, selling, and servicing the machines for impatient scientists, who were also undergoing a parallel crash course. The company had good reasons to buy Solexa. Although Illumina was a leader in chips, the new sequencers were breathing down the neck of the chip industry. With many scientists expecting ultrafast sequencers to edge out the chips, Illumina wanted a foot in both camps. It began pushing to transfer methods from one technology to the other.[29]

As happens so often in technology-based industries, the fast pace continually pushed companies to agglomerate—not only to obtain fresh capital but also to increase their strength in marketing and service. The pharmaceutical giant Roche bought 454 in 2007 for an estimated $150 million. The same year Roche acquired a chip company, NimbleGen of Madison, Wisconsin.[30]

A similar process happened at a company called Agencourt in Beverly, Massachusetts. Among other business activities, Agencourt was developing concepts from the irrepressible George Church at Harvard Medical School into a next-generation machine that became known as SOLiD, an acronym for Supported Oligonucleotide Ligation and Detection. Leading the design effort was Kevin McKernan, a cofounder of Agencourt. In the summer of 2006, Applied Biosystems leapt into the next-generation arena by buying SOLiD from Agencourt for $120 million. The process of consolidation marched on. In the summer of 2008, a company called Invitrogen, involved in many businesses associated with genomics, agreed to buy Applied Biosystems for $6.7 billion. The combined companies assumed a new name: Life Technologies, with the NASDAQ acronym LIFE.[31]

By early 2009, both Illumina and Life Technologies expressed confidence that they could soon deliver a complete human sequence for $10,000. Only two years earlier, Jim Watson's sequence had cost $1 million. Hence, they aspired to a price a hundredfold, or two orders of magnitude, cheaper. At the tenth annual Advances in Genomic Biology and Technology conference on Marco Island, Florida, John McPherson of the Ontario Institute for Cancer Research, said it was now possible to sequence large numbers of tumors (each with a complete human genome), going far deeper into the differences between one tumor and another and between tumors and normal tissue. "We really can ask the questions now that we've wanted to ask all along and just were limited by technology." It was not just a matter of speed or numbers of DNA subunits per second. Stan Nelson of the University of California, Los Angeles, said, "The quality has gone up shockingly in this last year." McKernan said the market was growing so "disruptively" that there would be "lots of fragmentation and differentiation."[32]

At Marco Island, as 454, Illumina, and Life Technologies trumpeted their progress, they were feeling the hot breath of a third, even faster generation of sequencing. A California company called Pacific Biosciences, which had emerged from what is called "stealth mode" at Marco Island a year earlier, announced it had used its single-molecule technique to sequence the classical lab bacterium *Escherichia coli* sixty-eight-fold, with DNA chunks, or "reads," averaging nearly 500 bases in length. The company planned to launch its sequencing machine commercially in the second half of 2010. Another new company, Complete Genomics, detailed its plans to start up a service that would offer human genome sequences for $5,000; the company expected to sequence 1,000 people the first year and then soar to 20,000 the next. As a test, Complete Genomics announced plans to sequence five human genomes for the Broad Institute.[33]

Another feisty newcomer was Helicos, based in Cambridge, Massachusetts. Its single-molecule sequencer began undergoing tests at laboratories like those of the Broad Institute and Stanford University. In 2009, as a bootleg project sandwiched between other research, three people using a Helicos machine succeeded in sequencing the DNA of Stanford professor Stephen Quake, the machine's inventor. This feat, Quake said, showed that big sequencing centers were not the only place for deciphering whole human sequences.[34]

Under this pressure, both Illumina and Life Technologies were pushing into the third generation. Illumina invested in companies like Avantome and Oxford Nanopore. Life Technologies, which had also acquired a third-generation Texas company called Visigen, announced it was spending more than $100 million on advanced sequencing. The "great vitality, dynamism, and inventiveness" of the sequencing industry, one observer wrote, made it difficult to predict which companies and technologies would dominate the next few years with the cost of a human genome sinking to $5,000 and less. The already profitable leaders—Illumina, 454, and Applied Biosystems—were selling more machines and related "consumables," not only to big centers but also to smaller labs. Even if the newcomers thrust their way into the market, the outlook was bright: "The next several years are likely to prove a Golden Age for next-generation sequencing."[35]

Rich and Poor Await the Payoff

As this narrative has emphasized, parallel changes in biological science, genetics, sanitation, crop breeding, pharmaceuticals, and medical practice have all led to rapid changes in human health. What is medically possible increases from year to year. This fact breeds hope that more can be done to palliate, postpone, or prevent painful and frightening diseases. The hope also breeds impatience. One question is always asked: When am I going to see the payoff from DNA science?

This question applies not only to nations where illnesses related to aging dominate medical spending but also to those still ravaged by infectious disease. A frightening example is malaria, the insect-borne disease that strikes hard in the less affluent nations of the tropics and subtropics, each year infecting hundreds of millions of people, sapping the energy of whole nations, and killing millions, many of them children.

The main line of defense always has been to shield people, particularly at night, from the *Anopheles* mosquitoes that draw blood for their meals and thereby infect people with the major malaria parasite *Plas-*

modium falciparum. The battle against malaria has always, first of all, been a matter of using bed nets, then, after steel wire could be woven cheaply, putting screens on millions of windows and porches.

A second line of defense has been to drain the bodies of water, however small, where the mosquitoes lay their eggs. A third line is to poison the insects' nervous systems with insecticides. Since the 1940s, a succession of chlorinated hydrocarbons, starting with DDT, has been used to kill mosquitoes, inside houses and outdoors, on such a large scale that the insects developed resistance within years.

A fourth line is to give drugs to those already suffering from malaria to beat back the disease. When the *Plasmodium* parasites evolve resistance to traditional medicines like quinine, yet others enter the battle. One of these is the artemisinin from the sweet wormwood plant long used as an herbal remedy in China—and now the target of a global push for mass manufacture. As is the case with the worldwide struggles against pests that threaten the major food crops, the fight against malaria is a constant battle against rapid evolution.

About the time that the genome project began to scale up in the mid-1990s, geneticists across the world sought ways to attack both malaria parasites and the mosquitoes that carry them. The central idea was to find both points of vulnerability and the sources of resistance to chemicals. Years of effort culminated in October 2002 with publication of the parasite's genome in *Nature* and the insect's in *Science*. Soon, sequences from human victims opened the way to study the genetics of people's susceptibility and resistance to malaria.[36]

Parallel efforts began with human immunodeficiency virus, or HIV, which leads to AIDS, afflicting tens of millions of people in poor countries and rich. The genomics of AIDS did not just focus on the almost incredible versatility of the virus in getting around the body's natural defenses and the "cocktails" of drugs that allow victims to live many years after infection. Of equal interest was the genetics of people's resistance to AIDS. Early in 2009, the sequencing machine manufacturer Illumina announced a collaboration with Duke University to sequence the genomes of fifty people known to be infected with the virus but possessing factors repressing the disease. Genomics was having an impact on diseases beyond those of rich countries.[37]

People's consciousness of the science and uses of genomics is growing rapidly, as is impatience for bigger payoffs. But will genomics benefit poor nations as well as rich? Determined people are working to harness genomics to worldwide public health, as in the combat against infectious diseases like AIDS or malaria. Since 1995, dozens of microbes causing major diseases such as cholera and tuberculosis have been sequenced.

The Influence of Forces Outside Science and Medicine

As 2010 approached, the tasks requiring genomic technologies and insights were clearly even vaster than they had appeared ten years earlier. In the absolute sense, the medical utility of genomics depends on understanding systems of enormous complexity, those governing normal and abnormal development and physiology. Ahead lie the mysteries of the brain, the immune system, and cancer.

It was also clear that the hope of making use of a person's genetic endowment to guide both human behavior and medical intervention depended upon forces far beyond the scope of scientific inquiry.

One sign of this was James Watson's proposal in August 2009 for a crash program by the NCI to develop multidrug approaches to cancer. The widely held idea, which Watson expressed, was that cancers should be attacked as AIDS has been, by a cocktail of chemicals to suppress recurrence of the disease that so often follows treatment with one drug at a time.

Such an approach, Watson wrote in the *New York Times*, required not only a mission beyond NCI's role in supporting research but also major changes in the FDA's system for approving new drugs and new uses for them. NCI needed to take the lead in drug development because the economic incentives were not strong enough for private industry. The FDA needed to change from its approach of approving one drug or device at a time.[38] Analogies to other grand-scale government-led enterprises that Watson could have drawn included American river-basin electricity projects and the creation in World War II of entire industries to make aviation gasoline, synthetic rubber, and penicillin.

The often-stated practical aim of genomics was to develop a medicine targeted at the causes rather than symptoms of disease with treatments far more closely tailored to individuals. People would avoid unnecessary treatments and be spared side effects from drugs that would not help them. With new, more complete, more specific information—universally recorded electronically and available to any physician wherever someone fell sick—mistakes in treatment, including many leading to death, would be fewer.

With more precise knowledge of predispositions to disease, obtained before and just after birth and later, physicians could advise patients about how to manage their environments and lifestyles to postpone or avoid disease and disability. In a time of intense worry about medical costs rising far faster than normal economic inflation, it was hoped that we could avoid a substantial fraction of today's medical expenses.

This hope of "rational" drug development benefiting from genetic knowledge has not disappeared. In principle, the number of targets will increase. Drug treatment will be tailored to individuals. Many drugs are laid aside after clinical trials because a substantial portion of patients suffer dangerous side effects or are not helped. But could such drugs be taken down from the shelf and given to the minority they did help? To be sure, drug companies are highly motivated to produce "blockbuster" drugs taken by millions of people and generating billions of dollars in annual sales. Drugs that would be used by people with little or no ability to pay, no matter how numerous, have been neglected. Conceivably, however, the "rescue" of failed drugs—to say nothing of a host of new, targeted drugs—could help the economics of drug companies. Whether the pharmaceutical firms could enter such realms without philanthropic or government aid remains to be determined.

The vision of a genomics-aided healthy society implies vast social change, inside and outside of the practice of medicine. Such change was foreshadowed by the loud debate in the United States over health-care economics in 2009 and 2010, and by the outbursts of anger after two studies appearing within days of each other in November 2009 urged less frequent screening for breast and cervical cancer. The need for evidence-based medical care, which genomics is contributing to now and will continue to do even more in the future, collides with many people's fear that their access to particular types of care would be rationed. Would every citizen, without regard to income, have access to, and regularly visit, medical professionals who could warn them of genetic susceptibility and prescribe preventive steps? Regular examinations could catch many illnesses, such as diabetes, at early stages. If they did, many fantastically expensive visits to emergency rooms could be avoided. But will people seek to know about their genetic risks and act on what they learn—or will they shrink from it? Studies to find out began at such places as the Scripps Institute in La Jolla, California.[39]

The questions are encompassing and unanswered. Will doctors and other medical professionals become and remain better informed about genetics as a factor in diagnosis and treatment? Will they collaborate to share records and insights and to avoid unnecessary interventions? Will it be necessary to escape from a fee-for-service system and rely instead on salaried doctors and universally nonprofit hospitals? That such systems are possible has already been demonstrated on a large scale in many places across the United States. An example is the Kaiser-Permanente health system, born during World War II, which has 2 million members in northern California alone.

Outside medicine, much vaster efforts would be needed. They would have to persist for many years, as has the movement to reduce the proportion of people who smoke. The efforts would include a mass movement—beginning in childhood—to control food intake and increase physical activity. Such programs would be a struggle in societies where food prices absorb only a modest share of personal income, and jobs are increasingly sedentary. Any suggestion of constraint would arouse resistance from at least some people.

It is possible that such a mass movement could arise from the small beginnings already evident in the nascent industry for personal genetic testing. Some of the customers share their information with others with a similar medical condition or ancestry, and others do so in the already widespread practice of social openness typified by networks like YouTube, Facebook, or Twitter. It is probably no accident that several direct-to-consumer testing companies arose in Silicon Valley in California. They share the ethos that has profoundly influenced electronic computation and communication over the last half century. The direct-to-consumer testing industry pitches the idea of people having a right to their complete health information—not merely to a tabulation of their genetic risks but to all medical records. When customers pay for the information, they may grow more interested in using it.[40]

Agriculture, Energy, and the Environment

The promise of genomics, of course, transcends medical care to embrace other fields of vast importance to human health—agriculture in particular. Since the 1940s, a "Green Revolution" in plant genetics has transformed the yields of such food plants as wheat and rice in nations where humanity is most thickly gathered. These include India, China, and Indonesia. Total world food production is estimated to have increased about 2.5-fold in the last fifty years. But now, with a world population approaching 7 billion, the momentous issues of global "food security" have become even more urgent. By mid-century, the world's population is expected to reach 9 billion. The proportion of the global population able to afford a richer diet will grow, while the proportion raising food will continue to diminish. As cities spread, the area for growing food is shrinking. Agriculture also is threatened by deforestation, which has led to widespread erosion and flooding.

Confronting the imminent, enormous demand for swift increases in the productivity of the world's farms, plant geneticists in many nations have gone beyond their model organism, a fast-growing thale cress called *Arabidopsis* that was sequenced in 2000, to decode the genomes of major food grains like rice and maize and even one chromosome of the much larger wheat genome.[41] Among the aims are development of plants resistant to heat, cold, pests, diseases, salinity, drought, and flooding and that will eventually produce their own fertilizer nitrogen, last for more than one crop season, and use sunlight more efficiently.

With progress in plant genetics in mind, David Baulcombe of Britain's Royal Society said in October 2009, "We need to take action now to stave off food shortages. If we wait even five to ten years, it may be too late." Baulcombe, a plant geneticist and pioneer in studies of micro RNAs, chaired the society's report urging a global program that would cost more than $3 billion. An inevitable outcome of such work, the report noted, will be a great expansion of genetic engineering of plants.[42] While genetically modified crops are grown widely in the

United States, their use is strongly resisted in Europe. As a consequence, farmers in food-exporting regions of Africa fear they will not be able to sell "GM crops."

The evident, apparently accelerating warming of the earth's atmosphere also creates potential threats to human health. Hence, the linked fields of energy and pollution have also become domains for genomics. These may one day even dwarf the impact of genomics on medical care. A central focus in the energy and environment crises lies in what is called synthetic biology, that is, the harnessing of microbes. By digesting grasses, fast-growing trees, and water-dwelling algae, genetically engineered microbes may provide new and "sustainable" sources of liquid fuel and help clean up the effluents of the vast energy industry. They may reduce the conflict between growing corn for fuel and growing it for food. Could such microbes not only clean up oil spills or toxic wastes from mines but also pull carbon dioxide from the atmosphere?

Hundreds of millions of dollars of public and private money are going into synthetic biology, encouraging divergent approaches and much controversy. The main difference appears to lie in which method researchers hope to use in "reprogramming" living cells to do new work. One is tailoring existing microorganisms by using what have been dubbed "Bio-bricks." The other is creating entirely new organisms. Typifying the first approach is the work of companies like DuPont and DOE's Lawrence Berkeley National Laboratory. There, Jay Keasling, who found genomic methods to speed the manufacture of the antimalarial drug artemisinin a millionfold, heads a newly created Joint BioEnergy Institute. Venter's company, San Diego–based Synthetic Genomics, is committed to the more radical pathway of artificial life. On their way to synthesizing an artificial bacterium, Nobel Prize winner Hamilton Smith, Clyde Hutchison, and others at Venter's non-profit institute stitched together the complete DNA of the smallest of bacteria. A Spanish newspaper called the feat a work of titans, "Un obra de titanes," and *Time* put it on its list of fifty best inventions of 2008. In the summer of 2009, they announced an advance in reprogramming the principal, extremely simple microbe with which they had been working since the 1990s, *Mycoplasma genitalium*. This was the progress that led the oil giant ExxonMobil, hitherto highly skeptical of biofuels, to back Venter in a ten-year quest to develop liquid fuel from algae.[43]

Changing Perceptions

Like the shock wave that ripples out from a supernova in all directions, the big-scale science of genomics represents an explosion in knowledge that shows no sign of contracting. Since 2000, as we have seen, the worldwide genomics team has been plunging into paradoxes—a far smaller number of genes than expected and a dramatic new focus on the "noncoding" regions of the human genome. Scientific surprises and frustrating new complexities constantly challenge the accuracy of the human sequence.[44]

Although many of the promised economic and medical benefits of the revolutionary enterprise of genomics still lie in the future, the study of genomes in ever-greater detail adds up to an increasingly effective system for discovering how life processes work. The discoveries are rapidly changing perceptions about being human amid a host of other species. The curtain has been drawn back on a genomic age.

ACRONYMS

ABI	Applied Biosystems, Inc.
AIDS	acquired immune deficiency syndrome
AML	acute myeloid leukemia
BAC	bacterial artificial chromosome
cDNA	complementary DNA
CEPH	Center for Study of Human Polymorphism
CF	cystic fibrosis
CNV	copy-number variation
CSHL	Cold Spring Harbor Laboratory
DMD	Duchenne muscular dystrophy
DNA	deoxyribonucleic acid
DOE	Department of Energy
ENCODE	"The Encyclopedia of DNA Elements"
EST	expressed sequence tag
FDA	Food and Drug Administration
GWA	genome-wide association
HGS	Human Genome Sciences
HHMI	Howard Hughes Medical Institute
HLA	human leukocyte antigen
IPO	initial public offering
LOD	logarithmic odds
MGH	Massachusetts General Hospital
MIT	Massachusetts Institute of Technology
MRC	Medical Research Council
NCI	National Cancer Institute
NEJM	*New England Journal of Medicine*
NHGRI	National Human Genome Research Institute
NIH	National Institutes of Health
NRC	National Research Council

OTA	Office of Technology Assessment
PAC	artificial chromosome made from a virus called P1
PCR	polymerase chain reaction
RFLP	restriction fragment length polymorphism
RNA	ribonucleic acid
siRNA	small interfering RNA
SNP	single-nucleotide polymorphism
STS	sequence-tagged site
TCGA	The Cancer Genome Atlas
TIGR	The Institute for Genomic Research
UC	University of California
UCSC	University of California, Santa Cruz
Wash U	Washington University
YAC	yeast artificial chromosome

NOTES

Chapter 1: Building the Toolbox

1. **Nobel autobiography:** Hamilton O. Smith, 1978, Nobel Foundation, http://nobelprize.org/nobel_prizes/medicine/laureates/1978/smith-autobio.html.

Nobel lecture: Hamilton O. Smith, "Nucleotide Sequence Specificity of Restriction Endonucleases," December 8, 1978, Nobel Foundation, http://nobelprize.org/nobel_prizes/medicine/laureates/1978/smith-lecture.html.

Newspapers: B. D. Colen, "American, Swiss Scientists Share Prize in Medicine; 3 DNA Researchers Win Nobel Prize," *Washington Post*, October 13, 1978, A1; Douglas Birch, "Hamilton Smith's Second Chance; Scientist's Journey: He Won the Nobel, but Lost His Way. Could He Put His Family Together? And Could He Crack One of Life's Great Puzzles?" *Baltimore Sun*, April 11–13, 1999.

Magazine: "Chemical Scalpel," *Newsweek*, October 23, 1978, 98.

2. **Books:** John Cairns, Gunther S. Stent, and James D. Watson, eds., *Phage and the Origins of Molecular Biology*, exp. ed. (Cold Spring Harbor, NY: Cold Spring Harbor Laboratory Press, 1992); Robert Olby, *The Path to the Double Helix* (Seattle: University of Washington Press, 1974); James D. Watson, *The Double Helix*, 1st ed. (New York: Atheneum, 1968); Francois Jacob, *The Statue Within: An Autobiography*, trans. Franklin Philip (Cold Spring Harbor, NY: Cold Spring Harbor Laboratory Press, 1995); Francis Crick, *What Mad Pursuit: A Personal View of Scientific Discovery* (New York, Basic Books, 1988); Victor K. McElheny, *Watson and DNA: Making a Scientific Revolution* (Cambridge, MA: Perseus, 2003); Brenda Maddox, *Rosalind Franklin: The Dark Lady of DNA* (New York: HarperCollins, 2002); Robert Olby, *Francis Crick: Hunter of Life's Secrets* (Cold Spring Harbor, NY: Cold Spring Harbor Laboratory Press, 2009).

3. **Magazine:** Victor K. McElheny, "Laboratory of Molecular Biology, Cambridge," *Science* 144 (April 24, 1964): 399–400.

Books: Arthur Kornberg, *For the Love of Enzymes: The Odyssey of a Biochemist* (Cambridge, MA: Harvard University Press, 1989); Georgina Ferry, *Max Perutz* *and the Secret of Life* (Cold Spring Harbor, NY: Cold Spring Harbor Laboratory Press, 2007).

4. **Book:** Horace Freeland Judson, *The Eighth Day of Creation: Makers of the Revolution in Biology*, exp. ed. (Cold Spring Harbor, NY: Cold Spring Harbor Laboratory Press, 1996).

5. **Newspapers:** Victor K. McElheny, "Repressor of Genes Isolated," *Boston Globe*, April 9, 1967, 20; Victor K. McElheny, "Cell Control Advances Are Claimed," *Washington Post*, May 1, 1967; Victor K. McElheny, "Russian-Born Magasanik Studies Cell Controls," *Boston Sunday Globe*, May 28, 1967, 8A.

6. **Scientific article:** Matthew Meselson and Robert Yuan, "DNA Restriction Enzyme from *E. coli*," *Nature* 217 (March 23, 1968): 1110; "News and Views," ibid., 1097–1098.

7. **Scientific articles:** H. O. Smith and K. W. Wilcox, *Journal of Molecular Biology* 51 (1970): 379; T. J. Kelly and H. O. Smith, *Journal of Molecular Biology* 51 (1970): 393.

8. **Interview:** Leroy Hood, by the author, Seattle, Washington, February 14, 2004, in *Mapping Life Notebook* 2, 12–15.

9. **Scientific article:** John F. Morrow, Stanley N. Cohen, Annie C. Y. Chang, Herbert W. Boyer, Howard M. Goodman, and Robert B. Helling, "Replication and Transcription of Eukaryotic DNA to *Escherichia coli*," *Proceedings of the National Academy of Sciences of the United States of America* 71, no. 5 (May 1974): 1743–1747.

Newspapers: Victor K. McElheny, "New Clues Found in Genetic Study," *New York Times*, April 11, 1974, 7; Victor K. McElheny, "Animal Gene Shifted to Bacteria; Aid Seen to Medicine and Farm," *New York Times*, May 20, 1974, 61; Victor K. McElheny, "Genetic Tests Renounced over Possible Hazards," *New York Times*, July 18, 1974, 1.

10. **Scientific articles:** Frederick Sanger et al., "Nucleotide Sequence of Bacteriophage Phi X 174 DNA," *Nature* 265 (February 24, 1977): 687, with commentary, 685; F. Sanger, S. Nicklen, and A. R. Coulson, "DNA Sequencing with Chain-Terminating Inhibitors," *Proceedings of the National Academy of Sciences USA* 74 (December 1977): 5463.

Magazine: John Rogers, "Atomic Structure of a Living Organism," *New Scientist*, March 10, 1977, 583.

11. **Magazine:** *ASM News* [American Society of Microbiology] 65, no. 5 (May 1999).

12. **Book:** See "'Odd Man Out'—Recombinant DNA," chapter 11 in McElheny, *Watson and DNA*, and endnotes.

13. **Newspapers:** Victor K. McElheny, "Safeguard Urged in Gene Transfer," *New York Times*, January 22, 1975, 9; Victor K. McElheny, "World Biologists Tighten Rules on 'Genetic Engineering' Work," *New York Times*, February 28, 1975, 1; Victor K. McElheny, "Science Seeks 'Safe Bugs' for Genetic Engineering," *New York Times*, March 3, 1975, 28.

14. **Nobel autobiography:** David Baltimore, 1975, Nobel Foundation, www.nobelprize.org/nobel_prizes/medicine/laureates/1975/baltimore-autobio.html.

Nobel lecture: David Baltimore, "Viruses, Polymerases and Cancer," December 12, 1975, Nobel Foundation, http://nobelprize.org/nobel_prizes/medicine/laureates/1975/baltimore-lecture.pdf.

Newspapers: Victor K. McElheny, "MIT Biologist Probes Virus-Cancer Link," *Boston Globe*, June 12, 1970, 10; Victor K. McElheny, "Winners of Nobel Prize in Medicine," *New York Times*, October 17, 1975, 12.

15. **Interview:** David Botstein, by the author, Stanford, California, February 12, 2001, transcript.

Book: David Botstein, "A Phage Geneticist Turns to Yeast," in *The Early Days of Yeast Genetics*, ed. Michael N. Hall and Patrick Linder (Cold Spring Harbor, NY: Cold Spring Harbor Laboratory Press, 1993), 361–373.

16. **Author's notes.**

Nobel banquet speech: Sydney Brenner, December 10, 2002, Nobel Foundation, http://nobelprize.org/nobel_prizes/medicine/laureates/2002/brenner-speech-text.html.

Magazine: Victor K. McElheny, "Three Nobelists Ask: Are We Ready for the Next Frontier?" *Cerebrum* (fall 2003), www.dana.org/news/cerebrum/detail.aspx?id=2966.

17. **Interview:** Sydney Brenner, by the author, La Jolla, California, February 14, 2001, transcript.

Nobel autobiography: Sydney Brenner, 2002, Nobel Foundation, http://nobelprize.org/nobel_prizes/medicine/laureates/2002/brenner-autobio.html.

18. **Newspaper:** Jack Hee, "The Funniest Scientist Who Ever Lived," *Straits Times* (Singapore), November 10, 2000, 14.

19. **Nobel lecture:** Sydney Brenner, "Nature's Gift to Science," December 8, 2002, Nobel Foundation, http://nobelprize.org/nobel_prizes/medicine/laureates/2002/brenner-lecture.pdf.

20. **News release:** Nobel Foundation, "The Nobel Prize in Physiology or Medicine 2002," October 7, 2002, Nobel Foundation, http://nobelprize.org/nobel_prizes/medicine/laureates/2002/press.html.

Newspaper: Thomas Maugh II, "'Suicide Cell' Work Earns Trio a Nobel; Science: Biologist at Salk Institute in La Jolla Shares Prize in Medicine/Physiology," *Los Angeles Times*, October 8, 2002, 1.

Books: Sydney Brenner, *My Life in Science* (as told to Lewis Wolpert) (London: Science Archive Limited, 2001); Andrew Brown, *In the Beginning Was the Worm: Finding the Secrets of Life in a Tiny Hermaphrodite* (New York: Columbia, 2003).

21. **Nobel lecture:** Frederick Sanger, "Determination of Nucleotide Sequences in DNA," December 8, 1980, Nobel Foundation, http://nobelprize.org/nobel_prizes/chemistry/laureates/1980/sanger-lecture.html.

Nobel autobiography: Frederick Sanger, 1980, Nobel Foundation, http://nobelprize.org/nobel_prizes/chemistry/laureates/1980/sanger-autobio.html.

Scientific article: Frederick Sanger, "Sequences, Sequences, and Sequences," *Annual Review of Biochemistry* 57 (1988): 1–28.

News release: Nobel Foundation, "The Nobel Prize in Chemistry 1980," Nobel Foundation, October 14, 1980, http://nobelprize.org/nobel_prizes/chemistry/laureates/1980/press.html.

Magazines: "Success in the Fens," *The Economist*, October 25, 1980, 116; Glyn Jones and Kate Douglas, "The Quiet Genius Who Decoded Life: The Techniques to Unravel the Human Genome Were Pioneered by Frederick Sanger Decades Ago," *New Scientist*, October 8, 1994, 33.

22. **Scientific articles:** Walter Gilbert and Allan Maxam, "The Nucleotide Sequence of the *lac* Operator," *Proceedings of the National Academy of Sciences USA* 70 (December 1973): 3581; Walter Gilbert and Allan Maxam, "A New Method for Sequencing DNA," *Proceedings of the National Academy of Sciences USA* 74 (February 1977): 560.

Nobel lecture: Walter Gilbert, "DNA Sequencing and Gene Structure," December 8, 1980, Nobel Foundation, http://nobelprize.org/nobel_prizes/chemistry/laureates/1980/gilbert-lecture.html.

Nobel autobiography: Walter Gilbert, 1980, Nobel Foundation, http://nobelprize.org/nobel_prizes/chemistry/laureates/1980/gilbert-autobio.html.

Speech: Walter Gilbert, "My Life in Molecular Biology," speech at Tokyo Institute of Technology, January 25, 2001, typescript, Gilbert files, Harvard University archives.

Newspaper: Pete Earley, "Shy Bookworm at Sidwell Friends Gains International Attention," *Washington Post*, October 16, 1980, A23.

Magazine: Gina Bari Kolata, "Lac System: New Research on How a Protein Binds to DNA," *Science* 184 (April 5, 1974): 52.

Book: McElheny, *Watson and DNA*, 136–139 (account of repressor isolation).

23. **Interview:** Leroy Hood, by the author, Seattle, Washington, February 14, 2004, in *Mapping Life Notebook* 2, 12–15.

Speeches: Leroy Hood, "My Life and Adventures Integrating Biology and Technology: A Commemorative Lecture for the 2002 Kyoto Prize in Advance Technologies," Systems Biology, www.systemsbiology.org/download/2002Kyoto.pdf; Leroy Hood, "New Ideas Thrive on New Organizational Structures," Batten Briefings, Darden Graduate School of Business, spring 2004, www.darden.virginia.edu/batten/pdf/BB_Hood_SP04.pdf.

Newspapers: Harold M. Schmeck Jr., "New 'Gene Machines' Speeding Up the Pace of Biological Revolution," *New York Times*, March 24, 1981, 1; Peter Eisler, "Was the U.S. Government Cheated on DNA Decoder?" *USA Today*, April 18, 2002, 1A. In re disputed credit for the invention of fluorescent DNA labeling for sequencing: Mary Ann Benitez, "HK-Born Scientist at Centre of Lawsuit over DNA Map Device," *South China Morning Post* (Hong Kong), April 20, 2002, 3; Tina Hesman "DNA Machine Generates Sequences, Controversy," *St. Louis Post-Dispatch*, May 29, 2002, A1.

Magazines: "Caltech and R+D Company Build New DNA Synthesizer," McGraw-Hill's *Biotechnology Newswatch* 1 (October 19, 1981): 3; Antoine Danchin, "Une histoire intense, presque violente," *La Recherche*, June 2000, 27. In re disputed credit: John Dodge, "An Uncivil Action," *Bio-IT World*, June 12, 2002; Cindy Coty, "Confusion and Litigation over Rights to Basic Sequencing Technology," *Genomics and Proteomics*, September 1, 2002, 54.

Books: Robert Cook-Deegan, *The Gene Wars: Science, Politics, and the Human Genome* (New York: W. W. Norton, 1994), 64–71; Kevin Davies, *Cracking the Genome: Inside the Race to Unlock Human DNA* (New York: Free Press, 2001), 143–144.

24. **Scientific article:** Lloyd M. Smith et al., including Leroy Hood, "Fluorescence Detection in Automated DNA Sequence Analysis," *Nature* 321 (June 12, 1986): 674.

Interview: Michael Hunkapiller, by the author, Palo Alto, California, February 22, 2007, transcript.

News releases, all from Business Wire: October 27, 1983, reporting quarterly sales of $3.259 million and net income of $865,000; April 16, 1984, announcing first stock trades on NASDAQ; February 12, 1985, announcing two-for-one stock split; March 1985, announcing an $18.4 million stock sale to Becton Dickinson, bringing cash reserves of $53 million; October 2, 1985, reporting a $3.1 million research agreement with Rothschild, Inc.; December 16, 1987, noting annual sales of $84.4 million and net income of $9.4 million.

Wire service: Lee Siegel, "Machine Pushes 'Biology into the Big Time,'" Associated Press, June 12, 1986.

Newspapers: "Applied Biosystems Reports Earnings for Qtr to July 1," *New York Times*, August 18, 1983, D5; "Briefs, Debt Issues," *New York Times*, June 30, 1983, D11; Thomas H. Maugh II, "Caltech scientists Develop Super-Fast DNA Analyzer," *Los Angeles Times*, June 12, 1986, Part I, 3; Jay Mathews, "Caltech's New DNA-Analysis Machine Expected to Speed Cancer Research," *Washington Post*, June 12, 1986, A26; Keith Schneider, "Technology; Gene Mapping Is Improved," *New York Times*, June 26, 1986, D2; David Fishlock, "Finding the Beads in Life's Genetic Strings; Out of the Backroom," *Financial Times*, July 10, 1986, Section I, 24; Jeannine Stein, "Superscientist Balances Home Life and Lab Life," *Los Angeles Times*, July 14, 1986, Part V, 1.

Magazines: "Machines to Read Genes," *The Economist* (U.S. edition), August 24, 1985, 79; Ricki Lewis, "Computerizing Gene Analysis: The 'Sequenator,' Capable of Deciphering Every Human Gene, Will Benefit Medicine and Biotechnology Alike," *High Technology* (December 1986): 46; Leslie Roberts, "New Sequencers to Take on the Genome," *Science* 238 (October 16, 1987): 271; Reginald Rhein, "Speeding Up the Search for Genetic Secrets," *Business Week*, November 9, 1987, 106D, noting sales to date of fifty machines at $90,000 apiece; Frederic Golden and Michael D. Lemonick, reported by Dick Thompson, "Mapping the Genome: The Race Is Over," *Time*, July 3, 2000, 18–23.

Book: J. Craig Venter, *A Life Decoded: My Genome, My Life* (New York: Viking Penguin, 2007), 100–102.

25. **Scientific articles:** Randall K. Saiki et al., including Kary B. Mullis, "Enzymatic Amplification of Beta-Globin Genomic Sequences and Restriction Site Analysis for Diagnosis of Sickle Cell Anemia," *Science* 230 (December 20, 1985): 1350; Randall K. Saiki et al., including Kary B. Mullis, "Primer-Directed Enzymatic Amplification of DNA with a Thermostable DNA Polymerase," *Science* 239 (January 29, 1988): 487.

Nobel lecture: Kary B. Mullis, "Polymerase Chain Reaction," December 8, 1993, Nobel Foundation, http://nobelprize.org/nobel_prizes/chemistry/laureates/1993/mullis-lecture.html.

Nobel autobiography: Kary B. Mullis, 1993, Nobel Foundation, http://nobelprize.org/nobel_prizes/chemistry/laureates/1993/mullis-autobio.html.

News release: Nobel Foundation, "The Nobel Prize in Chemistry 1993," Nobel Foundation, October 13, 1993, http://nobelprize.org/nobel_prizes/chemistry/laureates/1993/press.html.

Nobel presentation speech: Carl-Ivar Brändén, "The Nobel Prize in Chemistry 1993," presentation speech, Nobel Foundation, 1993, http://nobelprize.org/nobel_prizes/chemistry/laureates/1993/presentation-speech.html.

Newspapers: Andrew Pollack, "Two Biotech Pioneers to Merge," *New York Times*, July 23, 1991, D1 (reporting Hoffmann-LaRoche's purchase of PCR patent rights for $300 million); Malcolm W. Browne, "Four North Americans Win Nobel Prizes for Science," *New York Times*, October 14, 1993, A6.

Magazines: Jean L. Marx, "Multiplying Genes by Leaps and Bounds," *Science* 240 (June 10, 1988): 1408; Norman Arnheim and Corey H. Levenson, "Polymerase Chain Reaction," *Chemical and Engineering News*, October 1, 1990, 36; Marcia Barinaga, "Biotech Nightmare: Does Cetus Own PCR?" *Science* 251 (February 15, 1991): 739–740; Elizabeth Schaefer, "Cetus Retains PCR Patents," *Nature* 350 (March 7, 1991): 6; Marcia Barinaga, "And the Winner: Cetus Does Own PCR," *Science* 251 (March 8, 1991): 1174; Tim Appenzeller, "Democratizing the DNA Sequence," *Science* 247 (March 2, 1990): 1030; Kary B. Mullis, "The Unusual Origin of the Polymerase Chain Reaction," *Scientific American* 262 (April 1990): 56; Jeremy Cherfas, "Genes Unlimited," *New Scientist*, April 14, 1990, 29; William F. Allman, "Horizons: The Amazing Gene Machine, Biology, a Revolutionary Process for Rapidly Copying DNA," *Business Week*, July 16, 1990, 53.

Book: Paul Rabinow, *Making PCR: A Story of Biotechnology* (Chicago: University of Chicago Press, 1996).

26. **Scientific articles:** David Botstein et al., "Construction of a Genetic Linkage Map in Man Using Restriction Fragment Length Polymorphisms," *American Journal of Human Genetics* 32 (1980): 314; Arlene R. Wyman and Ray L. White, "A Highly Polymorphic Locus in Human DNA," *Proceedings of the National Academy of Sciences USA* 77 (1980): 6754.

Interview: David Botstein, by the author, Princeton, New Jersey, January 25–26, 2007, transcript.

Newspapers: Harold M. Schmeck Jr., "Fetal Tests Can Now Find Many More Genetic Flaws," *New York Times*, March 11, 1986, C1; Byron Spice, "Piecing Together Our Future; the New Map of the Human Genome Will Change Our Lives Forever," *Pittsburgh Post-Gazette*, June 25, 2000, A1 (story of hemachromatosis patient).

Magazines: Eric S. Lander, "The New Human Genetics," *Princeton Alumni Weekly*, March 25, 1987, cover, 1, 10–16; Jane Gitschier, "Willing to Do the Math: An Interview with David Botstein," *PLoS Genetics* 2, no. 5 (2006): e79.

Books: Jerry E. Bishop and Michael Waldholz, "Alta," in *Genome: The Story of the Most Astonishing Scientific Adventure of Our Time—the Attempt to Map All the Genes in the Human Body* (San Jose, CA: iUniverse.com, 1999), 49–68; Stephen S. Hall, "A Most Unsatisfactory Organism, 'Riflips' and Maps of Human Genes," in *Mapping the Next Millennium: The Discovery of New Geographies* (New York: Random House, 1992), 174–192; Cook-Deegan, "Mapping Our Genes," in *The Gene Wars*, 28–47.

Chapter 2: Getting Started

1. **Scientific articles:** James F. Gusella et al., "A Polymorphic DNA Marker Genetically Linked to Huntington's Disease," *Nature* 306 (1983): 234; Neil Risch, "The SNP Endgame: A Multidisciplinary Approach," *American Journal of Human Genetics* 76 (February 2005): 221.

Wire services: Paul Raeburn, "Studies in Remote Village Lead to Test for Huntington's Disease," Associated Press, November 8, 1983; Gino del Guercio, "Science Horizons: Geneticists Nearer to Cause of Huntington's; Hope to Solve Riddle of Other Genetic Diseases," United Press International, December 12, 1983.

Newspapers: Lawrence K. Altman, "Researchers Report Genetic Test Detects Huntington's Disease," *New York Times*, November 9, 1983, A1; Philip J. Hilts, "Genetic Disease Tests Seen Likely," *Washington Post*, November 9, 1983, A12; Richard Saltus, "For One Scientist, Genetic Debate Has Personal Sting," *Boston Globe*, October 30, 1989, 29.

Magazines: Miranda Robertson, "Huntington's Disease: The Beginning of the End of a Dilemma?" *Nature* 306 (November 17, 1983): 222; Matt Clark with Marsha Zabarsky, "Decoding a Killer Disease," *Newsweek*, November 21, 1983, 107; "Unlocking the Secrets of Genetic Disease," *The Economist* (U.S. edition), December 17,

1983, 77; Eric S. Lander, "The New Human Genetics: Mapping Inherited Diseases; Using DNA Markers, It Has Become Possible in Principle to Locate the Gene for Almost Any Simple Hereditary Disorder, a Feat with Revolutionary Implications," *Princeton Alumni Weekly*, March 25, 1987, 13.

Book: Jerry E. Bishop and Michael Waldholz, "Alta," in *Genome: The Story of the Most Astonishing Scientific Adventure of Our Time—the Attempt to Map All the Genes in the Human Body* (San Jose, CA: iUniverse.com, 1999), 49–68.

2. **Interview:** David Botstein, by the author, Lewis-Sigler Institute, Princeton University, Princeton, New Jersey, January 25, 2007, transcript.

3. **Interview:** Victor McKusick, by the author, Johns Hopkins University, Baltimore, Maryland, April 9, 2007, in *Mapping Life Notebook* 5, 339–345.

Newspaper: Harold M. Schmeck Jr., "Burst of Discoveries Reveals Genetic Basis for Many Diseases," *New York Times*, March 31, 1987, C1.

4. **Scientific article:** D. W. Russell et al., including Michael S. Brown and Joseph L. Goldstein, "The LDL Receptor in Familial Hypercholesterolemia: Use of Human Mutations to Dissect a Membrane Protein," *Cold Spring Harbor Symposia on Quantitative Biology* 51 (1986): 811.

Nobel lecture: Michael S. Brown and Joseph L. Goldstein, "A Receptor-Mediated Pathway for Cholesterol Homeostasis," Museum.com, December 9, 1985, http://nobelprize.virtual.museum/nobel_prizes/medicine/laureates/1985/brown-goldstein-lecture.pdf.

Nobel autobiographies: Michael S. Brown, 1985, Nobel Foundation, http://nobelprize.org/nobel_prizes/medicine/laureates/1985/brown-bio.html; Joseph L. Goldstein, 1985, Nobel Foundation, http://nobelprize.org/nobel_prizes/medicine/laureates/1985/goldstein-bio.html.

Nobel presentation speech: Viktor Mutt, "The Nobel Prize in Physiology or Medicine 1985," presentation speech, Nobel Foundation, December 10, 1985, http://nobelprize.org/nobel_prizes/medicine/laureates/1985/presentation-speech.html.

News release: Nobel Foundation, "Nobel Prize in Medicine or Physiology 1985," Nobel Foundation, http://nobelprize:org/nobel_prizes/medicine/laureates/1985/press.html.

Newspapers: William A. Knaus, "What Cholesterol Does," *Washington Post*, July 17, 1977, B3; Jane E. Brody, "Fatal Genetic Flaw Offers 'Human Lab' to Test Theories on Cholesterol," *New York Times*, November 10, 1981, C1.

5. **Scientific articles:** Louis M. Kunkel et al., "Analysis of Deletions in DNA from Patients with Becker and Duchenne Muscular Dystrophy," *Nature* 322 (July 3, 1986): 73; Anthony P. Monaco et al., including Louis M. Kunkel, "Isolation of Candidate cDNAs for Portions of the Duchenne Muscular Dystrophy Gene," *Nature* 323 (October 16, 1986): 646; Eric P. Hoffman, Robert H. Brown Jr., and Louis M. Kunkel, "Dystrophin, the Protein Product of the Duchenne Muscular Dystrophy Locus," *Cell* 51 (December 24, 1987): 919; Louis M. Kunkel, "2004 William Allan

Award Address, Cloning of the DMD Gene," *American Journal of Human Genetics* 76 (2005): 205.

Speeches: Louis M. Kunkel, seminars, Knight Science Journalism Fellowships, Massachusetts Institute of Technology, March 10, 1987, February 27, 1991, author's KSJF notebook 43, 87–97; April 15, 1992, April 28, 1993, April 26, 1995, notebook 63, 80–85; November 8, 1995, notebook 65, 128–135; December 4, 1996, notebook 70, 35–46; March 4, 1998, notebook 74, 152–166.

Wire services: Paul Raeburn, "'Historic Discovery' of Muscular Dystrophy Gene Praised," Associated Press, October 16, 1986; United Press International, "Scientists Trace a Gene Causing Muscle Disease," *New York Times*, October 16, 1986, A20.

Broadcast: David Nathan, interview by Charlayne Hunter-Gault, *MacNeil/Lehrer NewsHour*, PBS, October 16, 1986.

Newspapers: Andrew Veitch, "Suspect Gene Is Traced, Gene Responsible for Muscular Dystrophy Believed Isolated," *Guardian*, October 17, 1986; Harold M. Schmeck Jr., "Defect in Muscle Disease Pinpointed," *New York Times*, December 23, 1987, B12; Pearce Wright and Robert Matthews, "Scientists Identify Cause of Muscular Dystrophy," *Times*, December 24, 1987; Larry Thompson, "New Understanding of Muscular Dystrophy; Scientists Identify the Errant Protein," *Washington Post*, January 5, 1988, Z6.

Magazines: Peter N. Goodfellow, "Duchenne Muscular Dystrophy: Collaboration and Progress," *Nature* 322 (July 3, 1986): 12; Peter N. Goodfellow, "First Glimpse of the *DMD* Gene?" *Nature* 323 (October 16, 1986): 583; Gina Kolata, "Two Disease-Causing Genes Found," *Science* 234 (November 7, 1986): 669.

6. **Scientific articles:** Lap-Chee Tsui et al., "Cystic Fibrosis Locus Defined by a Genetically Linked Polymorphic DNA Marker," *Science* 230 (November 29, 1985): 1054; John R. Riordan et al., including Francis Collins and Lap-Chee Tsui, "Identification of the Cystic Fibrosis Gene: Cloning and Characterization of Complementary DNA," *Science* 245 (September 8, 1989): 1066.

Wire service: Rebecca Kolberg, "General Use of Cystic Fibrosis Test Opposed," United Press International, November 14, 1989.

Newspapers: Richard Saltus, "Gain Seen on Cystic Fibrosis," *Boston Globe*, October 10, 1985, 1; Jerry E. Bishop, "Scientists Closer to Developing Test for Cystic Fibrosis," *Wall Street Journal*, October 10, 1985, 17; Harold M. Schmeck Jr., "Genetic Marker for Cystic Fibrosis Reported Found," *New York Times*, October 11, 1985, A17; Harold M. Schmeck Jr., "Genetic Sign Found for Cystic Fibrosis: Scientists Discover Markers Linked to Faulty Gene—Step Called Significant," *New York Times*, November 28, 1985, A17; Richard Saltus, "Finding Stirs Hope for Prenatal Cystic Fibrosis Test," *Boston Globe*, December 10, 1985, 71; Reuters, "Scientists Identify Gene Causing Cystic Fibrosis," *Toronto Star*, August 23, 1989, A4; Jerry E. Bishop, "Scientists to Unveil Gene That Causes Cystic Fibrosis," *Wall Street Journal*, August 24, 1989, B5; Sandra Blakeslee, "Scientists Find Hope for Victims of Cystic

Fibrosis by Discovering Its Gene," *New York Times*, August 24, 1989, B13; Richard Saltus, "Gene Discovery Fuels Hope for Cure of Cystic Fibrosis," *Boston Globe*, August 25, 1989, 1; Richard Saltus, "Race Is on to Apply Discovery; Cystic Fibrosis Gene's First Payoff Likely to Be in Prevention, Not Treatment," *Boston Globe*, August 28, 1989, 25; Marilyn Dunlop, "'Curtain of Silence' Hid Hunt for Deadly Cystic Fibrosis Gene," *Toronto Star*, September 8, 1989, A20.

Magazine: Peter N. Goodfellow, "Cystic Fibrosis: Classical and Reverse Genetics," *Nature* 326 (April 30, 1987): 824.

7. **Scientific articles:** Manuel Perucho et al., "Human-Tumor-Derived Cell Lines Contain Common and Different Transforming Genes," *Cell* 27, Part 2 (December 1981): 467–476; Chiaho Shih and Robert A.Weinberg, "Isolation of a Transforming Sequence from a Human Bladder Carcinoma Cell Line," *Cell* 29 (May 1982): 161–169.

Wire service: Discovery of oncogenes in humans is reported in Associated Press, "Scientists Report Isolating Cancer Genes," citing *Newsday*, September 15, 1981.

Newspapers: Lasker awards to Michael Bishop and Harold Varmus are reported in Harold M. Schmeck Jr., "7 Medical Experts Win Lasker Prizes," *New York Times*, November 18, 1982, A21; Harold M. Schmeck Jr., "Some Genes Linked to Cancer Process," *New York Times*, September 19, 1981, 1.

Magazines: "Rooting Out Cancer," *The Economist* (U.S. edition), September 19, 1981, 103; Sharon McAuliffe and Kathleen McAuliffe, "The Genetic Assault on Cancer," *New York Times Magazine*, October 24, 1982, 39.

Books: Natalie Angier, *Natural Obsessions: The Search for the Oncogene* (Boston: Houghton Mifflin, 1988); Harold Varmus, *Genes and the Biology of Cancer* (New York: Scientific American Library, W. H. Freeman, 1993); Robert A. Weinberg, *One Renegade Cell: How Cancer Begins* (New York: Basic Books, 1998); Robert A. Weinberg, *Racing to the Beginning of the Road: The Search for the Origin of Cancer* (New York, Harmony Books, 1996).

8. **Testimony:** James D. Watson, prepared testimony, House Appropriations Committee, Labor/HHS/Education Subcommittee [John Porter, chairman], February 28, 1995, Federal News Service.

9. **Scientific article:** T. J. Ley et al., *New England Journal of Medicine* 307 (December 9, 1982): 1469.

Wire service: Daniel Haney, "Genetic Treatment Reverses Sickle Cell Anemia," Associated Press, December 8, 1982.

Newspapers: Harold M. Schmeck Jr., "Researchers Correct Defect in a Human Gene," *New York Times*, April 26, 1982, A1; Harold M. Schmeck Jr., "Activity of Genes Reported Altered in Treating Man," *New York Times*, December 9, 1982, 1; Victor Cohn, "Doctors Activate Dormant Gene to Treat Blood Disease Victims," *Washington Post*, December 10, 1982, A16.

Injection of a normal ADA gene is described in Richard Saltus, "4-Year-Old Gets Historic Gene Implant," *Boston Globe*, September 15, 1990, 1; Larry Thompson,

"Medicine's 4-Year-Old Pioneer; First Gene-Therapy Patient Opens Door to Treating 4,000 Inherited Diseases," *Washington Post*, September 25, 1990, Z8. Ashanti DaSilva was reported as alive in Ramez Naam, *More Than Human* (New York: Broadway Books, 2005).

10. **News release:** Kathleen Klein, "When Scientists at the UW and Penn Found a Way to Make 'Super Mice,' They Advanced the Genetic Engineering of Animals—and Humans," University of Washington, December 1995, www.washington.edu/alumni/columns/dec95/mice.html.

Wire services: Al Rossiter Jr., "Gene Transfer Produces 'Super Mice,'" United Press International, December 15, 1982; Warren E. Leary, "Researchers Transfer Genetic Information," Associated Press, December 15, 1982.

Newspapers: Victor Cohn, "Day of the 'Super-Mice,'" *Washington Post*, December 16, 1982, A1; Harold M. Schmeck Jr., "Rat Gene Implant in Mice Reported," *New York Times*, December 16, 1982, A1; editorial, "A Mouse That Can Roar," *Washington Post*, December 18, 1982, A18; Robert C. Cowen, "Experiment on Mice Highlights Concern over Genetic Engineering," *Christian Science Monitor*, December 20, 1982.

Magazine: Sharon Begley, "The Making of a Mighty Mouse," *Newsweek* (U.S. edition), December 27, 1982, 67.

11. **Website:** "In the United States, Howard and Georgianna Jones' IVF Program in Norfolk, VA Produced This Country's First IVF Baby, Born December 28, 1981," IVF1.com, June 19, 2009, www.ivf1.com/ivf (accessed November 14, 2009).

12. **Wire service:** Patricia McCormack, "Humulin, Manmade Insulin for Diabetics, Opening New Field of Medical Discovery," United Press International, November 12, 1982.

Newspapers: Cristine Russell, "FDA Approves Insulin Made by Splicing Genes," *Washington Post*, October 30, 1982, A6; Lawrence K. Altman, "A New Insulin Given Approval for Use in U.S.," *New York Times*, October 30, 1982, 1.

13. **Wire service:** Warren E. Leary, "Research Guidelines Relaxed," Associated Press, January 29, 1980.

Newspapers: Victor Cohn, "U.S. Lowers Bars to Gene Engineering," *Washington Post*, January 30, 1980, A1; Philip J. Hilts, "Brave New Science; Where Will It Lead? Part 4, Gene Splicers' Fear Recedes, Awe Remains," *Washington Post*, November 4, 1981, A1; Philip J. Hilts, "NIH Unit Votes to Ease but Retain Federal Rules on Gene-Splicing," *Washington Post*, February 9, 1982, A7; "U.S. Panel Urges Further Relaxing of Restrictions on Gene-Splicing," *New York Times*, February 9, 1982, C1.

14. **Newspapers:** Robert C. Cowen, "Gene-Splicing Opens New World for Agriculture," *Christian Science Monitor* (Midwestern edition), July 7, 1981, 3; Harold M. Schmeck Jr., "Revolution in Plant Genetics Forecast," *New York Times*, May 18, 1982, C7.

15. **Magazine:** "Gazing into the Crystal Ball of Genetic Research," *The Economist* (U.S. edition), April 23, 1983, 97.

16. **News release:** David Smith, "Evolution of a Vision: Genome Project Origins, Present and Future Challenges, and Far-Reaching Benefits," *Human Genome News* 7 (September–December 1995): 3.

Newspaper: Natalie Angier, "Great 15-Year Project to Decipher Genes Stirs Opposition," *New York Times*, June 5, 1990, C1.

Magazines: B. D. Davis, "Sequencing the Human Genome: A Faded Goal," *Bulletin of the New York Academy of Medicine* 68 (January–February 1992): 115–125; Robert Mullan Cook-Deegan, "Historical Sketch: The Alta Summit, December 1984," *Genomics* 5 (1989): 661–663.

Book: Christopher Wills, *Exons, Introns, and Talking Genes: The Science Behind the Human Genome Project* (New York: Basic Books, 1991), 71–75.

17. **Newspapers:** Victor Cohn, "Scientists Making a Copy of Disease-Fighting Gene," *Washington Post*, January 17, 1980, A7; Harold M. Schmeck Jr., "Natural Virus-Fighting Substance Is Reported Made by Gene Splicing; Scientists Say It Has Big Potential in Curing a Variety of Diseases, Including Colds and Other Infections," *New York Times*, January 17, 1980, 1; Jessica Tuchman Mathews, "Factories Too Tiny to See," *Washington Post*, January 23, 1980, A23; Lee Lescaze, "World Labs Racing to Develop and Produce New Cancer Drug," *Washington Post*, June 15, 1980, A8.

Magazines: "Biogen Claims to Have Won in Interferon Race," *Chemical Week*, January 23, 1980, 19; Matt Clark with Sharon Begley, "The Making of a Miracle Drug," *Newsweek* (U.S. edition), January 28, 1980, 82; "Gene Feat Spurs Interferon Race," *Chemical Week*, February 6, 1980, 43.

18. **Wire service:** Warren E. Leary, "Supreme Court Ruling Expected to Spur Genetic Engineering Work," Associated Press, June 17, 1980.

Newspapers: Fred Barbash, "Laboratory Life Forms Patentable," *Washington Post*, June 17, 1980, A1; Linda Greenhouse, "Science May Patent New Forms of Life, Justices Rule, 5 To 4: Dispute on Bacteria: Decision Assists Industry in Bioengineering in a Variety of Projects," *New York Times*, June 17, 1980, A1; "Excerpts from Supreme Court Opinions on Man-Made Life Forms," *New York Times*, June 17, 1980, D17; Harold M. Schmeck Jr., "Justices' Ruling Recognizes Gains in the Manipulation of Life Forms," *New York Times*, June 17, 1980, D16; Anthony J. Parisi, "Gene Engineering Industry Hails Court Ruling As Spur to Growth," *New York Times*, June 17, 1980, D16; Robert C. Cowen, "High Court: New Life Forms Can Be Patented," *Christian Science Monitor*, June 17, 1980, 3; Nathaniel Sheppard Jr., "Man in the News; Developer of a New Life Form: Ananda Mohan Chakrabarty," *New York Times*, June 18, 1980, A22; Thomas O'Toole, "In the Lab: Bugs to Grow Wheat, Eat Metal; Ruling on New Life Forms Clears Way for Industry," *Washington Post*, June 18, 1980, A1; Harold M. Schmeck Jr., "U.S. to Process 100 Applications for Patents on Living Organisms," *New York Times*, June 18, 1980, A22.

Magazines: "A Patent on Life," *The Economist*, June 21, 1980, 18; "Patent Ruling Won't Shift Gene R&D Goals," *Chemical Week*, June 25, 1980, 57; Aric Press et al., "The Right to Patent Life," *Newsweek*, June 30, 1980, 74.

19. **Wire service:** Issuance of Patent 4,237,224, "Process for Producing Biologically Functional Molecular Chimeras," to Stanley N. Cohen and Herbert W. Boyer and assigned to Stanford University and the University of California is reported in Warren E. Leary, "First Patent Granted for Gene-Splicing Technique," Associated Press, December 4, 1980.

Newspapers: Wallace Immen, "Magic in Lab: Turning Cells into Gold Mines," *Globe and Mail* (Canada), October 2, 1980; "Genentech Priced at 35," *New York Times*, October 14, 1980, D4; Robert J. Cole, "Genentech, New Issue, Up Sharply," *New York Times*, October 15, 1980, D1; James L. Rowe Jr., "Designer Genes Are Snapped Up; Gene-Splitting Genentech Stock: Wall Street's Hottest Offering," *Washington Post*, October 15, 1980, E1; "The Designer Gene Trend Continues," *Washington Post*, October16, 1980, D1; Karen W. Arenson, "A 'Hot' Offering Retrospective," *New York Times*, December 30, 1980, D1.

20. **Newspaper:** Stacy V. Jones, "Patents," *New York Times*, December 6, 1980, 22.

21. **Magazines:** "Finally, a Streamlining for the Patent Office," *Business Week* (industrial edition), December 22, 1980, 23; "Patent Bill Gets Warm Reception," *Chemical Week*, December 10, 1980, 15.

22. **Newspapers:** "Harvard Considers Commercial Role in DNA Research," *New York Times*, October 27, 1980, A1; Philip J. Hilts, "Ivy-Covered Capitalism; Colleges Weigh Spinoffs to Plug the Profits Leak," *Washington Post*, November 10, 1980, A1; "Harvard Rules Out Genes Business," *Christian Science Monitor* (Midwestern edition), November 19, 1980, 2.

Magazine: "Harvard; Clean Genes," *The Economist*, November 22, 1980, 35.

23. **News release:** Biogen, N.V., PR Newswire, September 30, 1981.

Newspapers: Leonard Sloane, "Business People; Professor Takes Leave for Year to Run Biogen," *New York Times*, October 2, 1981, D2; "Harvard Drafts Rules on Conflicts of Interest," *New York Times*, October 4, 1981, Part I, 29; Fox Butterfield, "Colleges Are Uneasy Over Faculties' Outside Jobs," *New York Times*, November 16, 1981, B11.

Magazine: An article titled "The Tempest Raging over Profit-Minded Professors," *Business Week*, November 7, 1983, 86, quotes Walter Gilbert about his resignation in 1982: "Harvard wanted its faculty to devote 80% of their time to the university. I clearly was planning to spend 80% to 90% of my time somewhere else."

24. **Interview:** Robert L. Sinsheimer, by Shelley Erwin, May 30–31, 1990, and March 26, 1991. Archives, California Institute of Technology, Session 3, 34–38.

Magazines: Robert L. Sinsheimer, "Historical Sketch: The Santa Cruz Workshop—May 1985," *Genomics* 5 (1989): 954–956; Roger Lewin, "In the Beginning Was the Genome," *New Scientist*, July 21, 1990, 34–38.

Books: Robert L. Sinsheimer, *The Strands of a Life: The Science of DNA and the Art of Education* (Berkeley: University of California Press, 1994), 257–270; Robert Mullan Cook-Deegan, *The Gene Wars: Science, Politics, and the Human Genome* (New York: W. W. Norton, 1994), 79–85, 353; Andrew Brown, *In the Beginning Was the Worm* (New York: Columbia University Press, 2003), 183–189.

25. **Interview:** Leroy Hood, by the author, Seattle, Washington, February 14, 2004, transcript.

Magazine: Leon Jaroff et al., "The Gene Hunt," *Time*, March 30, 1989, 62–67.

Speech: Leroy Hood, "New Ideas Thrive on New Organizational Structures," Batten Briefings, Darden Graduate School of Business, spring 2004, www.darden.virginia.edu/batten.

26. **Interview:** David Botstein, by the author, Princeton University, Princeton, New Jersey, January 25–26, 2007, transcript.

Magazine: Stephen S. Hall, "Botstein's Caveat," *Technology Review* (September–October 2000): 115.

27. **Speeches:** Leroy Hood, "New Ideas Thrive on New Organizational Structures," Batten Briefings, Darden Graduate School of Busines, spring 2004, www.darden.virginia.edu/batten.

Magazine: Leon Jaroff et al., "The Gene Hunt," 62–67.

28. **Magazine:** Sinsheimer, "Historical Sketch."

29. **Magazine:** Hall, "Botstein's Caveat."

30. **Interview:** Sinsheimer, by Erwin.

31. **Interview:** Hood, by the author.

32. **Interview:** Botstein, by the author, Princeton University, Princeton, New Jersey, January 25–26, 2007, transcript.

33. Walter Gilbert, "Memo About the Human Genome Institute," February 9, 1986, typescript, Gilbert files. Derived from letter to Robert Edgar, May 27, 1985, cited in Cook-Deegan, *The Gene Wars*, 87–88, 373; Walter Gilbert to Robert Moyzis, Los Alamos National Laboratory, February 10, 1986, typescript, Gilbert files.

34. **Magazine:** Bruce Alberts, "Limits to Growth: In Biology, Small Science Is Good Science," *Cell* 41 (June 1985): 337.

35. **Magazines:** Joseph Palca, "Human Genome: Department of Energy on the Map," *Nature* 321 (May 22, 1986): 371; Joseph Palca, "The Numbers Game," *Nature* 321 (May 22, 1986): 371 (quoting Frederick Blattner of the University of Wisconsin).

Book: James B. Wyngaarden, "Jim Watson and the Human Genome Project," in John Inglis, Joseph Sambrook, and Jan Witkowski, eds. *Inspiring Science: Jim Watson and the Age of DNA* (Cold Spring Harbor, NY: Cold Spring Harbor Laboratory Press, 2003), 377–385.

36. David E. Comings, City of Hope Medical Center, running notes, Workshop 1: Technology, Appendix IV to typescript, "Genome Sequencing Workshop, March 3 and 4, 1986, Santa Fe, New Mexico," in author's 86 Santa Fe file.

37. **Newspaper:** Douglas Birch, "Daring Sprint to the Summit; The Quest: A Determined Hamilton Smith Attempts to Scale a Scientific Pinnacle—and Reconcile with Family," *Baltimore Sun*, April 13, 1999, 1A.

Magazine: Roger Lewin, "Proposal to Sequence the Human Genome Stirs Debate: Some Molecular Biologists Fear That the Proposal to Sequence the Entire Human Genome Is Developing Unstoppable Momentum, Despite Its Uncertain Merits," *Science* 232 (June 27, 1986): 1598.

Books: Wills, *Exons, Introns, and Talking Genes*, 76; Cook-Deegan, *The Gene Wars*; Kevin Davies, *Cracking the Genome: Inside the Race to Unlock Human DNA* (New York: Free Press 2001), 15.

38. Mark W. Bitensky, Los Alamos National Laboratory, to Charles DeLisi, April 2, 1986, author's 86 Santa Fe file.

39. **Scientific article:** Renato Dulbecco, "Perspective: A Turning Point in Cancer Research: Sequencing the Human Genome," *Science* 231 (March 7, 1986): 1055.

Interview: Heiner Westphal, by the author, National Institutes of Health, Bethesda, Maryland, March 28, 2001, transcript.

Books: Victor K. McElheny, *Watson and DNA, Making a Scientific Revolution* (New York: Perseus Books, 2003), 242–243; Cook-Deegan, *The Gene Wars*, 105, 107–108; Davies, *Cracking the Genome*, 11; Wills, *Exons, Introns, and Talking Genes*, 16.

Annual report: James D. Watson, "Director's Report," in Cold Spring Harbor Laboratory's 1988 annual report, 1–11 (dated August 2, 1989).

40. **Author's notes:** James D. Watson, remarks, Baylor College of Medicine, Houston, Texas, May 31, 2007, in *Mapping Life Notebook* 7, 36; James D. Watson, remarks opening Cold Spring Harbor conference, "Personal Genomes," October 9, 2008, in *Mapping Life Notebook* 8, 46.

41. **Interviews:** Eric Lander, by the author, Broad Institute, Cambridge, Massachusetts, June 28, 2006, transcript; David Botstein, by the author, Princeton University, Princeton, New Jersey, January 25–26, 2007, transcript.

Magazine: Aaron Zitner, "The DNA Detective; MIT Mathematician Eric Lander Took a Detour into Biology—and Found Himself Leading the Biggest Biological Research Project in History," *Boston Globe Sunday Magazine*, October 10, 1999, 17.

42. **Scientific article:** James D. Watson, "The Human Genome Project: Past, Present, and Future," *Science* 248 (April 6, 1990): 44.

Interview: Paul Berg, by the author, Stanford University, Palo Alto, California, February 13, 2001, transcript.

Newspapers: Richard Saltus, "Genetics: Crash Effort to Map Human Genes Urged; Massive Project Could Open Window on Heredity, Wide Range of Diseases; Critics Fearful of Abuses," *Boston Globe*, June 9, 1986; Robert Cooke, Newsday, "Scientists Reach Genetics Milestone," *Toronto Star*, June 15, 1986, B7; Philip M.

Boffey, "Rapid Advances Point to the Mapping of All Human Genes. The Debate: Would Accelerated Project Distort Usual Methods of Biology?" *New York Times*, July 15, 1986, C1.

Magazines: Lewin, "Proposal"; Miranda Robertson, "The Proper Study of Mankind: Molecular Biology Has Made the Human Genome Accessible to Laboratory Investigation. But Does That Mean That the Sequence of the Human Genome Would Be Worth the Effort?" *Nature* 322 (July 3, 1986): 11; Robert Kanigel, "The Genome Project: An Audacious Effort to Decipher the Entire Genetic Code of a Human Being Could Revolutionize Medicine," *New York Times Magazine*, December 13, 1987, 44; James D. Watson and Robert Mullan Cook-Deegan, "Research Communications: Origins of the Human Genome Project," *FASEB Journal* 5 (1991): 8; David Botstein, remarks in "An Invitation to Genetics in the 21st Century," *Los Alamos Science* 20 (1992): 314–329; Leslie Roberts, "The Gene Hunters," *U.S. News & World Report* 128, January 3, 2000, 34; Leslie Roberts, "Controversial from the Start," *Science* 291 (February 16, 2001): 1182.

Annual Report: Watson, "Director's Report."

Books: Wills, *Exons, Introns, and Talking Genes*, 78–79; Cook-Deegan, *The Gene Wars*, 109–113; James D. Watson with Andrew Berry, *DNA: The Secret of Life* (New York: Knopf, 2003), 168; James D. Watson, Amy A. Caudy, Richard M. Myers, and Jan A. Witkowski, *Recombinant DNA*, 3rd ed. (New York: W. H. Freeman and Company, 2007), 275–277.

43. **Author's notes:** James D. Watson, remarks at Baylor College of Medicine, May 31, 2007; James D. Watson, remarks at Cold Spring Harbor, October 9, 2008, in *Mapping Life Notebook* 8, 46–52.

Wire services: "Son Is Missing," United Press International, May 28, 1986; Marlene Aig, "Nobel Prize Winner's Son Found," Associated Press, May 30, 1986.

44. **Interview:** David Botstein, by the author, Stanford University, February 12, 2001, transcript.

45. **Newspaper:** Saltus, "Genetics."

46. **Broadcast:** Eric Lander, interview, CNN, June 1, 2000, www.cnn.com/SPECIALS/2000/genome/story/interviews/lander.html.

47. **Author's notes:** James D. Watson, remarks at Cold Spring Harbor, October 9, 2008, in *Mapping Life Notebook* 8, 46–52.

48. **Magazine:** Leslie Roberts, "Human Genome: Questions of Cost," *Science* 237 (September 18, 1987): 1411.

49. **Archives:** John Burris, memorandum to Philip M. Smith, executive officer, April 25, 1986, National Academy of Sciences archives.

50. **Archives:** Frank Press to James B. Wyngaarden, June 16, 1986, and James B. Wyngaarden to Frank Press, June 26, 1986, National Academy of Sciences archives.

51. **Wire service:** Rob Stein, "Mapping Genes; Mapping the Human Genetic Code," United Press International, August 16, 1986.

52. **Archives:** Ruth Kirschstein notes, August 5, 1986, History Associates, Rockville, Maryland; John E. Burris, memorandum to participants, July 30, 1986, National Academy Archives; Notes, Board on Basic Biology, Commission on Life Sciences, National Research Council, National Academy Archives.

53. **Interviews:** Bruce Alberts, by the author, National Academy of Sciences, Washington, D.C., March 28, 2001, author's notes; David Botstein, by the author, 2001; David Botstein, by the author, 2007.

Book: Committee on Mapping and Sequencing the Human Genome, Board on Basic Biology, Commission on Life Sciences, National Research Council, *Mapping and Sequencing the Human Genome* (Washington, D.C.: National Academy Press, 1988).

54. **Interview:** Maynard Olson, by the author, University of Washington, Seattle, February 18, 2004, transcript.

55. **Archives:** Ruth Kirschstein, Box 102, Folder 8 (Dingell hearing), History Associates, Rockville, Maryland, photocopied April 13, 2007.

Documents: "OTA Report on the Human Genome Project," hearing before the Subcommittee on Oversight and Investigations of the Committee on Energy and Commerce, House of Representatives, 100th Congress, Second Session, April 27, 1988, Serial No. 100–123, U.S. Government Printing Office, 1988, 136 pp. In author's April 27, 1988, folder.

56. **Magazine:** Leslie Roberts, "Academy Backs Genome Project: The Nation Should Embark on an All-Out Effort to Map and Sequence the Genome, Says a National Academy of Sciences Panel, but How the Project Will Be Managed Remains Unclear," *Science* 239 (February 12, 1988): 725.

57. **Archives:** Howard Hughes Medical Institute, tentative agenda dated July 1, 1986, author's typed notes; "HHMI Meeting," dated July 23, 1986, handwritten notes, 24 pp; Ruth Kirschstein, MD, director, National Institute of General Medical Sciences, "The Activities of NIH in Characterizing the Human Genome," presented at the informational forum on the human genome, sponsored by the Howard Hughes Medical Institute, July 23, 1986, ms. 5 pp., 2 tables, chart of GenBank entries. All from History Associates, Rockville, Maryland, Box 103, Ruth Kirschstein, Folder 2. 86 HHMI Conference, July folder; photocopied April 13, 2007, in author's folder: 86 HHMI conference, July.

Newspaper: Sally Squires, "Putting Genes on a Map; World's Scientists Begin Planning a Detailed Description of Human DNA," *Washington Post*, July 30, 1986, Health section, 7.

Magazines: Lewin, "Shifting Sentiments over Sequencing the Human Genome; the Latest Gathering to Discuss the Prospects for Sequencing the Entire Human Genome Experienced an Important Change of Priorities; Mapping Is Now Seen As the Preferred Goal," *Science* 233 (August 8, 1986): 620; Robert L. Sinsheimer, "Human Genome Sequencing," *Science* 233 (September 19, 1986): 1246.

58. **Archives:** All from Ruth Kirschstein, Box 103, Folder 4, History Associates, Rockville, Maryland: "The Human Genome: Proceedings of the 54th Meeting of the Advisory Committee to the Director," National Institutes of Health, October 16–17, 1986; "Meeting to Plan the DAC," date uncertain; handwritten notes, October 16, 1986; "Alternative Uses of Funds and Personnel," typescript.

59. **Book:** Wyngaarden, "Jim Watson and the Human Genome Project," 380.

60. **Book:** Bradie Metheny, "Dr. Watson Goes to Washington," in John Inglis, Joseph Sambrook, and Jan Witkowski, eds. *Inspiring Science: Jim Watson and the Age of DNA* (Cold Spring Harbor, NY: Cold Spring Harbor Laboratory Press, 2003), 369–375.

61. **Archives:** James D. Watson press conference, October 3, 1988, transcript, James D. Watson Archives, Cold Spring Harbor Laboratory.

Newspapers: Harold M. Schmeck Jr., "DNA Pioneer to Tackle Biggest Gene Project Ever," *New York Times*, October 4, 1988, C1; Nicholas Wade, "Double Landmarks for Watson: Helix and Genome," *New York Times*, June 27, 2000, D5.

Books: Daniel J. Kevles, "Out of Eugenics: The Historical Politics of the Human Genome," in *The Code of Codes: Scientific and Social Issues in the Human Genome Project*, ed. Daniel J. Kevles and Leroy Hood (Cambridge, MA: Harvard University Press, 1993), 35; Cook-Deegan, *The Gene Wars*, 163; McElheny, *Watson and DNA*, 238–239.

Chapter 3: Scaling Up

1. **Documents:** "OTA Report on the Human Genome Project," hearing before the Subcommittee on Oversight and Investigations, Committee on Energy and Commerce, House of Representatives, 100th Congress, Second Session, April 27, 1988, committee print, Serial Number 100–123, U.S. Government Printing Office, 1988, 136 pp. In author's April 27, 1988, folder.

2. **Books:** Examples of histories of large systems: Louis C. Hunter, *Steamboats on the Western Rivers: An Economic and Technological History* (Cambridge, MA: Harvard University Press, 1949); Richard G. Hewlett and Oscar E. Anderson, *The New World*, vol. 1, *A History of the United States Atomic Energy Commission* (University Park: Pennsylvania State University Press, 1962); Margaret Gowing, *Britain and Atomic Energy*, 1939–1945 (New York: St. Martin's Press, 1964); Pierre Berton, *The Impossible Railway* (New York: Knopf, 1972); Thomas P. Hughes, *Networks of Power: Electrification in Western Society, 1880–1930* (Baltimore: Johns Hopkins University Press, 1983); Richard Rhodes, *The Making of the Atomic Bomb* (New York: Simon & Schuster, 1986); David Holloway, *Stalin and the Bomb: The Soviet Union and Atomic Energy, 1939–1956* (New Haven, CT: Yale University Press, 1994); Richard Rhodes, *Dark Sun: The Making of the Hydrogen Bomb* (New York: Simon & Schuster, 1995); Charles Murray and Catherine Bly Cox, *Apollo: The Race to the Moon* (New York:

Simon & Schuster, 1989); Stephen Ambrose, *Nothing Like It in the World: The Men Who Built the Transcontinental Railway, 1863–69* (New York: Simon & Schuster, 2000); David Haward Bain, *Empire Express: Building the First Transcontinental Railroad* (New York: Penguin Books, 2000); David A. Mindell, *Digital Apollo: Human and Machine in Space Flight* (Cambridge, MA: MIT Press, 2008).

3. **Interview:** Ronald Davis, by the author, Palo Alto, California, February 27, 2007, transcript.

4. **Archives:** James D. Watson, Center for Human Genome Research, National Institutes of Health, "The Human Genome Project Is Finally Underway," Bethesda, Maryland, November 28, 1989, typescript, 20 pp, James D. Watson Archives, Cold Spring Harbor Laboratory.

Newspapers: Larry Thompson, "A Big-Ticket, Low-Profile Science Venture: $3 Billion Project Aims to Decipher All 50,000-Plus Human Genes," *Washington Post*, January 4, 1989, A17; Larry Thompson, "$100 Million Sought for Study of Genes; White House Boosts Controversial Project," *Washington Post*, January 10, 1989, Z9; Robert Cooke, "Dean of DNA: Cold Spring Harbor Laboratory Director James Watson, Co-Discoverer of DNA's Chemical Structure, Will Soon Lead Another DNA Quest," *Newsday*, January 24, 1989, Part III, 1; Harold M. Schmeck Jr., "New Methods Fuel Efforts to Decode Human Genes; Improved Machines Can Write Messages in the Language of Heredity Itself," *New York Times*, May 9, 1989, C1; Larry Thompson, "The Man Behind the Double Helix; Gene-Buster James Watson Moves on to Biology's Biggest Challenge—Mapping Heredity," *Washington Post*, September 12, 1989, Z12.

Magazines: Joseph Palca, "Genome Projects Are Growing Like Weeds: NIH and DOE Continue to Move Ahead with Their Plans, but Other Agencies Are Looking to Get Involved," *Science* 245 (July 14, 1989): 131 (reporting establishment of the National Center for Human Genome Research, with authority to make grants directly); Jeffrey L. Fox, "Directing the Genome Project," *Nature Biotechnology* 7 (March 1989): 223; "Back to the Bases: Watson Slips into the Driver's Seat of the Genome Project," *Scientific American* 260 (January 1989): 8; David Dickson, "Watson Floats a Plan to Carve Up the Genome: The Head of the U.S. Genome Project Suggests Giving Individual Countries Responsibility for Specific Chromosomes," *Science* 244 (May 5, 1989): 521; Jeffrey Fox, "Genome Initiative: Watson's Great Human Chromosome Auction," *Nature Biotechnology* 7 (June 1989): 559; Leslie Roberts, "Genome Mapping Goal Now in Reach, James Watson Has Promised to Complete a Map of the Human Genome Within 5 Years; Now It Looks Like It Might Be Doable," *Science* 244 (April 28, 1989): 424; Leslie Roberts, "Plan for Genome Centers Sparks a Controversy," *Science* 246 (October 13, 1989): 204; James D. Watson and Elke Jordan, "Program Description: The Human Genome Project at the National Institutes of Health," *Genomics* 5 (1989): 654; Stu Borman, "Human Genome Project: Five-Year Plan Taking Shape," *Chemical and Engineering News*, December 11, 1989, 4.

Book: Victor K. McElheny, "Genome: 'It Is So Obvious,'" in *Watson and DNA: Making a Scientific Revolution* (New York: Perseus, 2003), 239–269.

5. **Newspaper:** William Bown, "Technology: American Know-How for Paid-Up Members Only," *Independent*, December 11, 1989, 13.

Magazines: Leslie Roberts, "Watson versus Japan: James Watson Is Threatening to Deny Japanese Scientists Access to DNA Sequence Databases Unless Japan Pays Its 'Fair Share' for the Genome Project," *Science* 246 (November 3, 1989): 576; "The Latecomers," *The Economist*, December 16, 1989, 84.

6. **Interview:** William Haseltine, by the author, Washington, D.C., March 25, 2001, transcript.

Wire service: Associated Press, "DNA Pioneer Quits Gene Map Project: Watson Resigns After Federal Review of His Holdings in Biology Companies," *New York Times*, April 11, 1992, 12.

Newspapers: Philip J. Hilts, "Head of Gene Map Threatens to Quit: Watson, Who Helped Discover DNA, Now Faces Pressure on Financial Holdings," *New York Times*, April 9, 1992, A26; Hilary Stout, "Watson, Head of U.S. Genome Project, Faces Questions over Stock Holdings," *Wall Street Journal*, April 9, 1992, B8; David Brown and Malcolm Gladwell, "Nobel Prize Biologist Watson Plans to Resign U.S. Position," *Washington Post*, April 9, 1992, A3; Hilary Stout, "Watson Resigns As Head of U.S. Gene-Mapping Project: Pioneering DNA Researcher Faced Probe on Holdings in Biotechnology Firms," *Wall Street Journal*, April 13, 1992, B9; Jerry E. Bishop, "Opposition to Businessman's Worm Genome Project Led to Conflict Charges," *Wall Street Journal*, April 13, 1992, B9; "Patents Pending; James Watson Was Right About Patenting Genes; His Resignation Is a Loss to Science," *Newsday*, May 2, 1992, 16.

Magazines: Leslie Roberts, "Friends Say Jim Watson Will Resign Soon," *Science* 256 (April 10, 1992): 171; editorial, "Healy in a Hurry: Dr. Bernadine Healy Has Rid the Human Genome Project of Dr. J. D. Watson in a Distasteful Way," *Nature* 356 (April 16, 1992): 547; Christopher Anderson, "Watson Resigns; Genome Project Open to Change: Will Leave Immediately, Acting Head Named; Smaller Centres, More Research Anticipated," *Nature* 356 (April 16, 1992): 549.

Annual report: James D. Watson, "Director's Report: Highlights of the Year," in Cold Spring Harbor Laboratory's 1991 annual report, 23–24 (dated May 14, 1992).

Book: Robert Cook-Deegan, *The Gene Wars: Science, Politics, and the Human Genome* (New York: W. W. Norton, 1994), 326–327.

7. **Interview:** David Botstein, by the author, Princeton, New Jersey, January 25, 2007, transcript.

Archives: "Program Advisory Committee on the Human Genome, First Meeting," Box 1211, Folder 9, History Associates, Rockville, Maryland, photocopied April 13, 2007.

Newspaper: Thompson, "$100 Million Sought."

Magazines: Leon Jaroff, "The Gene Hunt," *Time*, March 20, 1989, 62–67; "Genome Mapping Efforts Turning from Men and Mice to Microbes, Molds," McGraw Hill's *Biotechnology Newswatch* 9 (January 16, 1989): 6; Leslie Roberts, "Genome Project Under Way, At Last: With Many of the Major Questions Settled, NIH Is Trying to Figure Out Exactly What the New Genome Project Will Entail," *Science* 243 (January 13, 1989): 167.

8. **Magazine:** Roger Lewin, "Genome Planners Fear Avalanche of Red Tape," *Science* 244 (June 30, 1989): 1543.

Books: John Sulston and Georgina Ferry, *The Common Thread: A Story of Science, Politics, Ethics, and the Human Genome* (Washington, D.C.: Joseph Henry Press, 2002), 65–69; McElheny, *Watson and DNA*, 262–263, and endnote, 341.

9. **Testimony:** "Human Genome Project," Committee on Energy and Natural Resources, Senate, 100th Congress, Second Session, Hearing, July 11, 1990, 90 CIS S 31158.

Newspapers: William Allen, "Genome Project: 'Source Book for Biology' or 'Expensive Tinker Toy'?" *St. Louis Post-Dispatch*, February 5, 1989, 4B (reporting remarks at the American Association for the Advancement of Science annual meeting by Victor McKusick and Philip Bereano); Natalie Angier, "Great 15-Year Project to Decipher Genes Stirs Opposition," *New York Times*, June 5,1990, C1.

Magazine: Jeffrey Mervis, "On Capitol Hill: One Day in the Hard Life of the Genome Project, a Routine Hearing Turns Explosive When a Critic of the Program Finds Himself Under Attack by a Prominent Senator," *Scientist* 4 (August 20, 1990): 1.

10. **Magazine:** Leslie Roberts, "Tough Times Ahead for the Genome Project; As the Genome Project Comes Under Increasing Scrutiny, Congress Is Asking How Much and How Fast It Should Grow," *Science* 248 (June 29, 1990): 1601.

11. **Magazine:** Roberts, "Tough Times Ahead."

12. **Magazine:** "Genome Critics Charge: Feds Spend Too Much, Achieve Little," *McGraw Hill's Biotechnology Newswatch* 10 (November 5, 1990): 1.

13. **Magazines:** "Genome Critics Charge"; Leslie Roberts, "A Meeting of the Minds on the Genome Project? In Spite of Some Sharp Exchanges, There Was a Lot of Common Ground When the Project's Organizers Met Their Critics," *Science* 250 (November 9, 1990): 756.

14. **Interview:** David Lipman, by the author, National Center for Biotechnology Information, Bethesda, Maryland, April 5, 2007, in *Mapping Life Notebook* 5, 295–310.

Magazine: "The Geography of Genes," *The Economist*, December 16, 1989, 81.

Books: James Shreeve, *The Genome War: How Craig Venter Tried to Capture the Code of Life and Save the World* (New York: Knopf, 2004), 78–79; J. Craig Venter, "Scientific Heaven, Bureaucratic Hell," in *A Life Decoded: My Genome, My Life* (New York: Viking, 2007), 89–124.

15. **Interview:** Mark Guyer and Jane Peterson, by the author, National Human Genome Research Institute, Rockville, Maryland, April 13, 2007, in *Mapping Life Notebook* 5, 362–367.

16. **Scientific article:** Mark D. Adams et al., "Complementary DNA Sequencing: Expressed Sequence Tags and Human Genome Project," *Science* 252 (June 21, 1991): 1651.

Interview: Haseltine, by the author.

Magazines: Leslie Roberts, "Gambling on a Shortcut to Genome Sequencing: Craig Venter Says He Can Find All the Human Genes for a Fraction of the Cost of the Human Genome Project," *Science* 252 (June 21, 1991): 1618; "Strategies Shift to Targeting Genes," *New Scientist*, August 17, 1991, 14; C. Anderson, "U.S. Patent Application Stirs Up Gene Hunters: NIH Researcher Files for 337 Genes at One Time; When Should a Gene Be Patentable?" *Nature* 353 (October 10, 1991): 485.

Books: Kevin Davies, *Cracking the Genome: Inside the Race to Unlock Human DNA* (New York: Free Press, 2001), 59; James D. Watson, Amy A. Caudy, Richard M. Myers, and Jan A. Witkowski, *Recombinant DNA*, 3rd ed. (New York: W. H. Freeman and Company, 2007), 277.

17. **Wire service:** Catherine Arnst, "Gene Patent Bid by U.S. Agency Sparks Huge Row," *Toronto Star*, Reuters, October 27, 1991.

Newspapers: Edmund L. Andrews, "U.S. Seeks Patent on Genetic Codes, Setting Off Furor," *New York Times*, October 21, 1991, A1; Steve Connor, "Americans Try to Patent the Human Brain," *Independent*, October 20, 1991, 1; Nora Zamichow, "Scientists Question U.S. Plan to Patent Genes," *Los Angeles Times* (San Diego County edition), October 20, 1991, B2; Roger Highfield, "Scientists to Fight Gene 'Gold Rush,'" *Daily Telegraph*, October 26, 1991, 14; Larry Thompson, "NIH's Rush to Patent Human Genes," *Washington Post*, October 28, 1991, A3; Jessica Mathews, "The Race to Claim the Gene," *Washington Post*, November 17, 1991, C7; Richard Saltus, "Gene Patents: Weighing Protection vs. Secrecy; Outcome of Broad U.S. Application Could Sway the Course of Research Commitment," *Boston Globe*, December 2, 1991, 25.

Magazines: Anderson, "U.S. Patent Application Stirs Up Gene Hunters," 485; Leslie Roberts, "Genome Patent Fight Erupts: An NIH Plan to Patent Thousands of Random DNA Sequences Will Discourage Industrial Investment and Undercut the Genome Project Itself, the Plan's Critics Charge," *Science* 254 (October 11, 1991): 184; C. Anderson and Peter Aldhous, "Genome Project: Secrecy and the Bottom Line," *Nature* 354 (November 14, 1991): 96; editorial, "Free Trade in Human Sequence Data?" *Nature* 354 (November 21, 1991): 171–172; David L. Wheeler, "Britain and Congress Respond to Controversy Sparked by NIH Plan to Patent Genes," *Chronicle of Higher Education*, November 27, 1991, A29; "NIH Patent Fight 'Fascinates,' Fair Price Dominates House Subcommittee," *McGraw Hill's Biotechnology Newswatch* 11 (December 2, 1991): 4.

18. **Wire service:** Paul Recer, "British Agency Joins U.S. Decision Not to Seek Gene Fragment Patents," Associated Press, February 11, 1994.

Newspapers: Warren E. Leary, "U.S. Scientists to Seek Patent on 2,375 More Genes," *New York Times*, February 13, 1992, B16; "Biotechnology: U.S. Pursuit of Gene Patents Riles Industry," *Wall Street Journal*, February 13, 1992, B1; Malcolm Gladwell, "NIH Seeks Patent Protection for Human Genes; Move Renews Controversy over Control of Commercial Uses of Key Biotech Research," *Washington Post*, February 13, 1992, A16; Richard Saltus, "Scientists Criticize NIH Bid for Patent on Gene Fragments," *Boston Globe*, February 13, 1992, 26; Michael Unger, "No Dice to Spliced Genes: Patent Office," *Newsday*, September 24, 1992, 49; Natalie Angier, "U.S. Dropping Its Efforts to Patent Thousands of Genetic Fragments," *New York Times*, February 11, 1994, A16; Rick Weiss, "NIH Abandons Bid to Patent DNA Fragments; Exclusivity Could Impede Important Genetic Research, Director Says," *Washington Post*, February 11, 1994, A1; Michael Waldholz, "NIH Gives Up Effort to Patent Pieces of Genes," *Wall Street Journal*, February 11, 1994, B1.

Magazines: Jeffrey L. Fox, "Defying Its Parent, NIH Against Gene Patents," *Nature Biotechnology* 10 (February 1992): 120; "Intellectual Property; Obviously Not," *The Economist*, October 3, 1992, 90; Jeffrey L. Fox, "NIH Nixes DNA Patents: What Next?" *Nature Biotechnology* 12 (April 1994): 348.

19. **Magazine:** Alun Anderson, "Interview: Craig Venter," *Prospect* 121 (April 23, 2006), www.prospectmagazine.co.uk/2006/04/craigventer.

20. **Newspapers:** Nicholas Wade, "Scientist at Work: J. Craig Venter, a Bold Short Cut to Human Genes," *New York Times*, February 22, 1994, C1; Jerry E. Bishop and Hilary Stout, "Gene Scientist to Leave NIH, Form Institute," *Wall Street Journal*, July 7, 1992, B1; Robin Herman, "NIH Genes Researcher Is Leaving for His Own Lab," *Washington Post*, July 7, 1992, Z4; Alex Barnum, "Genes Scientist Starts Private Institute," *San Francisco Chronicle*, July 10, 1992, B2; Gina Kolata, "Biologist's Speedy Gene Method Scares Peers but Gains Backer," *New York Times*, July 28, 1992, C1; Sandra Sugawara, "A Healthy Vision; Investment Group Injects Millions into Maryland's Biotech Dream," *Washington Post*, November 16, 1992, F1; Gina Kolata, "Wallace Steinberg Dies at 61; Backed Health Care Ventures," *New York Times*, July 29, 1995, 27.

Magazines: John Carey et al., "The Gene Kings: Two Scientists Are Changing How DNA Is Mined—and Drugs Are Discovered," *Business Week*, May 8, 1995, 76; Mara Bovsun, "Venter, 30 Scientists to Leave NIH for Venture-Funded Lab," *McGraw Hill's Biotechnology Newswatch* 12 (July 20, 1992): 1.

Books: Venter, *A Life Decoded*, 149–158; Shreeve, *The Genome War*, 85–87.

21. **Broadcast:** James D. Watson, interview by Charlie Rose, *Charlie Rose*, PBS, June 21, 2000, transcript from 800-AllNews.

22. **Interviews:** Francis Collins, by the author, Bethesda, Maryland, April 11, 2007, in *Mapping Life Notebook* 5, 346–360; Francis Collins, by the Academy of

Achievement, Jackson Hole, Wyoming, May 23, 1998, www.achievement.org/autodoc/page/col1int-1.

Newspaper: Tim Friend, "Scientist on a Roll; Riding High in His Successful Gene Hunt; the Chief of Genetic Detectives," *USA Today*, July 24, 1990.

Magazines: Francis Collins, "No Longer Just Looking Under the Lamp Post," William Allen Award address, *American Journal of Human Genetics* 79 (September 2006): 426; Larry Thompson, "Healy and Collins Strike a Deal," *Science* 259 (January 1, 1993): 22.

23. **Interview:** Collins, by the author.

24. **Interview:** Bruce Birren, by the author, Broad Institute, Cambridge, Massachusetts, October 4, 2006, transcript.

25. **Interview:** Daniel Cohen, "Entretien avec Daniel Cohen," HistRecMed, January 21, 2002, http://picardp1.ivry.cnrs.fr/CohenD.html (accessed January 14, 2005).

Newspapers: Virginia Hamill, "2 U.S. Scientists Split Nobel Prize with Frenchman; 3 Scientists Share Nobel Prize in Medicine," *Washington Post*, October 11, 1980, A1; William Borders, "3 Cell Researchers Win Medicine Nobel," *New York Times*, October 11, 1980, Section 2, 1; "Sketches of Three Winners of Nobel Prize," *New York Times*, October 11, 1980, 7; Rita Reif, "6 Balthus Works Highlight Collection at Auction," *New York Times*, February 15, 1984, C17; "Record $1,185,800 Paid for a Balthus Painting," *New York Times*, March 28, 1984, C18.

Magazines: Antoine Danchin, "Une Histoire Intense, Presque Violente," *La Recherche* (June 2000): 27; Eldad Beck, "How Sweet It Is," *Jerusalem Report* (April 2, 1992): 31.

Book: Paul Rabinow, *French DNA: Trouble in Purgatory* (Chicago, University of Chicago Press, 1999).

26. **Scientific articles:** David T. Burke, Georges F. Carle, and Maynard V. Olson, "Cloning of Large Segments of Exogenous DNA into Yeast by Means of Artificial Chromosome Vectors," *Science* 236 (May 15, 1987): 806; H. Shizuya et al., including Melvin Simon, "Cloning and Stable Maintenance of 300-Kilobase-Pair Fragments of Human DNA in *Escherichia Coli* Using an F-Factor-Based Vector," *Proceedings of the National Academy of Sciences USA* 89 (September 1992): 8794; Panayiotis A. Ioannou et al., including Pieter J. de Jong, "A New Bacteriophage P1–Derived Vector for the Propagation of Large Human DNA Fragments," *Nature Genetics* 6 (January 1994): 84.

Interview: Pieter J. de Jong, by the author, Children's Hospital of Oakland Research Institute, Oakland, California, February 16, 2007, transcript.

27. **Scientific article:** David R. Cox et al., including Richard M. Myers, "Radiation Hybrid Mapping: A Somatic Cell Genetic Method for Constructing High-Resolution Maps of Mammalian Chromosomes," *Science* 250 (October 12, 1990): 245.

28. **Scientific article:** Maynard Olson, Leroy Hood, Charles Cantor, and David Botstein, "A Common Language for Physical Mapping of the Human Genome," *Science* 245 (September 29, 1989): 1434.

News release: "STS—New Strategy May Provide Common Link for Mapping," *Human Genome News* 2 (September 1990).

Newspapers: Larry Thompson, "New Gene-Mapping Plan May Speed Work, Cut Cost; Central Storage Facility May Not Be Needed," *Washington Post,* October 4, 1989, A3; Marilyn Chase, "Scientists Mapping Human Genes Seek New Ways to Express Findings," *Wall Street Journal*, October 5, 1989, B4; Sandra Blakeslee, "A Road Map for Genes," *New York Times*, October 10, 1989, C7; Knight Ridder, "Shortcut Found to Mapping Genetic Code," *Toronto Star*, October 13, 1989, C3.

29. **Scientific article:** Phil Green et al., including Helen Donis-Keller, "A Genetic Linkage Map of the Human Genome," *Cell* 51 (October 23, 1987): 319.

Archives: Eric Lander, "World's First Human Genetic Map—How It Will Be Useful," typescript, 5 pp., author's file, 87 Map: Collaborative.

Newspapers: Harold M. Schmeck Jr., "New Map of Genes May Aid in Fighting Hereditary Diseases," *New York Times*, October 8, 1987, A1; David Stipp, "Genetic Map That Could Speed Diagnosis of Inherited Diseases Touches Off Dispute," *Wall Street Journal*, October 8, 1987, 33; Richard Saltus, "Mass. Firm's Genetic Map Said to Aid Health Care," *Boston Globe*, October 8, 1987, 1.

Magazine: Leslie Roberts, "Flap Arises over Genetic Map; A Newly Published Genetic Linkage Map of the Human Genome Has Caused a Minor Furor over What Constitutes a Map and Who, If Anyone, Can Claim Scientific Precedence," *Science* 238 (November 6, 1987): 750.

30. **Scientific article:** D. Cohen, I. Chumakov, and J. Weissenbach, "A First-Generation Physical Map of the Human Genome," *Nature* 366 (December 16, 1993): 698.

Interview: Cohen, "Entretien avec Daniel Cohen."

Newspapers: Declan Butler, "Now Robots Have Joined the War Against Disease," *Independent*, November 9, 1992, 16; Robin McKie, "Science and Technology: French Map Path to Every Human Gene—Fine Art Has Funded Vital DNA Research," *Observer*, March 14, 1993, 61; Michael Waldholz, "French Produce First Map Ordering DNA Fragments of Human Genome," *Wall Street Journal*, December 16, 1993, B6; Natalie Angier, "Map of All Chromosomes to Guide Genetic Hunters," *New York Times*, December 16, 1993, A1; William Allen, "Scientists Here Aided French in Genetic 'Map,'" *St. Louis Post-Dispatch*, December 17, 1993, 1A; Clive Cookson, "A Spur for the Gene Hunters: The Mapping of Mankind's Genetic Makeup Sets a Medical Landmark," *Financial Times*, December 20, 1993, 12; Susan Watts, "The Gene Dilemma: French Steal Lead in DNA Research; Private and State Funds Help Project," *Independent*, January 6, 1994, 9.

Magazines: Tony Dajer, "Genetics 1992, the Genome Finds Its Henry Ford," *Discover*, January 1993, 86–87; "The Human Genome; Go Forth and Multiply," *The Economist*, March 13, 1993, 96; C. Anderson, "Genome Shortcut Leads to Problems," *Science* 259 (March 19, 1993): 1684; "French Win Race to Make Physical Map," *Applied Genetics News* 14, no. 6 (January 1994).

31. **Scientific article:** For illustrations showing the Genomatron, see Thomas J. Hudson et al., including Jean Weissenbach and Eric Lander, "An STS-Based Map of the Human Genome," *Science* 270 (December 22, 1995): 1947.

Newspapers: Laura Johannes, "Detailed Map of the Genome Is Now Ready," *Wall Street Journal*, December 22, 1995, B1; Richard Saltus, "In Effort to Map Human DNA, Scientists Find the Going Swift," *Boston Globe*, December 22, 1995, 23.

Magazine: Jean L. Marx, "A New Guide to the Human Genome," *Science* 270 (December 22, 1995): 1919.

32. **Scientific article:** Robert D. Fleischmann et al., including Claire M. Fraser, Hamilton O. Smith, and J. Craig Venter, "Whole-Genome Random Sequencing and Assembly of *Haemophilus influenzae*," *Science* 269 (July 28, 1995): 496.

News releases: "Two Bacterial Genomes Sequenced," *Human Genome News* 7 (May–June 1995); "Genetics and Genomics Timeline 1995," Genome News Network, obtained August 13, 2004.

Wire service: Paul Recer, "Sequencing of Bacteria Will Aid in Control of Disease, Expert Says," Associated Press, May 26, 1995.

Newspapers: Nicholas Wade, "Bacterium's Full Gene Makeup Is Decoded," *New York Times*, May 26, 1995, A16; Nigel Hawkes, "Germ Gene Sequence Revealed," *The Times* (of London), May 27, 1995, 16; Steve Connor, "Genetics Milestone May Be Key to the Code of Life," *Independent*, May 27, 1995, 3; Nicholas Wade, "First Sequencing of Cell's DNA Defines Basis of Life," *New York Times*, August 1, 1995, C1; Nigel Hawkes, "Cracking the Genetic Code," *The Times* (of London), July 31, 1995; Nicholas Wade, "Thinking Small Paying Off Big in Gene Quest," *New York Times*, February 3, 1997, A1.

Magazines: Rachel Nowak, "Bacterial Genome Sequence Bagged: The Publication of the First Complete Genome of a Free-Living Organism, the Bacterium *H. Influenzae*, Opens the Way to a Wealth of Fundamental and Practical Information," *Science* 269 (July 28, 1995): 468; "Genomic Wisdom," *The Economist* (U.S. edition), August 19, 1995, 73; Michael McCarthy, "Microbial Genome Sequencing Opens Door to New World," *Lancet* 349 (February 22, 1997): 548.

Books: Glyn Moody, *Digital Code of Life: How Bioinformatics Is Revolutionizing Science, Medicine, and Business* (Hoboken, NJ: Wiley, 2004), 321–323; Davies, *Cracking the Genome*, 104–111; Shreeve, *The Genome War*, 104–111; Venter, *A Life Decoded*, 193–211.

33. Yeast:

Scientific article: A. Goffeau et al., "Life with 6,000 Genes," *Science* 274 (October 25, 1996): 546.

News release: National Human Genome Research Institute (NHGRI), "International Team Completes DNA Sequence of Yeast," April 24, 1996, www.genome.gov/10000510.

Wire service: "Genetics and Genomics Timeline 1996," Genome News Network, obtained August 13, 2004.

Newspapers: Gina Kolata, "Map of Yeast Genes Promises Clues to Diseases of Humans," *New York Times*, April 25, 1996, B11; Rick Weiss, "Completion of Yeast Cells' Genetic Blueprint Is Hailed As Scientific 'Milestone,'" *Washington Post*, April 25, 1996, A10; Richard Saltus, "Codes of Yeast Genes Cracked; Information Seen of Value in Research on Human Diseases," *Boston Globe*, April 25, 1996, 29.

Magazines: Jane Bradbury, "Yeast Genome Sequenced Completely," *Lancet* 347 (April 27, 1996): 1175; Reginald Rhein, "Yeast Genome Holds Keys to Treatments for Human Ills," *McGraw Hill's Biotechnology Newswatch*, May 6, 1996, 1; Bob Kuska, "Yeast's Value Rises As Scientists Map Its 6,000 Genes," *Journal of the National Cancer Institute* 88 (June 1996): 706.

Syphilis:

Scientific article: C. M. Fraser et al., "Complete Genome Sequence of *Treponema Pallidum*, the Syphilis Spirochete," *Science* 281 (July 17, 1998): 375.

Newspaper: Nicholas Wade, "Genetic Map of Syphilis Is Decoded; Hope for Vaccine Is Raised," *New York Times*, July 17, 1998, A12.

Magazines: Elizabeth Pennisi, "Microbial Genetics: Genome Reveals Wiles and Weak Points of Syphilis," *Science* 281 (July 17, 1998): 324; Michael E. St. Louis and Judith N. Wasserheit, "Epidemiology: Elimination of Syphilis in the United States," *Science* 281 (July 17, 1998): 353.

34. **Newspaper:** Tina Hesman, "WU Center Loses a Leader, Not Its Vision, Staff Says," *St. Louis Post-Dispatch*, July 14, 2002, B1.

Magazine: Eliot Marshall, "A Strategy for Sequencing the Genome 5 Years Early; Human Genome Project," *Science* 267 (February 10, 1995): 783.

35. **Scientific article:** Maynard V. Olson, "A Time to Sequence," *Science* 270 (October 20, 1995): 394–396.

36. **Interview:** Collins, by the author.

News release: "International Large-Scale Sequencing Meeting," *Human Genome News* 7 (April–June 1996).

Magazine: David R. Bentley, "Policy Forum: Genomic Sequence Information Should Be Released Immediately and Freely in the Public Domain," *Science* 274 (October 25, 1996): 533.

Book: Shreeve, *The Genome War*, 46–47.

37. **Wire services:** Associated Press, "Dispute Arises over Patent for a Gene," *New York Times*, October 30, 1994, 32; Associated Press, "Gene Patent Dispute Is Settled," *New York Times*, February 16, 1995, A18.

Newspapers: Natalie Angier, "Scientists Identify a Mutant Gene Tied to Hereditary Breast Cancer," *New York Times*, September 15, 1994, A1; Natalie Angier, "Fierce Competition Marked Fervid Race for Cancer Gene," *New York Times*, September 20, 1994, C1; Steve Connor, "Concern over Cancer Gene Patent," *Independent*, September 15, 1994, 3; Michael Waldholz, "Feud Brewing over Breast-Cancer Gene Patent," *Wall Street Journal*, October 17, 1994, B1.

Magazine: Phyllida Brown and Kurt Kleiner, "Patent Row Splits Breast Cancer Researchers," *New Scientist*, September 24, 1994, 44.

38. **News releases:** "Merck to Make Comprehensive Database of Genetic Information Publicly Available," Business Wire, September 28, 1994; "Research Collaboration to Offer Access to World's Largest Human DNA-Sequence Database; Structural Details for 30,000 to 35,000 Human Genes," PR Newswire, October 3, 1994.

Newspapers: Jerry E. Bishop, "Plan May Blow Lid Off Secret Gene Research," *Wall Street Journal*, September 28, 1994, B1; Kathleen Day, "Merck to Reveal Findings on Human Genetic Code; Move Stirs Debate over Research Secrecy," *Washington Post*, September 29, 1994, B11; William Allen, "Pact Would Reveal Genetic Data; Washington U. Deal Would Speed Search for Disease-Causing Genes," *St. Louis Post-Dispatch*, September 29, 1994, 10A; Nigel Hawkes, "So Who Owns Genetic Data?" *The Times* (of London), October 3, 1994; Clive Cookson, "The Ultimate Privatisation: Who Should Own Human Genes," *Financial Times*, October 8, 1994, 9.

Magazines: Eliot Marshall, "HGS Opens Its Databanks—for a Price; Human Genome Sciences Wants Option to License Derivative Products," *Science* 266 (October 7, 1994): 25; Eliot Marshall, "A Showdown over Gene Fragments; Commercial Control of Genetic Data," *Science* 266 (October 14, 1994): 208; Kurt Kleiner, "Squabbling All the Way to the Genebank: Decoding the Human Genome Is Producing Some Extremely Useful—and Valuable—Information; the Conflict Between Private Profit and Public Interest," *New Scientist*, November 26, 1994, 1414.

Books: Davies, *Cracking the Genome*, 99–103; Shreeve, *The Genome War*, 88; Venter, *A Life Decoded*, 185–188.

On the "divorce" between The Institute for Genomic Research and Human Genome Sciences:

News release: "TIGR/HGS Funding Relationship Reaches Early Conclusion," PR Newswire, June 24, 1997.

Newspapers: Nicholas Wade, "Genome Project Partners Go Their Separate Ways," *New York Times*, June 24, 1997, C2; Tim Friend, "Gene Trailblazers Part Ways; Research Group, Biotech Firm Made Key Breakthroughs," *USA Today*, June 24, 1997, 4D; Beth Berselli, "Genetic Research Firm Severs Ties with Nonprofit," *Washington Post*, June 24, 1997, C4; Eric Fisher, "Biotech Firm, Institute Part Ways," *Washington Times*, June 24, 1997, B10.

Magazine: Carey et al., "The Gene Kings."

39. **Magazines:** Eliot Marshall, "Data Sharing: Genome Researchers Take the Pledge," *Science* 272 (April 26, 1996): 477; Eliot Marshall, "Bermuda Rules: Community Spirit, with Teeth," *Science* 291 (February 16, 2001): 1192.

Books: Sulston and Ferry, *The Common Thread*, 143–146; Jane Rogers, "The Finished Sequence of *Homo Sapiens*," in *Cold Spring Harbor Symposia on Quantitative Biology* 68 (2003): 1–2.

40. **Scientific articles:** James L. Weber and Eugene W. Myers, "Human Whole-Genome Shotgun Sequencing," *Genome Research* 7 (May 1, 1997): 401; Philip D. Green, "Against a Whole-Genome Shotgun," *Genome Research* 7 (May 1, 1997): 410.

Books: Davies, *Cracking the Genome*, 141–143; Shreeve, *The Genome War*, 45–47.

Chapter 4: Crisis

1. **Newspaper:** Douglas Birch, "Daring Sprint to the Summit. The Quest: A Determined Hamilton Smith Attempts to Scale a Scientific Pinnacle—and Reconcile with Family," *Baltimore Sun*, April 13, 1999, 1A.

Book: James Shreeve, *The Genome War: How Craig Venter Tried to Capture the Code of Life and Save the World* (New York: Knopf, 2004), 16–22.

2. **Interview:** Francis Collins, by the author, National Human Genome Research Institute (NHGRI), Bethesda, Maryland, April 11, 2007, in *Mapping Life Notebook* 5, 346–360.

News release: NHGRI, "Statement by the National Institutes of Health on the Perkin-Elmer-TIGR DNA Sequencing Proposal," May 1998, www.genome.gov/10000606.

Wire service: Paul Recer, "Gene Technology Must Prove Itself," Associated Press, May 11, 1998.

Newspapers: Nicholas Wade, "Scientist's Plan: Map All DNA Within 3 Years," *New York Times*, May 10, 1998, Section 1, 1; Bill Richards, "Perkin-Elmer Jumps into Race to Decode Genes," *Wall Street Journal*, May 11, 1998, B6; Justin Gillis and Rick Weiss, "Private Firm Aims to Beat Government to Gene Map," *Washington Post*, May 12, 1998, A1; Douglas Birch, "Md. Gene Researcher Again Takes on the Establishment; Venter Plans Fast Human Genome Map," *Baltimore Sun*, May 12, 1998, 1A; Paul Brown and Martin Walker, "U.S. Company Plans to Patent Key Gene Codes," *Guardian*, May 13, 1998, 3; Carl T. Hall, "New Fight in Gene Race," *San Francisco Chronicle*, May 13, 1998, D1; "Neues Unternehmen will menschliche Gene in Rekordzeit entziffern," *Frankfurter Allgemeine Zeitung*, May 15, 1998, 27; Fabrice Node-Langlois, "Coup d'accélérateur dans la course aux gènes; trois ans pour décrypter le génome humain," *Le Figaro*, May 16, 1998; Robin McKie, "I Asked the Big Questions, Like 'What Is Life, Anyway?'" *Observer*, May 17, 1998, 23; Charles Arthur, "Buccaneer Gene Scientist Hopes to Crack Secret of Life—and Wealth," *Independent*, May 17, 1998, 5.

Magazines: Eliot Marshall and Elizabeth Pennisi, "Hubris and the Human Genome," *Science* 280 (May 15, 1998): 994; Robert F. Service, "Picking Up the Pace of Sequencing," *Science* 280 (May 15, 1998): 995; "Genetic Warfare: Science Will Certainly Benefit from a Private Initiative to Sequence the Human Genome. Who Else Will Gain or Lose Remains to Be Seen," *The Economist* (U.S. edition), May 16,

1998, 87; Jonathan Knight and Kurt Kleiner, "Genome Project Goes into Overdrive," *New Scientist*, May 16, 1998, 11.

Books: Shreeve, *The Genome War*, 14; Kevin Davies, *Cracking the Genome: Inside the Race to Unlock Human DNA* (New York: Free Press, 2001), 147–150.

3. **Interview:** Eric S. Lander, by the author, Broad Institute, Cambridge, Massachusetts, August 28, 2006, transcript.

Book: Shreeve, *The Genome War*, 47–53, citing (on 379) interviews with Mark Adams, Francis Collins, David Cox, Richard Gibbs, Eric Lander, Francis Kalush, Ari Patrinos, and Craig Venter.

4. **Books:** Davies, *Cracking the Genome*, 158–159; Shreeve, *The Genome War*, 52–53.

5. **Magazine:** Richard Preston, "The Genome Warrior," *New Yorker*, June 12, 2000, 66.

6. **Book:** Shreeve, *The Genome War*, 51.

7. **Interview:** Collins, by the author.

Book: Craig Venter, *A Life Decoded: My Genome: My Life* (New York: Viking, 2007), 241.

8. **Newspapers:** Gillis and Weiss, "Private Firm Aims to Beat Government"; William F. Haseltine, "Life by Design; Gene-Mapping, Without Tax Money," *New York Times*, May 21, 1998, A33.

Magazines: John Carey, "The Duo Jolting the Gene Business," *Business Week*, May 25, 1998, 70; Marshall and Pennisi, "Hubris."

9. **Newspaper:** Hall, "New Fight in Gene Race."

10. **Magazine:** Marshall and Pennisi, "Hubris."

11. **Newspapers:** Node-Langlois, "Coup d'accélérateur"; Jean Yves Nau, "Jean Weissenbach," *Le Monde*, June 3, 1998.

12. **Newspaper:** Bill Lambrecht, "Maverick Scientist Could Threaten Funding for Washington U's Lab; For-Profit Firm Aims to Quickly Unravel Genetic Code; New Medicines Are at Stake," *St. Louis Post-Dispatch*, May 25, 1998, A1.

13. **Newspaper:** Nicholas Wade, "In Genome Race, Government Vows to Move Up Finish," *New York Times*, September 15, 1998, F3.

14. **Newspaper:** Birch, "Md. Gene Researcher."

15. **Scientific article:** Maynard Olson and Phil Green, "A 'Quality-First' Credo for the Human Genome Project," *Genome Research* 8 (May 1998): 414.

16. **Magazines:** Elizabeth Pennisi, "DNA Sequencers' Trial by Fire," *Science* 280 (May 8, 1998): 814; Preston, "The Genome Warrior."

17. **Newspaper:** Nicholas Wade, "Reading the Book of Life: A Historic Quest; Double Landmarks for Watson: Helix and Genome," *New York Times*, June 27, 2000, F5.

Magazine: Preston, "The Genome Warrior."

18. **News releases:** NHGRI, "Pilot Study Explores Feasibility of Sequencing Human DNA," April 1996, www.genome.gov/10000511; "NCHGR Initiates Se-

quencing Pilot Projects," *Human Genome News* 7 (April–June 1996); "Whitehead Receives $26 Million Grant to Begin Sequencing the Human Genome," Whitehead Institute, April 11, 1996, www.wi.mit.edu/news/archives/1996/el_0411.html.

Newspaper: "$60 Million Study Begins of DNA Sequencing," *Washington Post*, April 12, 1996, A3.

Magazine: Eliot Marshall and Elizabeth Pennisi, "NIH Launches the Final Push to Sequence the Genome," *Science* 272 (April 12, 1996): 188.

19. **Testimony:** "The Human Genome Project: Current Status and Benefits of Public and Private Efforts to Sequence the Human Genome," prepared testimony of Robert H. Waterston, MD, PhD, director of Washington University Genome Sequencing Center, before the House Committee on Science Subcommittee on Energy and the Environment, Federal News Service, April 6, 2000.

News release: NHGRI, "Faster, Cheaper Methods to Read DNA Focus on Systems Integration, Miniaturization," June 25, 1996, www.genome.gov/10000509.

Magazine: Tim Stevens, "Authors of the Book of Life," *Industry Week*, December 11, 2000, 47.

20. **Magazine:** Preston, "The Genome Warrior."

21. **Interviews:** Pieter J. de Jong, by the author, Children's Hospital of Oakland Research Institute, Oakland, California, February 16, 2007, in *Mapping Life Notebook* 5, 82–93, and transcript; Eric Lander, by the author, Broad Institute, July 12, 2006, transcript.

22. **Interview:** Lander, by the author, July 12, 2006.

23. **Interview:** Michael Hunkapiller, by the author, Alloy Ventures, Palo Alto, California, February 22, 2007, in *Mapping Life Notebook* 5, 129–143.

Magazine: Service, "Picking Up the Pace."

Book: Davies, *Cracking the Genome*, 143–146.

24. **Magazine:** Preston, "The Genome Warrior."

25. **Newspapers:** Aisling Irwin, "The Gene Genie Racing to Grab a Fast Billion," *Daily Telegraph*, May 14, 1998, 6; Arthur, "Buccaneer Gene Scientist."

Magazine: Elizabeth Pennisi, "Funders Reassure Genome Sequencers," *Science* 280 (May 22, 1998): 1185.

26. **Newspapers:** Irwin, "The Gene Genie"; Clive Cookson, "Wellcome Trust Puts £110m into Gene Race," *Financial Times*, May 14, 1998, 8; Nicholas Wade, "International Gene Project Gets Lift," *New York Times*, May 17, 1998, Section 1, 20.

Magazines: Preston, "The Genome Warrior"; Pennisi, "Funders Reassure Genome Sequencers."

Book: Shreeve, *The Genome War*, 54–55.

27. **Interview:** Lander, by the author, Broad Institute, Cambridge, Massachusetts, August 21, 2006, transcript.

28. **Magazine:** Meredith Wadman, "Rough Draft of Human Genome Wins Researchers' Backing," *Nature* 393 (June 4, 1998): 399.

29. **Newspapers:** Harold Varmus, "Progress on All Fronts in Race to Map Genes," *New York Times*, May 17, 1998, Section 4, 16; Vincent Tardieu, "Le décryptage du génome humain pourrait être achevé en 2001," *Le Monde*, June 3, 1998.

Magazine: David Dickson and Meredith Wadman, "Genome Effort 'Still in Need of Support,'" *Nature* 393 (May 21, 1998): 201.

30. **Magazine:** Dickson and Wadman, "Genome Effort."

31. **Testimony:** Federal News Service, June 16, 1998, prepared statements before the House Science Committee, Energy and Environment: Chairman Ken Calvert, R-California; J. Craig Venter, PhD, president and director of The Institute for Genomic Research; Maynard V. Olson; Dr. Ari Patrinos, associate director, Office of Biological and Environmental Research, Office of Energy Research, Department of Energy. Federal Document Clearing House Congressional Testimony, June 17, 1998: Francis S. Collins, director, NHGRI; David J. Galas, president, Chiroscience R&D, Inc.

Newspapers: Llewellyn H. Rockwell Jr., "Government and the Genome Project," *Washington Times*, June 16, 1998, A18; Nicholas Wade, "It's a Three-Legged Race to Decipher the Human Genome," *New York Times*, June 23, 1998, F3.

Magazines: "DOE Official Urges Congress to Keep Funding Human Genome Mapping," *Inside Energy/with Federal Lands*, June 22, 1998, 10; Daniel S. Greenberg, "Bad Blood in U.S. Genome Research," *Lancet* 351 (June 27, 1998): 1939.

Books: Kevin Davies, *Cracking the Genome: Inside the Race to Unlock Human DNA* (New York: Free Press, 2001), 153–154; Shreeve, *The Genome War*, 14.

32. **Newspaper:** Wade, "Three-Legged Race."

33. **News release:** NHGRI, "Human Genome Sequencing Projects Receive Third Year of Funding," July 1998, www.genome.gov/10000602.

34. **Magazine:** David L. Wheeler, "Will DNA Patents Hinder Research? Lawyers Say Not to Worry," *Chronicle of Higher Education*, July 16, 1999, A19.

35. **Scientific article:** R. Waterston and J. E. Sulston, "Genomics: The Human Genome Project—Reaching the Finish Line," *Science* 282 (October 2, 1998): 53–54.

36. **Interview:** Lander, by the author, August 28, 2006.

37. **Scientific article:** Francis S. Collins et al., "New Goals for the Human Genome Project: 1998–2003," *Science* 282 (October 23, 1998): 682.

News releases: NHGRI, "Genome Project Leaders Announce Intent to Finish Sequencing the Human Genome Two Years Early," September 1998, www.genome.gov/11508996; "U.S. HGP on Fast Track for Early Completion," *Human Genome News* 10, nos. 1 and 2 (February 1999).

Newspapers: Wade, "In Genome Race"; Justin Gillis, "Scientists Speed Up Timetable for Mapping Human Genes," *Washington Post*, September 15, 1998, A2; Usha Lee McFarling, Knight Ridder Tribune News, "Race to Read 'Book of Life' Spurs Modern Day Gold Rush; Companies Push to Map Human Genome Before Federal Project Makes It Public," *Houston Chronicle*, September 21, 1998, D7.

Magazines: Eliot Marshall, "Human Genome Project: NIH to Produce a 'Working Draft' of the Genome by 2001," *Science* 281 (September 18, 1998): 1774; David L. Wheeler, "Human Genome Project Plans to Speed Up Work on DNA Mapping," *Chronicle of Higher Education*, September 25, 1998, A20; Laurie Goodman, "The Human Genome Project Aims for 2003," *Genome Research* 8 (October 1998): 997; "New Goals for Genome Project," *Applied Genetics News* 19 (November 1998).

38. **Interview:** Jane Peterson and Mark Guyer, by the author, NHGRI, Rockville, Maryland, April 13, 2007, in *Mapping Life Notebook* 5, 362–367.

39. **News releases:** NHGRI, "Human Genome Project Announces Successful Completion of Pilot Project; Launches Large-Scale Effort to Sequence the Human Genome with New Awards, Accelerated Timetable," March 15, 1999, www.genome.gov/10002131; Massachusetts Institute of Technology News Office, "Whitehead Receives $35m for Genome Project's Next Phase," March 17, 1999.

Broadcast: Dan Charles, "Government and Private Company Working to Decode Human DNA Sequence," *Morning Edition*, National Public Radio (NPR), March 16, 1999.

Newspapers: Nicholas Wade, "One of 2 Teams in Genome-Map Race Sets an Earlier Deadline," *New York Times*, March 16, 2001, A21; Justin Gillis, "Gene Map Is on Fast Track; Millions in Grants Speed Up Project," *Washington Post*, March 16, 1999, A19; Robert Langreth, "Gene-Sequencing Race Between U.S. and Private Researchers Is Accelerating," *Wall Street Journal*, March 16, 1999, B5; William Allen, "Gene Project Wins $81 Million Infusion; Washington U Gets $38 Million for DNA Map Work; Largest 1-Year Grant in Its History," *St. Louis Post-Dispatch*, March 16, 1999, A1; Richard Saltus, "Cambridge Firm Joins Push to Finish Mapping Human Genome," *Boston Globe*, March 16, 1999, A10; Michael Lasalandra, "Cambridge Institute Given $35M for Gene Work," *Boston Herald*, March 16, 1999, 12; Clive Cookson, "Scientists Close in on Human Genetic Code," *Financial Times*, March 16, 1999, International section, 7.

Magazines: Meredith Wadman, "Human Genome Project Aims to Finish 'Working Draft' Next Year," *Nature* 398 (March 18, 1999): 177; Ehsan Masood, "Britain Sought to Speed Up Sequencing Efforts," *Nature* 398 (March 18, 1999): 177; Elizabeth Pennisi, "Academic Sequencers Challenge Celera in a Sprint to the Finish," *Science* 283 (March 19, 1999): 1822; John Carey, "Match Your Helix, Raise You a Genome," *Business Week*, March 29, 1999, 6; Marie-Clare McMenemy, "Human Genome Project Receives a Boost," *Lancet* 353 (March 27, 1999): 1078.

40. **News release:** "Celera Is Name of New Genomics Company Formed by Perkin-Elmer and Dr. J. Craig Venter; Members Chosen for Scientific Advisory Board," August 5, 1998.

41. **News release:** Incyte Pharmaceuticals, Inc., "Incyte Launches Pharmacogenetics Business Unit to Sequence and Map the Human Genome; New Business to

Commercialize Pharmaceutical Applications of Genetic Variation Data; New Series of Stock to Track Performance," August 17, 1998.

Wire service: "Palo Alto Company Will Map Most of Human Genome Within Year," Associated Press, August 18, 1998.

Newspapers: Nicholas Wade, "New Company Joins Race to Sequence Human Genome," *New York Times*, August 18, 1998, F6; Justin Gillis, "Calif. Firm Joins Race to Map Genes; Government, Rockville Company Already Seek Data Important to Drugmakers," *Washington Post*, August 18, 1998, C1.

Magazine: Eliot Marshall, "Genomics: A Second Private Genome Project," *Science* 281 (August 21, 1998): 1121.

42. **Scientific article:** *Science* : The *C. elegans* Sequencing Consortium, "Genome Sequence of the Nematode *C. Elegans*: A Platform for Investigating Biology," special issue, *Science* 282 (December 11, 1998): 2012.

News release: NHGRI, "International Genome Team Deciphers Genetic Instructions for a Complete Animal," December 11, 1998, www.genome.gov/10000570.

Wire services: John Sulston quote is from "Roundworm Genome Completely Sequenced," Reuters Health Medical News, December 11, 1998; Paul Recer, "Worm Has Entire Genetic Code Mapped," Associated Press, December 11, 1998; Lori Valigra, "Genetics of Complete Animal Deciphered," United Press International, December 10, 1998.

Broadcast: Joe Palca, "Earthworm Genetics," *All Things Considered*, National Public Radio, December 10, 1998.

Newspapers: Sue Goetinck, "Methods Used in Worm Study May Help Determine Genetic Blueprint of Humans," *Dallas Morning News*, December 10, 1998; Nicholas Wade, "Animal's Genetic Program Decoded, in a Science First," *New York Times*, December 11, 1998, A1; Rick Weiss, "Worm's Genetic Coding Detailed; 1st Total Map of an Animal May Yield Insights on Humans," *Washington Post*, December 11, 1998, A3; Tim Friend, "Blueprint for Human Genes Is Step Closer," *USA Today*, December 11, 1998, 1A; Monte Reel, "WU Scientists Help Make the First Map of an Animal's Genetic Structure; Milestone Could Lead to Human Illness Treatments; Team Deciphers Tiny Roundworm," *St. Louis Post-Dispatch*, December 11, 1998, A1; Robert Cooke, "Worm's Complex Genome Unraveled—'Map' Is Big Step Forward for Scientists," *Seattle Times*, December 11, 1998; Robert S. Boyd, "Genes Mapped for Creature Tiny, Slimy, and a Lot Like Us," *Philadelphia Inquirer*, December 11, 1998, A1; Tim Radford, "One Worm Plus £30m Equals a DNA Triumph," *Guardian*, December 11, 1998, 1; Tim Radford, "Analysis: The Genetic Code: A New Map of a (Quite) Close Cousin; British and American Scientists Make History Today," *Guardian*, December 11, 1998, 19; Clive Cookson, "Worms Help Scientists to Crack Animal Genetic Code," *Financial Times*, December 11, 1998, 7.

Magazines: Elizabeth Pennisi, "Genome Sequencing: Worming Secrets from the *C. elegans* Genome," *Science* 282 (December 11, 1998): 1972; Georgina Ferry, "The Human Worm," *New Scientist*, December 5, 1998, 32; "An Elegant Piece of Research," *The Economist* (U.S. edition), December 12, 1998, 83.

43. **Scientific article:** Neil Risch and Kathleen Merikangas, "The Future of Genetic Studies of Complex Human Diseases," *Science* 273 (September 13, 1996): 1516.

Magazine: Gary Taubes, "Epidemiology Faces Its Limits," *Science* 269 (July 14, 1995): 164.

44. **Scientific article:** D. G. Wang et al., "Large-Scale Identification, Mapping, and Genotyping of Single-Nucleotide Polymorphisms in the Human Genome," *Science* 280 (May 15, 1998): 1077.

Newspapers: Richard Saltus, "Top Scientist Urges Shortcut on Gene-Mapping Approach," *Boston Globe*, September 7, 1997, A8; Nicholas Wade, "In the Hunt for Useful Genes, a Lot Depends on 'Snips,'" *New York Times*, August 11, 1998, F1.

Magazines: Jennifer Fisher Wilson, "Hot Paper: SNP Potential: Variations Offer Clues to Heritable Diseases," *The Scientist* 14 (July 24, 2000): 29; Eliot Marshall, "Economics: Snipping Away at Genome Patenting," *Science* 277 (September 19, 1997): 1752; Francis Collins, Mark S. Guyer, and Aravinda Chakravarti, "Policy Forum: Variations on a Theme: Cataloging Human DNA Sequence Variation," *Science* 278 (November 28, 1997): 1580; Eliot Marshall, "Human Genome Project: 'Playing Chicken' over Gene Markers," *Science* 278 (December 19, 1997): 2046; Eliot Marshall, "The Hunting of the SNP," *Science* 278 (December 19, 1997): 2047.

45. **Newspapers:** Matthew Lynn, "Wellcome Seeks Alliance to Keep Genes Data Public," *Sunday Times*, August 30, 1998; Nicholas Wade, "10 Drug Makers Join in Drive to Find Diseases' Genetic Roots," *New York Times*, April 15, 1999, A27; David Pilling, "Scientists Combine in Bid to Crack Gene Code: Mapping Human DNA. Drug Companies, Government and Academics to Collaborate in $45m Project," *Financial Times* (U.S. edition), April 15, 1999, 5; Justin Gillis, "Drug Companies, Gene Labs to Join Forces; Collaboration Aims to Probe Genetic Differences—Without Proprietary Interests," *Washington Post*, April 15, 1999, E1; William Allen, "WU Scientists Will Play Key Role in Gene Project; Drug Companies Will Pitch in $45 Million; the Focus Is on Diseases," *St. Louis Post-Dispatch*, April 15, 1999, B1; Steve Connor, "Comment: Your Future Life Signposted," *Independent*, April 15, 1999, 5.

Magazines: Ehsan Masood, "Consortium Plans Free SNP Map of Human Genome," *Nature* 398 (April 15, 1999): 545; Eliot Marshall, "Genomics: Drug Firms to Create Public Database of Genetic Mutations," *Science* 284 (April 16, 1999): 406; Nicholas Wade, "Tailoring Drugs to Fit the Genes," *New York Times*, April 20, 1999, F9.

46. **Interview:** Lander, by the author, August 21, 2006.

47. **Newspapers:** Jules Crittenden, "Humanity's Map Unfolding—DNA Project Opens New Frontiers," *Boston Herald*, April 2, 2000, 6; Tom Spears, "Robots Are Running the Lab: A Laboratory Assembly Line Analyses Human Genetic Code," *Ottawa Citizen*, April 3, 2000, A4; Byron Spice, "Piecing Together Our Future; The New Map of the Human Genome Will Change Our Lives Forever," *Pittsburgh Post-Gazette*, June 25, 2000, A1.

48. **Scientific articles:** Mark D. Adams et al., including Gerald M. Rubin and J. Craig Venter, "The Genome Sequence of *Drosophila melanogaster*," *Science* 287 (March 24, 2000): 2185; Eugene M. Myers et al., including Mark D. Adams, Gerald M. Rubin and J. Craig Venter, "A Whole-Genome Assembly of *Drosophila*," *Science* 287 (March 24, 2000): 2196; Gerald M. Rubin et al., including Richard Gibbs, J. Craig Venter, Mark D. Adams, and Suzanne Lewis, "Comparative Genomics of the Eukaryotes," *Science* 287 (March 24, 2000): 2204.

News release: Celera Genomics, "Scientists at Celera Genomics and Berkeley *Drosophila* Genome Project Publish *Drosophila* Genome; Whole-Genome Shotgun Method Successful," March 23, 2000.

Wire service: Barbara J. Culliton, "The Humanized Fly: The Complete Genome Sequence of *Drosophila melanogaster* Will Be a Powerful Tool for Using Fly Biology to Study Human Medicine," Genome News Network, March 24, 2000.

Newspapers: "Scientists Break Genetic Code of Fruit Fly; Laboratory Insect Project Is Precursor to Human Genome," *Washington Post*, February 19, 2000, A3; Nicholas Wade, "On Road to Human Genome, a Milestone in the Fruit Fly," *New York Times*, March 24, 2000, A1; Nicholas Wade, "The Fly People Make History on the Frontiers of Genetics," *New York Times*, April 11, 2000, F1.

Magazines: Elizabeth Pennisi, "Fruit Fly Genome Yields Data and a Validation," *Science* 287 (February 25, 2000): 1374; "Fruit-Fly Genetics: Shotgun Success," *The Economist*, February 26, 2000, 93; Barbara R. Jasny and Floyd E. Bloom, "Flying to New Heights," *Science* 287 (March 24, 2000): 2157.

49. **Books:** Davies, *Cracking the Genome*, 145–147; Shreeve, *The Genome War*, 65–67.

50. **Interview:** Lander, by the author, August 21, 2006.

Newspapers: Justin Gillis, "Gene Map Alliance Hopes Fade; Commercial Usage Is Divisive Issue," *Washington Post*, March 6, 2000, A4; Justin Gillis, "Gene-Mapping Controversy Escalates; Rockville Firm Says Government Officials Seek to Undercut Its Effort," *Washington Post*, March 7, 2000, E1; Jane Martinson, "Protocol on Gene Research at Risk As Firm Demands Exclusive Rights," *Guardian*, March 7, 2000, 3; Justin Gillis, "Celera Leaves Door Open to Genetic Research Deal," *Washington Post*, March 8, 2000, E3; Nicholas Wade, "Genome Decoding Plan Is Derailed by Conflicts," *New York Times*, March 9, 2000, A20.

Magazines: Eliot Marshall, "Genome Sequencing: Talks of Public-Private Deal End in Acrimony," *Science* 287 (March 10, 2000): 1723; Eliot Marshall, Elizabeth

Pennisi, and Leslie Roberts, "In the Crossfire: Collins on Genomes, Patents, and 'Rivalry': As the Public Effort to Sequence the Human Genome Comes into the Final Stretch, Its Methods and Goals Are Increasingly Being Challenged by Private Firms," *Science* 287 (March 31, 2000): 2396.

Book: Shreeve, *The Genome War*, 292–296.

51. **News release:** "PE Corporation Announces Follow-On Public Stock Offering of 3.8 Million Shares of Celera Genomics Stock at $225 Per Share," Business Wire, March 1, 2000.

Newspaper: Jill J. Barshay, "Anatomy of a Stock Boom; a Sampling of Recent Equity Offerings in the Health Care Sector Indicates That Biotechnology Once Again Is Inspiring the Investment Community," *Star Tribune* (Minneapolis), March 13, 2000, 1D.

Magazine: Ellen Licking et al., "Move Over, Dot-Coms: Biotech Is Back," *Business Week*, March 6, 2000, 116.

52. **News releases:** White House [President Bill Clinton and Prime Minister Tony Blair], "The Human Genome Project: Benefiting All Humanity," FDCH Federal Department and Agency Documents, March 14, 2000 (also released by NHGRI); White House, "Transcript of White House Press Briefing March 14 by Joe Lockhart," U.S. Newswire, March 14, 2000; White House, "Transcript of Briefing by Directors of Office on Science and Technology Policy [Neil Lane] and the Human Genome Project [Francis Collins]," U.S. Newswire, March 14, 2000; "Celera Statement on the Policy Statement of President Clinton and Prime Minister Blair on the Availability of Genomic Information," Business Wire, March 14, 2000; U.S. Patent and Trademark Office, "U.S. Patent Policy Unaffected by U.S./U.K. Statement on Human Genome Sequence Data," release 00–17, March 16, 2000.

Wire services: Anne Gearan, "U.S., Britain Urge Open Access to Genetic Information," Associated Press, March 14, 2000; "U.S., U.K. Announce Gene Research Disclosure Pact," United Press International, March 14, 2000.

Newspapers: Justin Gillis, "Clinton, Blair Urge Open Access to Gene Data; President, Briton Step into Controversy on Code," *Washington Post*, March 15, 2000, E1; Sandra Sugawara and Ariana Eunjung Cha, "NASDAQ Swoons After Plea; Point Drop Is 2nd-Biggest," *Washington Post*, March 15, 2000, E1; Alex Berenson and Nicholas Wade, "A Call for Sharing of Research Causes Gene Stocks to Plunge," *New York Times*, March 15, 2000, A1; Ronald Rosenberg, "Biotechs, NASDAQ Falls Sharply; Plan for Free Access to Gene Data Blamed," *Boston Globe*, March 15, 2000, C1; Terence Corcoran, "Bill and Tony's $200-Billion Genome Expropriation," *Ottawa Citizen*, March 16, 2000, C4; editorial, "Free Genetic Data," *Financial Times*, March 16, 2000, 18; editorial, "Status Quo on the Genome," *Washington Post*, March 20, 2000, A16.

Magazines: Eliot Marshall, "Clinton and Blair Back Rapid Release of Data," *Science* 287 (March 17, 2000): 1903; "In-Gene-Uous; in Genomics, the British and

American Governments Are Meddling in Things That Do Not Concern Them and Neglecting Those That Do," *The Economist* (U.S. edition), March 18, 2000; Eliot Marshall, "Biotechnology: How a Bland Statement Sent Stocks Sprawling," *Science* 287 (March 24, 2000): 2127.

53. **News release:** "Celera Genomics to Acquire Paracel, Inc.," Business Wire, March 20, 2000.

54. **Magazine:** Preston, "The Genome Warrior."

55. **Magazine:** "Insight," special section, *Nature* 405 (June 15, 2000): 819–866.

56. **Scientific articles:** Editorial, "The Nature of the Number," *Nature Genetics* 25 (June 2000): 127; Samuel J. Aparicio, "News and Views: How to Count . . . Human Genes," *Nature Genetics* 25 (June 2000): 129; Brent Ewing and Phil Green, "Analysis of Expressed Sequence Tags Indicates 35,000 Human Genes," *Nature Genetics* 25 (June 2000): 232; Hugues Roest Crollius et al., including Jean Weissenbach, "Estimate of Human Gene Number Provided by Genome-Wide Analysis Using *Tetraodon Nigroviridis* DNA Sequence," *Nature Genetics* 25 (June 2000): 235; Peter Liang et al., including John Quackenbush, "Gene Index Analysis of the Human Genome Estimates 120,000," *Nature Genetics* 25 (June 2000): 239.

57. **Broadcast:** "Blueprint of the Body: Eric Lander: 'This Is Biology's Periodic Table,'" CNN, June 1, 2000, www.cnn.com/SPECIALS/2000/genome/story/interviews/lander.html.

58. **Testimony:** "Prepared Testimony of J. Craig Venter, PhD, President and Chief Scientific Officer, Celera Genomics, a PE Corporation Business, Before the Joint Economic Committee," Federal News Service, June 7, 2000.

59. **Broadcast:** *Charlie Rose*, PBS, June 19, 2000, transcript 2707.

60. **Newspaper:** Paul Jacobs and Aaron Zitner, "Genome Milestone: Cracking the Code; Scientists Reach Milestone in Mapping of Human Genome; Research: Private Firm and Public Project Finish Remarkable Achievement in a Dead Heat. It Opens the Door to Profound Breakthroughs in Medicine," *Los Angeles Times*, June 27, 2000, Part A; Part I, 1.

61. **Magazine:** Aaron Zitner, "MIT Mathematician Eric Lander Took a Detour into Biology—and Found Himself Leading the Biggest Biological Research Project in History," *Boston Globe Sunday Magazine*, October 10, 1999, 17.

62. **Broadcast:** Harold Varmus and Arnold Levine, interview, *Charlie Rose*, PBS, June 21, 2000, transcript 2709.

63. **Newspaper:** David Baltimore, "50,000 Genes, and We Know Them All (Almost)," *New York Times*, June 25, 2000, Section 4, 17.

64. **Magazine:** David Stipp, "He's Brilliant. He's Swaggering. And He May Soon Be Genomics' First Billionaire," *Fortune*, June 25, 2001, http://money.cnn.com/magazines/fortune_archive/2001/06/25/305422/index.htm.

65. **Interview:** Jim Kent, by the author, University of California, Santa Cruz (UCSC), February 28, 2007, in *Mapping Life Notebook* 5, 218–222.

Newspaper: Nicholas Wade, "Reading the Book of Life; Grad Student Becomes Gene Effort's Unlikely Hero," *New York Times*, February 13, 2001, F1.

66. **Transcripts:** White House, Office of the Press Secretary, "Remarks . . . on the Completion of the First Survey of the Entire Human Genome Project," The East Room, 10:19 a.m.; Press briefing by Neil Lane, Francis Collins, Craig Venter, and Ari Patrinos, in the James S. Brady Briefing Room, 11:08 a.m. EDT; Press briefing, Capitol Hilton, Francis Collins, Ari Patrinos, 12:30 p.m. EDT.

News releases: Celera Genomics, "Celera Genomics Completes the First Assembly of the Human Genome—Assembled Genome Has 3.12 Billion Base Pairs," June 26, 2000; NHGRI, "International Human Genome Sequencing Consortium Announces 'Working Draft' of Human Genome," June 2000, www.genome.gov/10001457; "Caltech and the Human Genome Project," June 26, 2000, http://pr.caltech.edu/media/PressReleases/PR12060.html; advertisement, Compaq Computer, *Wall Street Journal*, June 28, 2000, A21.

Newspapers: *New York Times* coverage, June 27, 2000: Nicholas Wade, "Reading the Book of Life: The Overview; Genetic Code of Human Life Is Cracked by Scientists," A1; Natalie Angier, "The Context, a Pearl and a Hodgepodge: Human DNA," A1; Nicholas Wade, "Now, the Hard Part: Putting the Genome to Work," F1; Nicholas Wade, "Tools Already in Use; Frustration and Rivalry Fueled a Passion for Genetic Discovery," F2; "Whose DNA Is It? In a Way, Nobody's," F2; Nicholas Wade, "The Public Consortium; an Adroit Director of an Unwieldy Team," F3; Nicholas Wade, "A Historic Quest; Double Landmarks for Watson: Helix and Genome," F5. *New York Times* coverage, June 28, 2000: editorial, "The Genome Announcement," A30.

Other newspapers, June 27, 2000: Rick Weiss and Justin Gillis, "Teams Finish Mapping Human DNA; Clinton, Scientists Celebrate 'Working Draft' of Human Genetic Blueprint," *Washington Post*, A1; Richard Saltus, "Decoding of Genome Declared Rivals Present Rough Findings," *Boston Globe*, A1; Ronald Rosenberg and Raja Mishra, "The Beginning of a New Science Mapping Nature's Blueprint for a Human Being Is One of the Biggest Scientific Achievements in a Century, Now Comes the Hard Part: Curing Diseases," *Boston Globe*, C1; Raja Mishra, "The Quest to Map the Human Genome Ends with a Truce," *Boston Globe*, C5; Tom Abate and Carolyn Lochhead, "Scientists Crack Human Gene Code," *San Francisco Chronicle*, A1; Robert Cooke, "Genome Project—a Human Blueprint—'Milestone for Humanity'—2 Research Teams Decode Human Genome, Unveiling New Possibilities for Medicine," *Newsday*, D2; Bryn Nelson, "Genome Project—a Human Blueprint—Moment Awaited for Past Ten Years," *Newsday*, D18; Byron Spice, "Genome Map Is Just a Start, Local Scientists Caution," *Pittsburgh Post-Gazette*, A3; James D. Watson, "Done in My Lifetime; James Watson, Who with Francis Crick Discovered the DNA Double Helix, Applauds Its Decoding 47 Years Later," *Guardian*, 18.

Magazines: Frederic Golden and Michael D. Lemonick, reported by Dick Thompson, "Mapping the Genome: The Race Is Over; the Great Genome Quest Is Officially a Tie, Thanks to a Round of Pizza Diplomacy. Yet Lead Researcher Craig Venter Still Draws Few Cheers from His Colleagues," *Time*, July 3, 2000, 18–23; James D. Watson, "The Double Helix Revisited: The Man Who Launched the Human Genome Project Celebrates Its Success," *Time*, July 3, 2000, 30; Colin MacIlwain, "World Leaders Heap Praise on Human Genome Landmark," *Nature* 405 (June 29, 2000): 983; Elizabeth Pennisi, "Finally, the Book of Life and Instructions for Navigating It. The Publicly and Privately Funded Teams Have Both Finished Drafts of the Human Genome. Now Comes the Daunting Task of Developing Tools to Figure Out Just What These Volumes Say," *Science* 288 (June 30, 2000): 2304; Eliot Marshall, "Human Genome: Rival Genome Sequencers Celebrate a Milestone Together," *Science* 288 (June 30, 2000): 2294; Sharon Begley, "Reality Check on the Genome Project," *Newsweek*, June 27, 2000; editorial, "The End of the Beginning," *Nature Genetics* 25 (August 2000): 363.

Books: Nicholas Wade, *Life Script: How the Human Genome Discoveries Will Transform Medicine and Enhance Your Health* (New York: Simon and Schuster, 2001), 22, 62; Victor K. McElheny, *Watson and DNA: Making a Scientific Revolution* (New York: Perseus, 2003), 268–269.

67. **Interview:** David Haussler, by the author, UCSC, February 28, 2007, in *Mapping Life Notebook* 5, 211–218.

News releases: UCSC, "Human Genome Project Draws on Expertise of UC Santa Cruz; Computer Scientists to Analyze, Assemble Genome Data," AScribe Newswire, June 26, 2000; UCSC, "Working Draft of Human Genome Sequence Available on UC Santa Cruz Web Site," AScribe Newswire, July 6, 2000.

Newspaper: Tom Abate, "UC Santa Cruz Puts Human Genome Online; Programming Wizard Does Job in 4 Weeks," *San Francisco Chronicle*, July 7, 2000, A2.

Chapter 5: Surprises

1. **News release:** Sarah H. Wright, "Lander Celebrates Genome Milestone in Heavily Attended Talk," *MIT Tech Talk*, February 28, 2001, 1.

2. **Scientific article:** The ENCODE Project Consortium, "The ENCODE (Encyclopedia of DNA Elements) Project," *Science* 306 (October 22, 2004): 636.

News releases: NHGRI, "ENCODE Consortium Publishes Scientific Strategy: New Technology Development Grants Will Aid Quest to Find All Functional Elements in Human DNA," October 21, 2004, www.genome.gov/12513444; "Workshop, the Comprehensive Extraction of Biological Information from Genomic Sequence, National Human Genome Research Institute [NHGRI], Bethesda, Md., July 23–24, 2002," www.genome.gov/10005568; NHGRI, "Launch of Pilot Project to Identify All Functional Elements in Human DNA; Encode Project Will

Clear Path for Mining Genomic Sequence," March 4, 2003, www.genome.gov/10506706.

Magazines: Elizabeth Pennisi, "Genome Institute Wrestles Mightily with Its Future," *Science* 298 (November 29, 2002): 1694; Ted Agres, "Navigating the Genomic Road Ahead," *Genomics and Proteomics*, December 1, 2002, 15; Elizabeth Pennisi, "Searching for the Genome's Second Code; the Genome Has More Than One Code for Specifying Life. The Hunt for the Various Types of Noncoding DNA That Control Gene Expression Is Heating Up," *Science* 306 (October 22, 2004): 632; Elizabeth Pennisi, "A Fast and Furious Hunt for Gene Regulators," *Science* 306 (October 22, 2004): 635; Jennifer Couzin, "Functional Genomics: How to Make Sense of Sequence," *Science* 299 (March 14, 2003): 1642.

3. **Scientific article:** "Human Genome," Nature Collections, Supplement to Nature Publishing Group, June 1, 2006, paperback, contains reports in *Nature* of detailed analyses of each human chromosome: X (March 17, 2005); Y (March 17, 2005); 22 (December 2, 1999); 21 (May 18, 2000); 20 (December 20/27, 2001); 19 (April 1, 2004); 18 (September 22, 2005); 17 (April 20, 2006); 16 (December 23/30, 2004); 15 (March 30, 2006); 14 (February 6, 2003); 13 (April 1, 2004); 12 (March 16, 2006); 11 (March 23, 2006); 10 (May 27, 2004); 9 (May 27, 2004); 8 (January 19, 2006); 7 (July 10, 2003); 6 (October 23, 2003); 5 (September 16, 2004); 4 (April 7, 2005); 3 (April 27, 2006); 2 (April 7, 2005); 1 (June 1, 2006).

4. **Magazine:** Elizabeth Pennisi, "Mouse Sequencers Take Up the Shotgun," *Science* 287 (February 18, 2000): 1179.

5. **Newspapers:** Justin Gillis and Terence Chea, "Celera Genomics's President Resigns; Tension with Parent Company's CEO May Have Led to Venter's Departure," *Washington Post*, January 23, 2002, E1; Justin Gillis, "New Chief of Celera to Weigh Overhaul; Review to Follow Ouster of President," *Washington Post*, January 24, 2002, E1; Justin Gillis, "Celera Changed, Venter Couldn't; Departure of Genomics Pioneer Was Sealed When Company Shifted Focus to Drugmaking," *Washington Post*, January 28, 2002, E1; Scott Hensley, "Venter Leaves Celera As Science, Business Clash," *Wall Street Journal*, January 23, 2002, B1; Andrew Pollack, "Scientist Quits the Company He Led in Quest for Genome," *New York Times*, January 23, 2002, C1; Andrew Pollack, "The Genome Is Mapped. Now He Wants Profit," *New York Times*, February 24, 2002, Section 3, 1; Naomi Aoki, "Venter Quits As President of Celera; Genetics Firm Says New Leader Must Focus on Developing Drugs," *Boston Globe*, January 23, 2002, C1; Tom Abate, "Celera Genomics President Steps Down," *San Francisco Chronicle*, January 23, 2002, B1.

Magazines: Carina Dennis, "Venter's Departure Sees Celera Seek Therapies," *Nature* 415 (January 31, 2002): 461; Vicki Brower, "Genomics 'Bad Boy' Movin' On, Venter Exits Firm He Helped Found," McGraw Hill's *Biotechnology Newswatch*, February 4, 2002, 1; John Carey, "Craig Venter: Decoupling a Gene Decoder," *Business Week*, February 4, 2002, 46.

Books: James Shreeve, *The Genome War: How Craig Venter Tried to Capture the Genome and Save the World* (New York: Knopf, 2004), 369–371; J. Craig Venter, *A Life Decoded, My Genome: My Life* (New York: Viking, 2007), 327–330.

6. **News releases:** ExxonMobil, "ExxonMobil to Launch Biofuels Program," Business Wire, July 14, 2009; Synthetic Genomics, Inc., "Synthetic Genomics Inc and ExxonMobil Research and Engineering Company Sign Exclusive, Multi-Year Agreement to Develop Next Generation Biofuels Using Photosynthetic Algae," July 14, 2009.

Wire service: "Synthetic Genomics' Exxon Biofuel Pact Worth Up to $300M," Genome Web, July 14, 2009.

Newspapers: Jad Mouawad, "A Biofuel Drop in the Bucket?" *New York Times*, July 14, 2009, B1; Carl Mortished, "Biggest Oil Company Puts Its Faith in Algae," *Times*, July 15, 2009, 3; Alok Jha, "Exxon Joins Gene Scientist to Create Algae Biofuel," *Guardian*, July 15, 2009, 24; Sheila McNulty and Ed Crooks, "News Analysis: Exxon Faces Questions over Goals," *Financial Times*, July 31, 2009, 16.

7. **Scientific articles:** F. Sanger et al., "Nucleotide Sequence of Bacteriophage Phi X 174 DNA," *Nature* 265 (February 24, 1977): 687; Susan M. Berget, Claire Moore, and Phillip A. Sharp, "Spliced Segments at the 5-Prime Terminus of Adenovirus 2 Late mRNA," *Proceedings of the National Academy of Sciences USA* 74 (August 1977): 3171; Richard E. Gelinas and Richard J. Roberts "One Predominant 5-Prime Undecanucleotide in Adenovirus 2 Late Messenger RNAs," *Cell* 11 (July 1977): 533–544; Louise T. Chow, James M. Roberts, James B. Lewis, and Thomas R. Broker, "A Map of Cytoplasmic RNA Transcripts from Lytic Adenovirus Type 2 Determined by Electron Microscopy of RNA-DNA Hybrids," *Cell* 11 (August 1977): 819; Walter Gilbert, "News and Views: Why Genes in Pieces?" *Nature* 271 (February 9, 1978): 501.

Nobel lectures: Phillip A. Sharp, "Split Genes and RNA Splicing," December 8, 1993, Nobel Foundation, http://nobelprize.org/nobel_prizes/medicine/laureates/1993/sharp-lecture.html; Richard J. Roberts, "An Amazing Distortion in DNA Induced by a Methyltransferase," December 8, 1993, Nobel Foundation, http://nobelprize.org/nobel_prizes/medicine/laureates/1993/roberts-lecture.html.

Magazines: Maria Szekely, "News and Views: Phi X 174 Sequenced," *Nature* 265 (February 24, 1977): 685; John Rogers, "Atomic Structure of a Living Organism," *New Scientist*, March 10, 1977, 583.

Book: James D. Watson, Amy A. Caudy, Richard M. Myers, and Jan A. Witkowski, *Recombinant DNA*, 3rd ed. (New York: W. H. Freeman and Company, 2007), 252.

8. **Scientific articles:** International Human Genome Sequencing Consortium et al., "Initial Sequencing and Analysis of the Human Genome," *Nature* 409 (February 15, 2001): 860; J. Craig Venter et al., "The Human Genome Sequence," *Science* 291 (February 16, 2001): 1304; Robert H. Waterston, Eric S. Lander, and John E. Sulston, "Genetics: On the Sequencing of the Human Genome," *Pro-*

ceedings of the National Academy of Sciences USA 99 (March 19, 2002): 3712; Phil Green, "Commentary: Whole Genome Disassembly," *Proceedings of the National Academy of Sciences USA* 99 (April 2, 2002): 4143; Eugene W. Myers et al., "Commentary: On the Sequencing and Assembly of the Human Genome," *Proceedings of the National Academy of Sciences USA* 99 (April 2, 2002): 4145; Robert H. Waterston, Eric S. Lander, and John E. Sulston, "More on the Sequencing of the Human Genome," *Proceedings of the National Academy of Sciences USA* 100 (March 18, 2003): 3022; Mark D. Adams et al., "The Independence of Our Genome Assemblies," *Proceedings of the National Academy of Sciences USA* 100 (March 18, 2003): 3025.

A human gene count of 20,500 is given in Michele Clamp et al., "Genetics: Distinguishing Protein-Coding and Noncoding Genes in the Human Genome," *Proceedings of the National Academy of Sciences USA* 104 (December 4, 2007): 19428.

News release: Broad Institute, "Human Gene Count Tumbles Again; New Analysis Reveals Several Thousand Genes to Be Spurious, Leads to Gene Count Revision," www.broadinstitute.org/news/102.

Wire service: Associated Press, "Scientists Say Decoders Needed Them," *New York Times*, March 5, 2002, A8.

Newspapers: Rick Weiss, "Restrictions on DNA Data Reignite Debate; Scientific Journal Lets For-Profit Firm Withhold Code, Angering Public-Sector Researchers," *Washington Post*, December 8, 2000, A2; Gina Kolata, "Celera to Charge Other Companies to Use Its Genome Data," *New York Times*, December 8, 2000, A26; Scott Hensley and Antonio Regalado, "Scientists Publish Critique of Celera's Work: Rivals Charge Firm Recycled Public Data in Genome Map," *Wall Street Journal*, March 5, 2002, A2; Justin Gillis, "Rivals Resume Battle of the Genome Map," *Washington Post*, March 5, 2002, E1; Steve Connor, "American Firm 'Cheated' in Race to Map Genome," *Independent*, March 5, 2002, 8.

Magazines: Eliot Marshall, "Human Genome: Storm Erupts over Terms for Publishing Celera's Sequence," *Science* 290 (December 15, 2000): 2042; Jennifer Couzin, "Genomics: Taking Aim at Celera's Shotgun," *Science* 295 (March 8, 2002): 1817.

Books: Shreeve, *The Genome War*, 363–367; Venter, *A Life Decoded*, 323–324.

9. **Magazine:** Nancy J. Nelson, "Draft Human Genome Sequence Yields Several Surprises," *Journal of the National Cancer Institute* 93 (April 4, 2001): 493.

10. **Interview:** Phil Green, by the author, University of Washington, Seattle, February 13, 2004.

11. **Scientific article:** David Baltimore, "News and Views: Our Genome Unveiled," *Nature* 409 (February 15, 2001): 814–816.

12. **News release:** NHGRI, "Mouse: Next in Line for DNA Sequencing," September 1999, www.genome.gov/10002108.

Magazines: Elizabeth Pennisi, "Mouse Genome Added to Sequencing Effort," *Science* 286 (October 8, 1999): 210; Elizabeth Pennisi, "Gene Skews Patterns of

Inheritance," *Science* 286 (November 12, 1999): 1269; Elizabeth Pennisi, "Mouse Economy: A Mouse Chronology," *Science* 288 (April 14, 2000): 248.

Book: Glyn Moody, *Digital Code of Life: How Bioinformatics Is Revolutionizing Science, Medicine, and Business* (Hoboken, NJ: Wiley, 2004), 130–131.

13. **News release:** NHGRI, "Public-Private Consortium to Accelerate Sequencing of Mouse Genome; Results Will Expedite Discovery of Human Genes," October 2000, www.genome.gov/10001343.

Newspapers: Byron Spice, "$58 Million Mouse Project Could Aid in Deciphering Human Genome," *Pittsburgh Post-Gazette*, October 6, 2000, A4; Justin Gillis, "Public-Private Group to Unravel Mouse Genome," *Washington Post*, October 6, 2000, E3; Justin Gillis and Terence Chea, "Gene Researcher Celera Prepares to Map a Mouse," *Washington Post*, October 13, 2000, E5.

Magazine: Eliot Marshall, "Public-Private Project to Deliver Mouse Genome in 6 Months," *Science* 290 (October 13, 2000): 242.

Book: Moody, *Digital Code*, 131.

14. **Newspapers:** Justin Gillis, "Celera Cites Progress on Mouse Genes; Similarity to Humans Makes Map Important," *Washington Post*, June 2, 2000, E3; Nicholas Wade, "Genetic Sequence of Mouse Is Also Decoded," *New York Times*, February 13, 2001, F5.

Wire Service: Malcolm Ritter, "Scientists Close to Deciphering Mouse Genetic Code," Associated Press, February 12, 2001.

15. **Magazine:** Eliot Marshall, "Genome Sequencing: Celera Assembles Mouse Genome; Public Labs Plan New Strategy," *Science* 292 (May 4, 2001): 822.

16. **Newspaper:** Justin Gillis, "Public-Private Group Finishes Rough Map of Mouse Genome," *Washington Post*, May 9, 2001, A7.

17. **News release:** NHGRI, "International Team of Researchers Assembles Draft Sequence of Mouse Genome: Ninety-Six Percent of Mouse Genome Sequenced; Results Freely Available on Public Databases on Internet," May 6, 2002, www.genome.gov/10002983.

Newspapers: Nicholas Wade, "Mouse Genome Is New Battleground for Project Rivals," *New York Times*, May 7, 2002, F2; Justin Gillis, "Scientists Compile Map of Mouse's Genome; Placed in Public Domain, Data on Key Lab Animal Has Implications for Human Medical Research," *Washington Post*, May 7, 2002, A6.

Magazines: Erika Check, "Draft Mouse Genome Makes Public Debut," *Nature* 417 (May 9, 2002): 106; Eliot Marshall, "Genome Sequencing: Public Group Completes Draft of the Mouse," *Science* 296 (May 10, 2002): 1005; Elizabeth Pennisi, "Charting a Genome's Hills and Valleys. Comparisons of the Sequences of the Mouse and Human Genomes Have Turned Up Unexpected Features," *Science* 296 (May 31, 2002): 1601.

18. **Scientific article:** Mouse Genome Sequencing Consortium, "Initial Sequencing and Comparative Analysis of the Mouse Genome," *Nature* 420 (December 5, 2002): 520.

Interview: David Haussler, by the author, University of California, Santa Cruz (UCSC), February 28, 2007, in *Mapping Life Notebook* 5, 218.

News releases: Tim Stephens, "Mouse Genome Sequence Published with First Comparative Analysis of Mouse and Human Genomes," UCSC, http://press.ucsc.edu/text.asp?pid=272; Tim Stephens, "Grad Student's Work on High-Profile Projects Started When He Was an Undergraduate," *Santa Cruz Currents*, January 19, 2004, http://currents.ucsc.edu/03–04/01–19/roskin_profile.html.

19. **Magazine:** Check, "Draft Mouse Genome."

20. **Magazine:** Pennisi, "Charting a Genome's Hills and Valleys," 1601.

21. **Website:** National Medals, www.nationalmedals.org/medals/1963.php.

22. The genetics award from the Gruber Foundation, begun in 2001, went to David Botstein in 2003, Mary-Claire King in 2004, Robert H. Waterston in 2005, and Maynard V. Olson in 2007. The Allan Prize Lectureship of the American Society of Human Genetics was awarded to Victor McKusick in 1977, Botstein and Ray White in 1989, Kary B. Mullis in 1990, Bert Vogelstein in 1998, Lou Kunkel in 2004, and Francis Collins in 2005. For a torrent of genome-related innovations, Lee Hood received MIT's Lemelson Prize in 2003. Sydney Brenner, Eric Lander, and Craig Venter shared the Novartis Drew biomedical research award in 2001. Winners of the 2002 Dan David Prize were John E. Sulston, Brenner, and Waterston.

23. **News releases:** "Gairdner Foundation Announces 2002 International Awards for Excellence in Medical Science," Canada News Wire, April 22, 2002; "Green and Olson to Receive Gairdner Awards," University of Washington, *University Week* 19, no. 24 (April 25, 2002).

24. **Newspapers:** J. Craig Venter, "We Are All Virtually Identical Twins," *Globe and Mail* (Canada), April 27, 2002, F6; Brad Evenson, "Nine Feuding DNA Pioneers Win Gairdner: Genome Scientists Continue to Argue over Sharing Credit," *National Post*, April 24, 2002, A7; Carolyn Abraham, "Human Genome Decoder Going Nonprofit; Company Founder to 'Educate the Public' on Ethical Issues," *Globe and Mail* (Canada), April 24, 2002, A8; Lynda Hurst, "Genome Just a Start for Maverick Scientist," *Toronto Star*, April 24, 2002, A2.

25. **Newspaper:** Tom Paulson, "Seattle Scientists at Forefront of Human Genome Project; Researchers Have Already Cracked the Human Genetic Code, but Now They Have a Better Description of It Than the One Made in 2001," *Seattle Post-Intelligencer*, April 15, 2003, A1.

26. **News release:** NHGRI, "International Consortium Completes Human Genome Project; All Goals Achieved; New Vision for Genome Research Unveiled," April 14, 2003, www.genome.gov/11006929.

Newspapers: Paulson, "Seattle Scientists"; Nicholas Wade, "Once Again, Scientists Say Human Genome Is Complete," *New York Times*, April 15, 2003, F1; Rick Weiss, "Genome Project Completed; Findings May Alter Humanity's Sense of Itself, Experts Predict," *Washington Post*, April 15, 2003, A6; Roxanne Roberts, "The

Double Helix's 50th: A Party with a Twist," *Washington Post*, April 15, 2003, C1; Steve Sternberg, "'First Edition' of Genome Goes to Press," *USA Today*, April 15, 2003, 8D; Earl Lane, "New Landmark for Genome Project," *Newsday*, April 15, 2003, A9; Bernadette Tansey, "Genome Project Completes Map of Human DNA; Celebrating Scientists Expect Medicine to Be Transformed As Result," *San Francisco Chronicle*, April 15, 2003, A2; Todd Ackerman, "Road Map to Human Life Complete," *Houston Chronicle*, April 15, 2003, A1; Francis Collins, "A Common Thread," *Washington Post*, April 22, 2003, A19.

Magazine: Elizabeth Pennisi, "Reaching Their Goal Early, Sequencing Labs Celebrate," *Science* 300 (April 18, 2003): 409.

27. **Scientific article:** International Human Genome Sequencing Consortium, "Finishing the Euchromatic Sequence of the Human Genome," *Nature* 431 (October 21, 2004): 931.

28. **News release:** NHGRI, "International Consortium Completes."

Newspapers: Peter N. Spotts, "Complete Genome Map Opens Roads for Science," *Christian Science Monitor*, April 15, 2003, 2; Roger Highfield, "Human Code Is Spelled Out in 3bn DNA 'Letters,'" *Daily Telegraph*, April 15, 2003, 13; Paulson, "Seattle Scientists"; Wade, "Once Again"; Sternberg, 'First Edition' of Genome Goes to Press"; Tom Paulson, "Gates Gives UW $70 Million; Grant Could Make Seattle a Leading Center of Human Genome Research," *Seattle Post-Intelligencer*, April 24, 2003, A1; Roberts, "The Double Helix's 50th."

Magazines: Pennisi, "Reaching Their Goal Early."

29. **News release:** NHGRI, "International Consortium Completes."

30. **Newspaper:** Wade, "Once Again."

31. **Newspaper:** Wade, "Once Again."

32. **Newspaper:** Wade, "Once Again."

33. **Newspaper:** Paulson, "Seattle Scientists."

34. **Magazine:** "This Is It. Honest," *The Economist* (U.S. edition), April 19, 2003.

35. **Newspaper:** Highfield, "Human Code."

36. **Scientific article:** Mitchell Guttman et al., including Aviv Regev, John L. Rinn, and Eric S. Lander, "Chromatin Signature Reveals over a Thousand Highly Conserved Large Non-Coding RNAs in Mammals," *Nature* 458 (March 12, 2009): 223.

News release: Leah Eisenstadt, "Missing Links of the Transcriptome: Clues in Chromatin Reveal New Class of RNA Molecules, Nearly 1,600 Strong," Broad Institute, February 1, 2009, www.broadinstitute.org/news/1213.

37. **Scientific articles:** The Chimpanzee Sequencing and Analysis Consortium, "Initial Sequence of the Chimpanzee Genome and Comparison with the Human Genome," *Nature* 437 (September 1, 2005): 69; Ze Cheng et al., including Evan E. Eichler, "A Genome-Wide Comparison of Recent Chimpanzee and Human Segmental Duplications," *Nature* 437 (September 1, 2005): 88; Wen-Hsiung Li and Matthew A. Saunders, "News and Views: The Chimpanzee and Us," *Nature* 437

(September 1, 2005): 50; Ewen F. Kirkness et al., including Clare M. Fraser and J. Craig Venter, "The Dog Genome: Survey Sequencing and Comparative Analysis," *Science* 301 (September 23, 2003): 1998; Kerstin Landblad-Toh et al., including Elaine A. Ostrander, "Genome Sequence, Comparative Analysis and Haplotype Structure of the Domestic Dog," *Nature* 438 (December 8, 2005): 803; Hans Ellegren, "The Dog Has Its Day," *Nature* 438 (December 8, 2005): 745.

News release: NHGRI, "Researchers Publish Dog Genome Sequence; Analysis Sheds Light on Human Disease, Differences Among Canine Breeds," December 7, 2005, www.genome.gov/17515860.

Newspapers: David Perlman, "Man, Chimps Share Genes; Comparison of Genomes Shows Thousands Are Virtually Identical," *San Francisco Chronicle*, September 1, 2005, A2; Tim Friend, "Dogs Share More Than Affection with Us," *USA Today*, September 26, 2003, 3A; Nicholas Wade, "Scientific Team Puts Together a Rough Draft of a Dog Genome," *New York Times*, September 26, 2003, 16; Nicholas Wade, "Dog's Genome Could Provide Clues to Disorders in Humans," *New York Times*, December 8, 2005, A37.

Magazine: Sylvia Pagan Westphal, "Genome of Man's Best Friend Is Still Riddled with Gaps," *New Scientist*, October 4, 2003, 14.

38. **News release:** "454 Life Sciences and Max Planck Institute Complete Draft 1X Sequence of Neanderthal Genome; Completion of More Than 3 Billion Bases of Sequenced Neanderthal DNA Brings Closure to the Sequencing Stage of an Ambitious Project Initiated Two and a Half Years Ago," Business Wire, February 12, 2009.

Newspapers: Nicholas Wade, "Neanderthal Genome Hints at Language Potential but Little Human Interbreeding," *New York Times*, February 13, 2009, A12; Steve Connor, "Science Unlocks Neanderthal Secrets; for the First Time, the Genetic Blueprint of an Extinct Human Species Has Been Discovered," *Independent*, February 13, 2009, 2.

Magazines: Elizabeth Pennisi, "Tales of a Prehistoric Human Genome," *Science* 323 (February 13, 2009): 866; Elizabeth Pennisi, "Wanted: Clean Neanderthal DNA," *Science* 323 (February 13, 2009): 868.

39. **Scientific articles:** Jonathan Sebat et al., including Michael Wigler, "Large-Scale Copy Number Polymorphism in the Human Genome," *Science* 305 (July 23, 2004): 525; Nigel P. Carter, "As Normal As Normal Can Be?" *Nature Genetics* 36 (September 2004): 931; A. John Iafrate et al., "Detection of Large-Scale Variation in the Human Genome," *Nature Genetics* 36 (September 2004): 949; Joseph H. Nadeau and Charles Lee, "Copies Count," *Nature* 439 (February 16, 2006): 798; Kevin V. Shianna and Huntington F. Willard, "Human Genomics: In Search of Normality," *Nature* 444 (November 23, 2006): 428; Richard Redon et al., including Stephen W. Scherer and Matthew E. Hurles, "Global Variation in Copy Number in the Human Genome," *Nature* 444 (November 23, 2006): 444; Barbara E. Stranger et al., including Emmanouil Dermitzakis, "Relative Impact of Nucleotide and Copy

Number Variation on Gene Expression Phenotypes," *Science* 315 (February 9, 2007): 848; Jonathan Sebat, "Commentary, Major Changes in Our DNA Lead to Major Changes in Our Thinking," *Nature Genetics Supplement* 39 (July 2007): S3–S5; Stephen W. Scherer et al., "Challenges and Standards in Integrating Surveys of Structural Variation," *Nature Genetics Supplement* 39 (July 2007): S7–S15; Nigel P. Carter, "Methods and Strategies for Analyzing Copy Number Variation Using DNA Microarrays," *Nature Genetics Supplement* 39 (July 2007): S16–S21; Gregory M. Cooper, Deborah A. Nickerson, and Evan E. Eichler, "Mutational and Selective Effects on Copy-Number Variants in the Human Genome," *Nature Genetics Supplement* 39 (July 2007): S22–S29; Donald F. Conrad and Matthew E. Hurles, "The Population Genetics of Structural Variation," *Nature Genetics Supplement* 39 (July 2007): S30–S36; Steven A. McCarroll and David M. Altshuler, "Copy-Number Variation and Association Studies of Human Disease," *Nature Genetics Supplement* 39 (July 2007): S37–S42; James R. Lupski, "Genomic Rearrangements and Sporadic Disease," *Nature Genetics Supplement* 39 (July 2007): S43–S47; Charles Lee, A. John Iafrate, and Arthur R. Brothman, "Copy Number Variations and Clinical Cytogenetic Diagnosis of Constitutional Disorders," *Nature Genetics Supplement* 39 (July 2007): S48–S54.

Magazine: Erika Check, "Large Genomic Differences Explain Our Little Quirks," *Nature* 435 (May 19, 2005): 252.

40. **Interview:** Thomas Gingeras, by the author, Affymetrix, Inc., Santa Clara, California, February 27, 2007, transcript and in *Mapping Life Notebook* 5, 197–205.

41. **Scientific articles:** Papers authored or coauthored by Thomas Gingeras: Robert J. Lipschutz et al., "High Density Synthetic Oligonucleotide Arrays," *Nature Genetics* 21 (January 1999): 20; Philipp Kapranov et al., "Large-Scale Transcriptional Activity in Chromosomes 21 and 22," *Science* 296 (May 3, 2002): 916.

Interview: Gingeras, by the author, February 27, 2007.

42. **Scientific articles:** Papers authored or coauthored by Thomas Gingeras: Philipp Kapranov et al., "Beyond Expression Profiling: Next Generation Uses of High Density Oligonucleotide Arrays," *Briefings in Functional Genomics and Proteomics* 2 (April 2005): 47; Simon Cawley et al., "Unbiased Mapping of Transcription Factor Binding Sites Along Human Chromosomes 21 and 22 Points to Widespread Regulation of Noncoding RNAs," *Cell* 116 (February 20, 2004): 499; Jill Cheng et al., "Transcriptional Maps of 10 Human Chromosomes at 5-Nucleotide Resolution," *Science* 308 (May 20, 2005): 1149; "The Multitasking Genome," *Nature Genetics* 38 (June 2006): 608; J. Robert Manak et al., "Biological Function of Unannotated Transcription During the Early Development of *Drosophila melanogaster*," *Nature Genetics* 38 (September 3, 2006): 1151; Philipp Kapranov et al., "Genome-Wide Transcription and the Implications for Genomic Organization," *Nature Reviews Genetics* 8 (June 2008): 413; Philipp Kapranov et al., "RNA Maps Reveal New RNA Classes and a Possible Function for Pervasive Transcription," *Science* 316 (June 8, 2007): 1484; "Implications of Chimaeric Non-Co-Linear Transcripts," *Na-*

ture 461 (September 10, 2009): 206. Paper by other authors: Michael Pheasant and John S. Mattick, "Raising the Estimate of Functional Human Sequences," *Genome Research* 17 (September 2007): 1245.

43. **Speech:** Thomas Gingeras, keynote address delivered at "The Biology of Genomes" meeting, Cold Spring Harbor Laboratory, May 14, 2005, in *Mapping Life Notebook* 3, 235–238.

44. **Scientific articles:** John M. Greally, "News and Views; Genomics: Encyclopaedia of Humble DNA," *Nature* 447 (June 14, 2007): 782; The ENCODE Project Consortium, "Identification and Analysis of Functional Elements in 1% of the Human Genome by the ENCODE Pilot Project," *Nature* 447 (June 14, 2007): 799; George M. Weinstock, "ENCODE: More Genomic Empowerment," *Genome Research* 17 (June 2007): 667; Mark B. Gerstein et al., including Sherman Weissman and Michael Snyder, "What Is a Gene, Post-ENCODE? History and Updated Definition," *Genome Research* 17 (June 2007): 669; Thomas Gingeras, "Perspective: Origin of Phenotypes: Genes and Transcripts," *Genome Research* 17 (June 2007): 682; Steven Henikoff, "News and Views; ENCODE and Our Very Busy Genome," *Nature Genetics* 39 (July 2007): 817.

News releases: "Highlights from the ENCODE-Related Papers in *Genome Research*," Genome Research, June 13, 2007, www.genome.org; other releases from NHGRI; European Bioinformatics Institute; Wellcome Trust Sanger Institute; University of California, Davis; University of Washington; University of Virginia; Affymetrix, Inc.; Yale University; Boston University; NimbleGen Systems, Inc.

Wire services: "Human Genome Not So Tidy After All, ENCODE Project Suggests," GenomeWeb.com, June 13, 2007; Richard Ingham, "Landmark Study Prompts Rethink of Genetic Code," Agence France Presse—English, June 13, 2007; John Lauerman, "'Junk DNA' Isn't Junk; Purpose Seen in Genetic Silent Majority," Bloomberg.com, June 13, 2007; Daniel J. DeNoon, "Genetics Revolution Arrives: First Chapter of Human DNA 'Encyclopedia' Rewrites Biology," WebMD.com, June 13, 2007.

Broadcast: Ira Flatow, "Reading Between the Genomes," *Talk of the Nation: Science Friday*, National Public Radio, June 14, 2007.

Newspapers: Rick Weiss, "Intricate Toiling Found in Nooks of DNA Once Believed to Stand Idle," *Washington Post*, June 14, 2007, A1; Joseph Hall, "DNA 'Junk' Appears to Have Uses," *Toronto Star*, June 14, 2007, A22; Colin Nickerson, "DNA Study Challenges Basic Ideas in Genetics; Genome 'Junk' Appears Essential," *Boston Globe*, June 14, 2007, A1; Roger Highfield, "Life Just Got a Lot More Complicated; the Deeper Scientists Look into the Human Genetic Code, the More Baffling It Becomes—But It Is Also Increasing Their Understanding of Disease," *Daily Telegraph*, June 19, 2007, 29.

News accounts and commentary also appeared in Switzerland, Italy, Spain, and Germany.

Magazines: Erika Check, "Genome Project Turns Up Evolutionary Surprises; Findings Reveal How DNA Is Conserved Across Animals," *Nature* 447 (June 14, 2007): 760; Elizabeth Pennisi, "DNA Study Forces Rethink of What It Means to Be a Gene," *Science* 316 (June 15, 2007): 1556; Melissa Lee Phillips, "First Pages of Regulation 'Encyclopedia,'" the-scientist.com, June 13, 2007, www.the-scientist .com/news/home/53280; Andy Coghlan, "'Junk' DNA Makes Compulsive Reading," *New Scientist*, June 16, 2007, 20.

45. **Interview:** Thomas Gingeras, by the author, Cold Spring Harbor Laboratory, Cold Spring Harbor, New York, May 6, 2009, in *Mapping Life Notebook* 8, 258–270.

Chapter 6: Human Variety

1. **Books:** John Cairns, Joseph L. Lyon, and Mark Skolnick, eds., *Cancer Incidence in Defined Populations*, Banbury Report 4 (Cold Spring Harbor, NY: Cold Spring Harbor Laboratory Press, 1980); John Cairns, *Cancer: Science and Society* (San Francisco: W. H. Freeman, 1978).

2. **News release:** "Millennium Evening at the White House, Informatics Meets Genomics," Clinton Presidential Materials Project, October 12, 1999, http://clinton6 .nara.gov/1999/10/1999–10–12-millennium-evening-transcript-informatics-meets -genomics.html (accessed January 9, 2009).

3. The author was present.

4. **Newspaper:** Keith O'Brien, "Will His Map Lead to Medical Treasure?" *Boston Globe*, November 14, 2005, C2 (interview with David Altshuler).

Magazine: Joannie Fischer, "Snipping Away at Human Disease," *U.S. News & World Report* 129 (October 23, 2000): 58.

5. **Interview:** David Altshuler, by the author, Broad Institute, Cambridge, Massachusetts, October 25, 2006, transcript and in *Mapping Life Notebook* 5, 18–25.

6. **Scientific article:** Eric S. Lander, "The New Genomics: Global Views of Biology," *Science* 274 (October 25, 1996): 536.

7. **Scientific articles:** Mark Daly et al., "High-Resolution Haplotype Structure in the Human Genome," *Nature Genetics* 29 (October 2001): 229; Stacey Gabriel et al., including Mark Daly and David Altshuler, "The Structure of Haplotype Blocks in the Human Genome," *Science* 296 (June 24, 2002): 2225 (published online May 23, 2002); Pui-Yan Kwok, "Genetic Association by Whole-Genome Analysis?" *Science* 294 (November 23, 2001): 1689; Nila Patil, "Blocks of Limited Haplotype Diversity Revealed by High-Resolution Scanning of Human Chromosome 21," *Science* 294 (November 23, 2001): 1719.

Newspapers: Geeta Anand, "U.S. to Help Fund Effort to Map Slight Variations in Human Genes," *Wall Street Journal*, June 28, 2001, B4; Nicholas Wade, "Gene-Mappers Take New Aim at Diseases," *New York Times*, October 30, 2002, A23.

Magazines: "Genome-Wide Survey Provides Support for Building Haplotype Map," *MIT Tech Talk*, June 12, 2002; David Adam, "Genetics Group Targets Disease Markers in the Human Sequence," *Nature* 411 (July 12, 2001): 105; Laura Helmuth, "Map of the Human Genome 3.0," *Science* 293 (July 27, 2001): 583 (reporting on the NIH "Developing a Haplotype Map of the Human Genome for Finding Genes Related to Health and Disease" meeting, July 18–19, 2001); Marina Chicurel, "Faster, Better, Cheaper Genotyping," *Nature* 412 (August 9, 2001): 580; Jennifer Couzin, "New Mapping Project Splits the Community; a New Type of Map, Known As the Haplotype Map, Promises to Speed the Search for Elusive Genes Involved in Complex Diseases. But Some Geneticists Question Whether It Will Work," *Science* 296 (May 24, 2002): 1391; Xavier Bosch, "Researchers Try to Unlock Human Genome Potential," *Lancet* 360 (November 9, 2002): 1481; Nell Boyce, "All the Difference in the World," *U.S. News & World Report* 133 (November 11, 2002): 80.

8. **Scientific articles:** Richard Gibbs, "Deeper into the Genome," *Nature* 437 (October 27, 2005): 1233; David B. Goldstein and Gianpiero L. Cavalleri, "News and Views: Genomics: Understanding Human Diversity," *Nature* 437 (October 27, 2005): 1241; International HapMap Consortium, "A Haplotype Map of the Human Genome," *Nature* 437 (October 27, 2005): 1299; International HapMap Consortium, "A Second Generation Human Haplotype Map of over 3.1 Million SNPs," *Nature* 449 (October 18, 2007): 851; Pardis C. Sabeti et al., "Genome-Wide Detection and Characterization of Positive Selection in Human Populations," *Nature* 449 (October 18, 2007): 913.

News releases: October 17, 2007: NHGRI, "Consortium Publishes Phase II Map of Human Genetic Variation; New Map Improves Power to Find Variants Involved in Common Diseases, Reveals More Signs of Adaptive Evolution"; Broad Institute, "New Chapter for HapMap: Phase 2 Effort Yields Dense Map of Human Variation, Enables Deeper Insights into Human Disease and Evolution"; Perlegen Sciences, Inc., "Perlegen's Contribution to Phase 2 HapMap of Human Genetic Variation Empowers Scientists Studying Common Diseases." October 26, 2005: NHGRI; Harvard University; Broad Institute; Wellcome Trust Sanger Institute; Baylor College of Medicine; Broad Institute, "David Altshuler Comments on the HapMap."

Broadcasts: Francis Collins, interview by Joe Palca, "Scientists Create New Gene Map," *Morning Edition*, National Public Radio, October 27, 2005; David Altshuler and David Goldstein of Duke University, interviews by Ira Flatow, "HapMap Project Catalogs Genetic Similarities and Differences in Human Beings," *Talk of the Nation: Science Friday*, National Public Radio, October 28, 2005.

Newspapers: Gareth Cook, "Geneticists Map What Makes Us Different: Advance May Aid Study of Diseases," *Boston Globe*, October 27, 2005; Nicholas Wade, "Genetic Catalog May Aid Search for Roots of Disease," *New York Times*, October 27, 2005, A20; Thomas H. Maugh II, "Disease Hunters Get a New Genetic Map; the

Blueprint Is Called a 'Phenomenal Tool' That Will Save Researchers Time and Money," *Los Angeles Times*, October 27, 2005, A20; Steve Connor, "Scientists Reveal Genetic Map That Will Unlock the Secrets of Human Diversity," *Independent*, October 27, 2005, 7; Margaret Munro, "Powerful New Tool Hunts Genes That Cause Illnesses: Toronto Discovery Hailed As Key to New Era in Diagnosis, Treatment," *Ottawa Citizen*, October 27, 2005, A1.

Magazines: Jennifer Couzin, "Genomics: New Haplotype Map May Overhaul Gene Hunting," *Science* 310 (October 28, 2005): 601; Elizabeth G. Phimister, "Genomic Cartography—Presenting the HapMap," *New England Journal of Medicine* 353 (October 27, 2005): 1766; "Finding Needles in a Haystack; Human Genetics," *The Economist*, October 29, 2005; Erika Check Hayden, "So Similar, Yet So Different: Tiny Pieces of the Genome Can Already Explain Many Human Characteristics," *Nature* 449 (October 18, 2007): 762.

9. **Scientific article:** International SNP Map Working Group, "A Map of Human Genome Sequence Variation Containing 1.42 Million Single Nucleotide Polymorphisms," *Nature* 409 (February 15, 2001): 928.

10. **Magazine:** Couzin, "New Mapping Project."

11. **News release:** "Affymetrix to Introduce 1 Million-SNP Product in Early 2007 and Single 500K Array Before End of 2006; Affymetrix and Broad Institute Collaborate to Increase Power of Whole-Genome Association Studies," Business Wire, July 18, 2006.

12. **Magazine:** Miranda Robertson writes, "There have been two outstanding contributions of medical genetics to science. One is the analysis of mutant haemoglobins; the other is the remarkable achievements of Michael Brown and Joseph Goldstein (University of Texas, Dallas) who uncovered the hypercholesterolemia that predisposes to heart disease" ("The Proper Study of Mankind. Molecular Biology Has Made the Human Genome Accessible to Laboratory Investigation. But Does That Mean That the Sequence of the Human Genome Would Be Worth the Effort?" *Nature* 322 [July 3, 1986]: 11).

13. **Website:** IVF1 at www.ivf1.com/ivf; Fergus Walsh, "30th Birthday for First Test-Tube Baby," British Broadcasting Corporation, July 14, 2008, http://news.bbc.co.uk/2/hi/health/7505635.stm.

Newspapers: John Naish, "The Hidden Fallout of Failed Fertility Treatment," *Times*, July 19, 2008, 4; Mark Henderson, "Race for the Secrets of IVF Success As the First Test-Tube Baby Turns 30," *The Times* (of London), July 19, 2008, 28.

Magazines: "Life After Superbabe," *Nature* 454 (July 17, 2008): 253; Helen Pearson, "Making Babies: The Next 30 Years," *Nature* 454 (July 17, 2008): 260; Ruth Deech, "30 Years: From IVF to Stem Cells," *Nature* 454 (July 17, 2008): 280; "Assisted Reproduction," *The Economist*, July 19, 2008; Peggy Orenstein, "In Vitro We Trust," *New York Times Magazine*, July 20, 2008, 11; "No IVF Please, We're British," *The Economist* (U.S. edition), July 19, 2008.

14. **Scientific articles:** Alec J. Jeffreys, Victoria Wilson, and Swee Lay Thein, "Hypervariable 'Minisatellite' Regions in Human DNA," *Nature* 314 (March 7, 1985): 67; C. Thomas Caskey, "Genetics: New Aid to Human Genome Mapping," *Nature* (March 7, 1985): 11; Peter Gill, Alec J. Jeffreys, and David J. Werrett, "Forensic Application of DNA 'Fingerprints,'" *Nature* 318 (December 12, 1985): 577; Barbara E. Dodd, "Forensic Science: DNA Fingerprinting in Matters of Family and Crime," *Nature* 318 (December 12, 1985): 506.

Newspapers: Andrew Veitch, "Genetic 'Print' That Could Point the Finger; British Scientists Discover Method of Identifying People by Their Genes," *Guardian*, March 11, 1985; Andrew Veitch, "Son Rejoins Mother As Genetic Test Ends Immigration Dispute; Ghanian Boy Allowed to Join Family in Britain," *Guardian*, October 31, 1985; Andrew Veitch, "Rapists May Be Trapped by Genes," *Guardian*, December 13, 1985; Peter Marsh, "Genetic 'Fingerprint' May Aid Crime Fighters," *Financial Times*, December 16, 1985, Section 2, 20; Boyce Rensberger, "Genealogical Fingerprints," *Washington Post*, December 23, 1985, A7; Lawrence K. Altman, "New DNA Test Offers Biological 'Fingerprints' for Crime Fight," *New York Times*, February 4, 1986, C1; "Invisible Print of a Killer: Joseph Wambaugh Discusses Death with Robert Winder," *Independent*, February 22, 1989, 21.

Magazine: Carla Straessle, "A Genetic 'Fingerprint,'" *Maclean's*, April 22, 1985, T8.

15. **Broadcast:** Deborah Roberts, "Families Cope with Not Having Body to Grieve for After Attacks on World Trade Center," *20/20*, ABC News, October 3, 2001.

Newspapers: Judy Peet, "These Rescuers Sniff, Search and Comfort," *Star-Ledger* (Newark, NJ), September 26, 2001, 8; Joe Bauman, "Utah Firm Helping ID Victims of Terror," *Deseret News* (Salt Lake City), October 8, 2001, B1.

16. **Wire service:** Caroline Wright, "U.K. Government U-Turn over DNA Database," PHG Foundation, October 26, 2009, www.phgfoundation.org.news.

Website: "Suspects' DNA Data Plans Changed," *BBC News*, October 19, 2009, http://news.bbc.co.uk/2/hi/uk_news/politics/8314973.stm.

Newspapers: Joel Rubin and Richard Winton, "L.A. Announces Plan to Reduce Backlog of Unexamined DNA Evidence from Violent Crimes," *Los Angeles Times*, October 20, 2008; Benjamin Protess and Joel Rubin, "As DNA Test Backlogs Soar, U.S. Cuts Funding," *Los Angeles Times*, November 9, 2008; Rubin, "L.A. County Sheriff's Officials Acknowledge That Genetic Evidence in 5,635 Rape Cases May Be Untested," *Los Angeles Times*, November 13, 2008; Maura Dolan and Jason Felch, "DNA: Genes As Evidence; Tracing a Crime Through a Relative," *Los Angeles Times*, November 25, 2008; Maura Dolan and Jason Felch, "The Danger of DNA: It Isn't Perfect," *Los Angeles Times*, December 26, 2008; Richard Winton and Joel Rubin, "LAPD, Union Tangle over Collection of Officers' DNA,"

Los Angeles Times, January 27, 2009; Joel Rubin, "Wider Scope of Backlog in L.A. County Sheriff's DNA Testing Is Revealed," *Los Angeles Times*, January 28, 2009; Solomon Moore, "Study Calls for Oversight of Forensics in Crime Labs," *New York Times*, February 19, 2009, A13; Solomon Moore, "FBI and States Vastly Expanding Databases of DNA," *New York Times*, April 19, 2009, A1; Solomon Moore, "In a Lab, an Ever-Growing Database of DNA Profiles," *New York Times*, May 12, 2009, D3; Jamie Doward, "Anger over DNA Tests on Asylum Seekers," *Observer*, September 20, 2009, 18; Alan Travis, "Plans to Retain DNA Records of Innocent People Dropped: Policing Bill Changed After European Court Ruling: Campaigners Claim Move a Victory for Human Rights," *Guardian*, October 20, 2009, 12.

Magazines: John Travis, "Forensic Science: Scientists Decry Isotope, DNA Testing of 'Nationality,'" *Science* 326 (October 2, 2009): 30; editorial, "Genetics Without Borders," *Nature* 461 (October 8, 2009): 697.

17. **Author's notes:** May 14, 2004, in *Mapping Life Notebook* 2, 177–185.

18. **Scientific articles:** International HapMap Consortium, "A Haplotype Map of the Human Genome"; Goldstein and Cavalleri, "News and Views: Genomics: Understanding Human Diversity," 1241; Richard Gibbs, "Deeper into the Genome," 1233.

Author's notes: Teleconference, October 26, 2005, in *Mapping Life Notebook* 4, 32–37.

News releases: National Human Genome Research Institute, "International Consortium Completes Map of Human Genetic Variation," October 26, 2005, www.genome.gov/17015412; others from Harvard University, the Wellcome Trust Sanger Institute, the Broad Institute, and Baylor College of Medicine.

Broadcast: Ira Flatow, "HapMap Project Catalogs Genetic Similarities and Differences in Human Beings."

Newspapers: Cook, "Geneticists Map What Makes Us Different"; Wade, "Genetic Catalog"; Maugh, "Disease Hunters"; Steve Sternberg, "Scientists Ready to 'Map' Gene Variations, Diseases," *USA Today*, October 27, 2005, 7D; Connor, "Scientists Reveal Genetic Map"; Munro, "Powerful New Tool Hunts Genes"; Sharon Begley, "Science Journal: Linking DNA Profiles to Diseases May Not Lead to Prevention," *Wall Street Journal*, November 4, 2005, B1.

Magazines: Couzin, "Genomics: New Haplotype Map"; "Finding Needles in a Haystack."

19. **Panel discussion:** "Ethical, Legal, and Social Implications of Direct-to-Consumer Genomic Test Marketing," Session 7, "The Biology of Genomes" meeting, Cold Spring Harbor Laboratory, May 8, 2008, author's notes, in *Mapping Life Notebook* 7, 195–207.

Wire service: Jesse J. Holland, "Congress Passes Anti-Genetic Discrimination Bill," Associated Press, May 2, 2008.

Newspaper: Amy Harmon, "Congress Passes Bill to Bar Bias Based on Genes," *New York Times*, May 2, 2008, A1.

20. **Scientific articles:** Stephen P. Daiger, "Genetics: Was the Human Genome Project Worth the Effort?" *Science* 308 (April 15, 2005): 362; Robert J. Klein et al., including Josephine Hoh, "Complement Factor H Polymorphism in Age-Related Macular Degeneration," *Science* 308 (April 15, 2005): 385; Jonathan L. Haines et al., including Margaret A. Pericak-Vance, "Complement Factor H Variant Increases the Risk of Age-Related Macular Degeneration," *Science* 308 (April 15, 2005): 419; Albert O. Edwards et al., including Lindsay A. Farrer, "Complement Factor H Polymorphism and Age-Related Macular Degeneration," *Science* 308 (April 15, 2005): 421.

News releases: University of Texas Southwestern Medical Center, Dallas, and Yale University, March 10–11, 2005.

Wire services: Randolph E. Schmid, "Genetic Mutation Linked to Illness That Blinds Millions," Associated Press, March 11, 2005; "Genetics Plays Role in AMD," United Press International, March 14, 2005.

Broadcast: Josephine Hoh, interview by host Ira Flatow, "Human Genome Project," *Talk of the Nation: Science Friday*, National Public Radio, March 11, 2005.

Newspapers: Andrew Pollack, "3 Studies Link Variant Gene to Risk of Severe Vision Loss," *New York Times*, March 11, 2005, A17; Clive Cookson, "Gene Sheds Light on Old-Age Blindness," *Financial Times*, March 11, 2005, 9; Roger Highfield, "Find Gives Hope to Elderly Blind," *Daily Telegraph*, March 11, 2005, 12.

Magazine: Elizabeth Tolchin, "A Common Haplotype of AMD Emerges," *Drug Discovery and Development*, June 1, 2005, 20.

21. **Magazines:** Elizabeth Pennisi, "Breakthrough of the Year: Human Genetic Variation; Equipped with Faster, Cheaper Technologies for Sequencing DNA and Assessing Variation in Genomes on Scales Ranging from One to Millions of Bases, Researchers Are Finding Out How Truly Different We Are from One Another," *Science* 318 (December 21, 2007): 1842; Donald Kennedy, "Breakthrough of the Year," *Science* 318 (December 21, 2007): 1833.

22. **Speech:** Kari Stefansson, Keynote Address 2, delivered at the Nature Publishing Group "Genomics of Common Diseases" conference, Broad Institute, Cambridge, Massachusetts, September 8, 2008, author's notes, in *Mapping Life Notebook* 7, 374–381.

Books: Kevin Davies, *Cracking the Genome: Inside the Race to Unlock Human DNA* (New York: Free Press, 2001), 130–140; James Watson, with Andrew Berry, *DNA: The Secret of Life* (New York: Alfred A. Knopf, 2003), 318–320; David Ewing Duncan, "Prelude" (interview with Kari Stefansson), in *The Geneticist Who Played Hoops with My DNA* (New York: William Morrow, 2005), 1–18.

A sampling of early coverage:

Newspapers: Clive Cookson, "Iceland Cashes in on Its Viking Gene Bank," *Financial Times*, April 25, 1997, 2; Stephen D. Moore, "In the Lab: Gene Hunter Targets

Isolated Iceland," *Wall Street Journal*, July 10, 1997, 1; "Tremor Gene Tracked Down," *Financial Times*, August 28, 1997, 24; Robin McKie, "Steam Rises over Iceland Gene Pool," *Observer*, November 8, 1998, 26; John Schwartz, "For Sale in Iceland: A Nation's Genetic Code; Deal with Research Firm Highlights Conflicting Views of Progress, Privacy and Ethics," *Washington Post*, January 12, 1999, A1; Sarah Lyall, "A Country Unveils Its Gene Pool and Debate Flares," *New York Times*, February 16, 1999, F1.

Magazines: Andy Coghlan, "Heirs of the Vikings Cash In on Their Heritage," *New Scientist*, May 3, 1997, 12; Martin Enserink, "Opponents Criticize Iceland's Database," *Science* 282 (October 30, 1998): 859; Ehsan Masood, "Iceland Poised to Sell Exclusive Rights to National Health Data," *Nature* 396 (December 3, 1998): 395; Andy Coghlan, "Selling the Family Secrets," *New Scientist*, December 5, 1998, 20; "Norse Code," *The Economist* (U.S. edition), December 5, 1998, 99; Martin Enserink, "Iceland OKs Private Health Databank," *Science* 283 (January 1, 1999): 13; Michael Specter, "Decoding Iceland: The Next Big Medical Breakthroughs May Result from One Scientist's Battle to Map the Viking Genome," *New Yorker* 74, January 18, 1999, 40.

Bankruptcy coverage:

News release: "deCODE Genetics, Inc. Files Voluntary Chapter 11 Petition to Facilitate Sale of Assets; Enters into Asset Purchase Agreement and Receives Commitment for Debtor-in-Possession Financing to Continue Operations During Chapter 11 Process," PR Newswire, November 17, 2009.

Wire services: "deCODE Genetics Files for Chapter 11, Seeks Sale of Assets," GenomeWeb.com, November 17, 2009; "Bankruptcy Court Allows deCODE Genetics to Continue Operations," GenomeWeb.com, November 24, 2009.

Newspapers: Nicholas Wade, "A Genetics Company Fails, Its Research Too Complex," *New York Times*, November 18, 2009, B2; Jeanne Whalen, "Genomics Pioneer deCODE Seeks Bankruptcy Protection," *Wall Street Journal*, November 18, 2009, B4; Jeremy Laurance, "Firm That Led the Way in DNA Testing Goes Bust; Icelandic Company Made Brilliant Discoveries but Was Unable to Profit from Them," *Independent*, November 18, 2009, 18.

Magazines: Erika Check Hayden, "Icelandic Genomics Firm Goes Bankrupt," *Nature* 462 (November 26, 2009): 401; Jocelyn Kaiser, "Bankruptcy Won't Stop deCODE, Says Its Founder, Stefansson," *Science* 326 (November 27, 2009): 1172; Kari Stefansson, "Icelandic Database Not at Risk from Bankruptcy," *Nature* 463 (January 7, 2010): 25.

23. **Scientific article:** Struan F. Grant et al., including Kari Stefansson, "Variant of Transcription Factor 7–Like 2 (*TCF7L2*) Gene Confers Risk of Type 2 Diabetes," *Nature Genetics* 38 (March 2006): 320 (published online January 15, 2006).

News release: "deCODE Discovers Major Genetic Risk Factor for Type 2 Diabetes; Most Significant Inherited Component of T2D Yet Found; Discovery May

Enable Development of Diagnostics and Drugs of Major Benefit to Public Health," PR Newswire, January 15, 2006.

Newspapers: Nicholas Wade, "Gene Increases Diabetes Risk, Scientists Find," *New York Times*, January 16, 2006, A1; David Brown, "Report Explores Cause of Diabetes; Study of Homogeneous Icelanders Links Gene to Disease," *Washington Post*, January 16, 2006, A4.

24. **Scientific articles:** Jose C. Florez et al., including David Altshuler, "*TCF7L2* Polymorphisms and Progression to Diabetes," *New England Journal of Medicine* 355 (July 20, 2006): 241; Stephen O'Rahilly and Nicholas J. Wareham, "Genetic Variants and Common Diseases—Better Late Than Never," *New England Journal of Medicine* 355 (July 20, 2006): 306.

News release: Nicole Davis, "Gene's Forecast of Diabetes Not Carved in Stone; Study Confirms the Link Between Type 2 Diabetes and a Recently Identified Gene Variant, but Finds That the Increased Genetic Risk Can Be Overcome by Lifestyle Changes," Broad Institute, July 19, 2006, www.broadinstitute.org/news/189.

25. **Scientific article:** Timothy M. Frayling et al., "A Common Variant in the *FTO* Gene Is Associated with Body Mass Index and Predisposes to Childhood and Adult Obesity," *Science* 316 (May 11, 2007): 889 (published online April 12, 2007).

News release: Wellcome Trust, "Major Genetic Study Identifies Clearest Link Yet to Obesity Risk," April 13, 2007.

Newspapers: Mark Henderson, "'Fat' Gene Found by Scientists," *Times*, April 13, 2007, 1; Steve Connor, "Scientists Find 'Fat' Gene That Leads to Weight Gain," *Independent*, April 13, 2007; James Randerson, "Obesity Is Not Just Gluttony—It May Be in Your Genes: Rogue Copy in DNA Can Lead to Weight Gain: Scientists Insist Diet and Exercise Still Important," *Guardian*, April 13, 2007, 6.

Magazines: Jocelyn Kaiser, "Mysterious, Widespread Obesity Gene Found Through Diabetes Study Genetics," *Science* 316 (April 13, 2007): 185; Jennifer Couzin and Jocelyn Kaiser, "Genome-Wide Association: Closing the Net on Common Disease Genes. Huge Data Sets and Lower Cost Analytical Methods Are Speeding Up the Search for DNA Variations That Confer an Increased Risk for Diabetes, Heart Disease, Cancer, and Other Common Ailments," *Science* 316 (May 11, 2006): 820.

26. **Scientific article:** Valgerdur Steinthorsdottir et al., including Kari Stefansson, "A Variant in *CDKAL1* Influences Insulin Response and Risk of Type 2 Diabetes," *Nature Genetics* 39 (June 2007): 770 (published online April 26, 2007).

News release: "deCODE Genetics Discovers Novel Gene Variant Conferring Risk of Type 2 Diabetes Through Decreased Insulin Secretion," deCODE Genetics, April 26, 2007.

27. **Scientific articles, all published online April 25, 2007:** Diabetes Genetics Initiative of Broad Institute of Harvard and MIT, Lund University, and Novartis Institutes for Biomedical Research, "Genome-Wide Association Analysis Identifies

Novel Loci for Type 2 Diabetes and Triglyceride Levels," *Science* 316 (June 1, 2007): 1331; Eleftheria Zeggini et al., including Mark McCarthy and Andrew T. Hattersley, "Replication of Genome-Wide Association Signals in U.K. Samples Reveals Risk Loci for Type 2 Diabetes," *Science* 316 (June 1, 2007): 1336; Laura J. Scott et al., including Francis S. Collins and Michael Boehnke, "A Genome-Wide Association Study of Type 2 Diabetes in Finns Detects Multiple Susceptibility Variants," *Science* 316 (June 1, 2007): 1341.

News release: Broad Institute, "Genomic Tour de Force for Diabetes," April 26, 2007.

Wire service: Lauran Neergaard, "Massive Study Finds New Genetic Risk Factors for Type 2 Diabetes," Associated Press Financial Wire, April 26, 2007.

Newspapers: Nicholas Wade, "Scientists Identify 7 New Diabetes Genes," *New York Times*, April 27, 2007, A22; Alice Dembner, "Genetic Flaws May Point Way on Diabetes; Scientists Find New Risk Factors for the Disease," *Boston Globe*, April 27, 2007, A2.

Magazine: Meredith Wadman, "Genome Miners Rush to Stake Claims; Race to Discover Disease-Linked Genes Reaches Fever Pitch," *Nature* 447 (June 7, 2007): 623.

28. **Scientific articles:** Wellcome Trust Case Control Consortium, "Genome-Wide Association Study of 14,000 Cases of Seven Common Diseases and 3,000 Shared Controls," *Nature* 447 (June 7, 2007): 661; Anne M. Bowcock, "Genomics: Guilt by Association," *Nature* 447 (June 7, 2007): 645; John A. Todd et al., including David G. Clayton, "Robust Associations of Four New Chromosome Regions from Genome-Wide Analyses of Type 1 Diabetes," *Nature Genetics* 39 (July 2007): 857; David Altshuler and Mark Daly, "Guilt Beyond a Reasonable Doubt," *Nature Genetics* 39 (July 2007): 813.

Website: "Serious Diseases Genes Revealed," *BBC News*, June 6, 2007, http://news.bbc.co.uk/2/hi/6724369.stm.

Newspapers: Nicholas Wade, "Researchers Detect Variations in DNA That Underlie Seven Common Diseases," *New York Times*, June 7, 2007, A32; Roger Highfield and Stephen Adams, "Gene Scientists Bring Hope of Cure for Seven Major Diseases," *Daily Telegraph*, June 7, 2007, 1; Clive Cookson, "Discovery of 24 Genes Boosts Disease Research," *Financial Times*, June 7, 2007, 4; Steve Connor, "Millions Offered Hope by Discovery of the Genes Behind Killer Diseases," *Independent*, June 7, 2007.

Magazines: Emily Singer, "Genes for Several Common Diseases Found. A Study of Seven Illnesses, Including Diabetes and Cardiovascular Disease, Identifies New Candidate Genes," Technologyreview.com, June 7, 2007, www.technologyreview.com/biomedicine/18834/?a=f; Brandon Keim, "Largest Gene-Disease Study Sets Record for Associations, Hype," Wired.com, June 7, 2007, www.wired.com/wiredscience/2007/06/largest_genedis/#previouspost; "Peekaboo; Molecular Medicine; Searching for Disease Genes," *The Economist* (U.S. edition), June 9, 2007.

29. **Speech:** Peter Donnelly, "Genome-Wide Association Studies and Their Follow-Up," Nature Publishing Group "Genomics of Common Diseases" conference, Broad Institute, Cambridge, Massachusetts, September 9, 2008, author's notes, in *Mapping Life Notebook* 8, 13–20.

30. **News release:** Institute for Systems Biology, "Leading U.S. Bioscience Pioneers Enter International Collaboration with Government of Luxembourg to Accelerate Biomedical Research; International Public-Private Initiative to Drive Innovation in the U.S. and Overseas," June 6, 2008, www.systemsbiology.org/Press_Release_60608.

Wire service: "U.S. Institutes to Help Luxembourg Build $200M-Plus Personalized Medicine Programs," GenomeWeb.com, June 6, 2008.

31. **News releases:** "U.K. Biobank Gets Unanimous Backing from International Experts After Piloting Phase; Multi-Million Pound Project to Go National from End of 2006," August 22, 2006; "U.K. Biobank Reaches Landmark Figure [100,000] In Wales," April 16, 2008; "U.K. Biobank Enlists 250,000th Participant," U.K. Biobank, January 30, 2009.

32. **Interview:** Neil Risch, by the author, Palo Alto, California, February 24, 2007, transcript.

News releases: "Kaiser Permanente Unveils Groundbreaking Genetic Research Program; Researchers Aim to Reveal the Genetic and Environmental Causes Behind Deadly and Disabling Diseases," PR Newswire, February 14, 2007; "Robert Wood Johnson Foundation Awards $8.6 Million Grant to Kaiser Permanente to Help Develop Largest, Most Diverse U.S. Biobank; Database Has Power to Reveal Environmental and Genetic Factors Behind Many Deadly Diseases; Groundbreaking Research Will Lead to Better Treatments," Robert Wood Johnson Foundation, December 17, 2008.

Wire services: Paul Elias, "Kaiser Launches Study into Genes, Environmental Disease Causes," Associated Press, February 14, 2007; "Kaiser Permanente to Study Genetic Causes of Diseases," GenomeWeb.com, March 1, 2007; "Robert Wood Johnson Foundation Gives $8.6M for Kaiser Permanente's West Coast Biobank," GenomeWeb.com, December 17, 2008; "National Institutes of Health awards more than $54 million to Kaiser Permanente to conduct health research," Reuters, October 12, 2009; "Kaiser Permanente, UCSF Land $25M Grant for Genotyping Effort," GenomeWeb.com, October 12, 2009.

Website: For information about the study, see www.dor.kaiser.org/studies/rpgeh.

Newspapers: Carl T. Hall, "Kaiser Starts Major Study of Members' Health; HMO Will Survey Health and Habits over 50 Years," *San Francisco Chronicle*, February 15, 2007, B3; Rhonda L. Rundle, "Kaiser Seeks Genetic Data in Effort to Build Database," *Wall Street Journal*, February 15, 2007, B3.

33. **Scientific article:** S. Romeo et al., "Genetic Variations in ANGPTL Family Members Implicate Nonredundant Roles in Triglyceride Metabolism in Humans," *J. Clin. Invest.* 119 (January 5, 2009): 70–79.

Speech: Helen Hobbs, Keynote Address 1, delivered at the Nature Publishing Group "Genomics of Common Diseases" conference, September 6, 2008, author's notes, in *Mapping Life Notebook* 7, 294–302, typed in Chronology, 2008 Genome September.

News releases: Nancy Ross-Flanigan, "Ace of Hearts," *HHMI Bulletin* 17 (spring 2004): 18–22; "Heart of the Matter: Helen Hobbs, Who Thrives in the Fast Pace and Intensity of Medicine, Did Not Settle Easily into the More Measured World of Research," *HHMI Bulletin* 22 (February 2009): 40–41; "Helen H. Hobbs, MD," Howard Hughes Medical Institute, www.hhmi.org/research/investigators/hobbs_bio.html (accessed August 9, 2009).

Newspaper: Robert Miller, "Dallas Heart Study Gets New Lease on Life," *Dallas Morning News*, September 3, 2006.

34. Presentation of hard drive containing DNA sequence of James D. Watson, Cullen Auditorium, Baylor College of Medicine, May 31, 2007, author's notes, in *Mapping Life Notebook* 7, 18–39, typed in Chronology, 2007 Genome May.

35. **Broadcast:** "Nobel Laureate James Watson's Reflections on Life," *Charlie Rose*, PBS, October 30, 2009, transcript.

36. **Wire service:** Denise Lavoie, "Researchers to Post Personal Genes on Web," Associated Press, October 20, 2008.

Newspapers: Ellen Nakashima, "Genome Database Will Link Genes, Traits in Public View," *Washington Post*, October 18, 2008, A1; Amy Harmon, "Project Lets Anyone Take a Peek at the Experts' Genetic Secrets," *New York Times*, October 20, 2008, A1; Sarah Rubenstein, "A New Sort of Facebook: Posting Your Genes on the Web," wsj.com, October 20, 2008, http://blogs.wsj.com/health/2008/10/20/a-new-sort-of-facebook-posting-your-genes-on-the-web/tab/article; Carey Goldberg, "Subjects' DNA Secrets to Be Revealed. Web Postings May Boost Research," *Boston Globe*, October 20, 2008, B1; Ellen Goodman, "A Molecular Full Monty," *Boston Globe*, October 24, 2008, A15.

Magazines: Emily Singer, "Genomes on Display," technologyreview.com, October 20, 2008, www.technologyreview.com/biomedicine/21579/?a=f; Nancy Shute, "Sharing Your Personal Genetic Map with the World," *U.S. News & World Report*, October 20, 2008, www.usnews.com/health/articles/2008/10/20/sharing-your-personal-genetic-map-with-the-world.html; Erika Check Hayden, "Accessible Genomes Move Closer," *Nature* 455 (October 23, 2008): 1014; editorial, "Getting Personal," *Nature* 455 (October 23, 2008): 1007; Steven Pinker, "My Genome, My Self," *New York Times Magazine*, January 11, 2009, 24.

37. **News releases:** "National Geographic and IBM Launch Landmark Project to Map How Humankind Populated Planet; Five-Year Genographic Project Allows Individuals to Trace Own Migratory History," National Geographic, April 13, 2005, http://press.nationalgeographic.com/pressroom/pressReleaseFiles/1113322781773/1113322781790/Geno_LeadRelease_Final%204.26.06.pdf; Applied Biosystems,

"DNA Analysis Technologies Aiding Genographic Project; Landmark Global Population Study Begins to Unlock Genetic History of Human Migration," Business Wire, May 5, 2008.

Newspapers: Nicholas Wade, "Geographic Society Is Seeking a Genealogy of Humankind," *New York Times*, April 13, 2005, A16; Charles Forelle, "Project Hopes to Trace Your Ancestors Back 10,000 Years," *Wall Street Journal*, April 13, 2005, B1; Thomas H. Maugh II, "Gene Project to Trace Migration in a Massive Effort, Researchers Will Collect and Analyze DNA from 100,000 Worldwide," *Los Angeles Times*, April 13, 2005, A24; Steve Connor, "Worldwide DNA Project to Map the Origins of Humanity," *Independent*, April 13, 2005, 8; Mike Toner, "Way Past the Family Tree; Science: National Geographic Project Hopes to Trace Genomes to Beginning of Species," *Atlanta Journal-Constitution*, April 17, 2005, 1B; Nicholas Wade, "Still Evolving, Human Genes Tell New Story," *New York Times*, March 7, 2006, A1.

Magazine: Graciela Flores, "New Genome Project, New Controversy," *The Scientist* 19 (May 9, 2005): 12.

38. **News releases:** "Navigenics Launches with Preeminent Team of Advisors, Partners, and Investors," PR Newswire, November 6, 2007; "deCODE Launches deCODEme; the Company That Has Led in the Discovery of Genes That Confer Risk of Common Diseases Is Empowering Individuals to Explore Their Own Genome; Webcast to Be Held at 9 A.M. EST," PR Newswire, November 16, 2007.

Wire services: David P. Hamilton, "Genetic Genealogy Hits the Big Time," venturebeat.com, October 17, 2007; David P. Hamilton, "Will 23andMe and Navigenics Lock Up Your Genome and Charge You for the Key?" venturebeat.com, October 23, 2007; David P. Hamilton, "Life Sciences: Navigenics Finally Offers You a Peek at Your Genome—Except Not Really, and Not Yet," venturebeat.com, November 7, 2007; Matt Jones, "HHS Group Recommends Policies for Oversight of Genetic Testing," GenomeWeb.com, November 8, 2007.

Newspapers: Ron Winslow, "Is There a Heart Attack In Your Future? Genetic Tests Promise to Map Your Personal Health Risks, but Some Question Usefulness," *Wall Street Journal*, November 6, 2007, D1; Ron Winslow, "Decode Me? Personal DNA Sleuthing on a Q-Tip," wsj.com, November 16, 2007, http://blogs.wsj.com/health/2007/11/16; Amy Harmon, "My Genome, Myself: Seeking Clues in DNA," *New York Times*, November 17, 2007, 1; "Where to Go for DNA Tests," *New York Times*, November 17, 2007, A16; Nicholas Wade, "Experts Advise a Grain of Salt with Mail-Order Genomes, at $1,000 a Pop," *New York Times*, November 17, 2007, A16; Clive Cookson, "DNA Test Brings Personal Genome Closer," *Financial Times*, November 17, 2007, 4; Andrew Clark, "Google-Backed Firm Offers DNA Testing for $999," *Guardian*, November 20, 2007, 29; Fred Reed, "Genetic Profiles: Right to Know?" *Washington Times*, November 24, 2007, C10; Eric J. Topol, Commentary, "What You Can Learn from a Gene Scan," *Wall Street Journal*, December 22, 2007, A10.

Magazines: David Ewing Duncan, "Welcome to the Future. With a Little Help from Google, Secretive Silicon Valley Start-Up 23andMe Is Betting You'll Want to Web-Surf Your Own DNA. But Is the Science Ready?" Portfolio.com, October 15, 2007, www.portfolio.com/news-markets/national-news/portfolio/2007/10/15/23andMe-Web-Site/index1.html; Robert Langreth, "Personal Genome Race Goes into Overdrive," Forbes.com, November 16, 2007, www.forbes.com/2007/11/16/personal-genome-services-tech-cx_rl_1116decode.html; Thomas Goetz, "The Age of the Genome," Wired.com, November 17, 2007, www.wired.com/medtech/genetics/magazine/15–12/ff_genomics_age; "Personal Genetics. Within Spitting Distance? The Era of Personalised Medicine Takes a Step Closer. But It Is Not Quite Here Yet," *The Economist*, November 24, 2007; editorial, "Risky Business," *Nature Genetics* 39 (December 2007): 1415.

39. **News releases:** "23andMe to Hold Press Meeting at World Economic Forum Annual Meeting in Davos, Switzerland, on Wednesday, January 23rd, 2008," Business Wire, January 18, 2008; Misha Angrist, "So You Want to Know Your Genome; In People's Desire for This Information, There's a Teachable Moment," Duke Today, February 8, 2008, http://news.duke.edu/2008/02/genome.html.

Newspapers: Rick Weiss, "Genetic Testing Gets Personal; Firms Sell Answers on Health, Even Love," *Washington Post*, March 25, 2008, A1; Allen Salkin, "When in Doubt, Spit It Out," *New York Times*, September 14, 2008, Style section, 1.

Magazines: Michael Shulman, "Talk of the Town, Double Helix Dept.," *New Yorker* 84 (September 22, 2008): 35; "The 50 Best Inventions of the Year," *Time* 172 (November 10, 2008): 68.

40. **News releases:** "Navigenics Proposes Standards for Personal Genomics Services, Coupled with Prospective Outcomes Studies, to Safeguard Consumers; Company Acts to 'Ensure the Integrity of This Critical Step Toward Personalized Health Care,'" PR Newswire, April 8, 2008; California Department of Public Health, "Cease and Desist" letter to such firms as 23andMe, Navigenics, and deCODEme, June 9, 2008, ww2.cdph.ca.gov/HealthInfo/news/Pages/LabTestingLandingPg.aspx.

Wire services: Edward Winnick, "New York State Warns Consumer Genomics Services Firms About Offering Tests," GenomeWeb.com, April 18, 2008; "California Sends Warning Letter to Consumer Genetic Testing Firms," GenomeWeb.com, June 16, 2008; Marcus Wohlsen, "Calif. Cracks Down on 13 Genetic Testing Start-Ups," Associated Press, June 16, 2008.

Newspapers: Gautam Naik, "As Gene Tests Spread, Questions Follow," *Wall Street Journal*, December 13, 2007, D1; "State Probes 6 Online Firms Testing DNA; Complaints Focus on Genetic Accuracy, Compliance with Law," *San Francisco Chronicle*, April 24, 2008, C1; Steve Johnson, "Gene-Testing Firms Under Fire," *San Jose Mercury News*, June 13, 2008; Bobbie Johnson, "California Clamps Down on Genetic Testing Industry," *Guardian*, June 19, 2008, 19; Bernadette Tansey,

"Genome Testing Firms Defy State Order to Stop," *San Francisco Chronicle*, June 25, 2008, C1; Andrew Pollack, "Gene Testing Questioned by Regulators," *New York Times*, June 26, 2008, C1; David Magnus, Mildred Cho, and Robert Cook-Deegan, "Genetic Test Firms Must Follow Law," *San Jose Mercury News*, July 11, 2008; Karen Ravn, "Medicine. A Closer Look: Gene Tests; DNA Firms Under Microscope; Confusion over Testing Policy Persists As California Orders Several Companies to 'Cease and Desist,'" *Los Angeles Times*, July 14, 2008, F3; Andrew Pollack, "California Licenses 2 Companies to Offer Gene Services," *New York Times*, August 20, 2008, C3.

Magazines: Robert Langreth and Matthew Herper, "States Crack Down on Online Gene Tests," Forbes.com, April 18, 2008, www.forbes.com/2008/04/17/genes-regulation-testing-biz-cx_mh_bl_0418genes.html; Victoria Colliver, Robert Langreth, and Matthew Herper, "I Want My DNA Test," Forbes.com, May 19, 2008, www.forbes.com/forbes/2008/0519/042a.html; Robert Langreth, "California Orders Stop to Gene Testing," Forbes.com, June 14, 2008, www.forbes.com/2008/06/14/stop-gene-testing-biz-healthcare-cz_rl_0614genetest.html; Johnson, "California Clamps Down," 19; Alexis Madrigal, "DNA Testers Reveal Cease-and-Desist Letter," Wired.com, June 18, 2008, www.wired.com/wiredscience/2008/06/exclusive-dna-t; Thomas Goetz, "Attention, California Health Dept.: My DNA Is My Data," Wired.com, June 17, 2008, www.wired.com/wiredscience/2008/06/attention-calif; Anna Gosline, "Genetics Get Personal; What's It Like to Get a Glimpse of Your Own Genome?" *New Scientist*, July 5, 2008, 36.

41. **Panel discussion:** "Ethical, Legal, and Social Implications of Direct-to-Consumer Genomic Test Marketing," Session 7, "The Biology of Genomes" meeting, Cold Spring Harbor Laboratory, May 8, 2008, author's notes, in *Mapping Life Notebook* 7, 195–207. Speakers: Kari Stefansson, Dietrich Stephan, Linda Avey, Kathy Hudson, Joseph McInerney.

42. **Magazine:** Nancy Shute, "Unraveling Your DNA's Secrets," *U.S. News & World Report* 142 (January 8, 2007): 50–58.

43. **Scientific articles:** David J. Hunter, Muin J. Khoury, and Jeffrey M. Drazen, "Letting the Genome Out of the Bottle—Will We Get Our Wish?" *New England Journal of Medicine* 358 (January 10, 2008): 105; "Susanne B. Haga, Huntington F. Willard," letter, *New England Journal of Medicine* 358 (May 15, 2008): 2184, with a reply from David J. Hunter, Muin J. Khoury, and Jeffrey M. Drazen.

44. **Magazine:** Editorial, "Positively Disruptive," *Nature Genetics* 40 (February 2008): 119.

Chapter 7: The Struggle for Medical Relevance

1. **Newspaper:** Stephen Smith, "MGH to Use Genetics to Personalize Cancer Care," *Boston Globe*, March 3, 2009, A1.

Magazine: Erika Check Hayden, "Personalized Cancer Therapy Gets Closer; Genetic Testing Allows Doctors to Select Best Treatment," *Nature* 458 (March 12, 2009): 131.

2. **Scientific articles:** Timothy J. Ley et al., including Richard K. Wilson, "DNA Sequencing of a Cytogenetically Normal Acute Myeloid Leukemia Genome," *Nature* 456 (November 6, 2008): 66–72; Elaine Mardis et al., "Comparing Acute Myeloid Leukemia Genomes with Different Outcomes Following Comprehensive Analysis of Whole Genome Re-Sequencing Data" (paper presented at Session 1, "The Biology of Genomes" meeting, Cold Spring Harbor Laboratory, May 5, 2009), in *Mapping Life Notebook* 8, 66–70, and Abstracts, Biology of Genomes 2009.

Speeches: Timothy J. Ley, "Sequencing Acute Myeloid Leukemia Genomes with 'Next Generation' Technologies," delivered at the Nature Publishing Group "Genomics of Common Diseases" conference, Broad Institute, Cambridge, Massachusetts, September 6–9, 2009, in *Mapping Life Notebook* 7, 357–361; Elaine Mardis, "Cancer Genomics via Whole Genome Sequencing" (paper presented at the "Personal Genomes Conference," Cold Spring Harbor Laboratory, October 10, 2008), in *Mapping Life Notebook* 8, 233–235; David Wheeler et al., including Richard Gibbs, "Analysis of the Cancer Genome by Second Generation Sequencers," Personal Genomes Conference, in *Mapping Life Notebook* 8, 240–242.

3. **News release:** "Illumina Announces Roadmap to Five Million Variants for Next-Generation Whole-Genome Genotyping," Business Wire, October 21, 2009.

Wire services: Dan Koboldt, "Back from Baylor," Mass Genomics, October 16, 2009, www.massgenomics.org/2009/10/back-from-baylor.html; Andrea Anderson, "Researchers Discuss 1000 Genomes Progress at ASHG," GenomeWeb.com, October 22, 2009.

Website: 1,000 Genomes at www.1000genomes.org.

4. **News release:** "Institute for Systems Biology to Work with Complete Genomics to Conduct Large-Scale Huntington's Disease Study; Project to Sequence an Unprecedented 100 Human Genomes in Just Six Months," Complete Genomics, Inc., November 2, 2009, www.completegenomics.com/pages/materials/ISBPR_FINAL_01Nov09.pdf.

5. **Scientific article:** Murim Choi et al., including Richard Lifton, "Genetic Diagnosis by Whole Exome Capture and Massively Parallel DNA Sequencing," *Proceedings of the National Academy of Sciences USA* 106 (November 10, 2009): 19096–19101.

News releases: "Diagnosis Emerges from Complete Sequencing of Patient's Genes," Howard Hughes Medical Institute, October 19, 2009; "Scan of Turkish Infant's Genome Yields a Surprise Diagnosis," Yale University, October 19, 2009.

Wire services: "International Research Team Demonstrates Dx Potential of Exome Sequencing," GenomeWeb.com, October 20, 2009; "Genome Analysis

Changes Diagnosis," *BBC News*, October 24, 2009, http://news.bbc.co.uk/2/hi/health/8315258.stm.

6. **Scientific articles:** H. Christina Fan et al., including Stephen Quake, "Noninvasive Diagnosis of Fetal Aneuploidy by Shotgun Sequencing DNA from Maternal Blood," *Proceedings of the National Academy of Sciences USA* 105 (October 6, 2008): 16266; Fiona M. F. Lun et al., including Y. M. Dennis Lo, "Noninvasive Prenatal Diagnosis of Monogenic Diseases by Digital Size Selection and Relative Mutation Dosage on DNA in Maternal Plasma," *Proceedings of the National Academy of Sciences USA* 105 (December 16, 2008): 19920; Rossa W. K. Chiu et al., including Y. M. Dennis Lo, "Noninvasive Prenatal Diagnosis of Fetal Chromosomal Aneuploidy by Massively Parallel Genomic Sequencing of DNA in Maternal Plasma," *Proceedings of the National Academy of Sciences USA* 105 (December 23, 2008): 20458.

Speech: Stephen Quake, "Single-Molecule and Single Cell DNA Sequencing," delivered at the Nature Publishing Group "Genomics of Common Diseases" conference, Broad Institute, Cambridge, Massachusetts, September 7, 2008, in *Mapping Life Notebook* 7, 323–327.

News releases: Howard Hughes Medical Institute, "New Blood Test for Down Syndrome," October 6, 2008; "New Prenatal Test for Down Syndrome Less Risky Than Amniocentesis, Stanford/Packard Scientists Say," Business Wire, October 6, 2008; Sequenom Corp., "New Noninvasive Prenatal Diagnostic Technology Allows Diagnosis of Inherited Diseases Such As Cystic Fibrosis Caused by Single Gene Mutations; Novel 'Mutation Counting' Approach Allows the Precise Measurement of Multiple Disease-Causing Mutations Inherited by a Fetus," Business Wire, November 24, 2008; Sequenom Corp., "Next-Generation Noninvasive Diagnostic Technology Shown to Accurately Detect Fetal Down Syndrome in First Trimester of Pregnancy," Business Wire, December 1, 2008.

Wire services: Andrea Anderson, "Stanford-Led Team Develops Sequencing-Based Maternal Blood Test for Fetal Chromosomal Abnormalities," GenomeWeb.com, October 7, 2008; "Sequenom Shares Fall on Potential Competition; Analyst Says Sell-Off Is Unwarranted," GenomeWeb.com, October 7, 2008; "Sequenom, Collaborators Demonstrate Efficacy of Non-Invasive Prenatal Monogenic Disease Test," GenomeWeb.com, November 25, 2008.

Newspapers: Lisa M. Krieger, "New Prenatal Test May Be Less Risky Than Amniocentesis," *Contra Costa Times* (California), October 6, 2008; Andrew Pollack, "Blood Tests Ease Search for Down Syndrome," *New York Times*, October 7, 2008, D5; Ishani Ganguli, "Blood Test Might Identify Down Syndrome," *Washington Post*, October 14, 2008, HE2; David Pallister, "Researchers Devise Safer Down's Syndrome Test," *Guardian*, October 7, 2008, 8; Gautam Naik, "Prenatal Test Has Promise, Study Shows," *Wall Street Journal*, November 25, 2008, D2.

Magazine: Kathryn B. Garber, "This Month in Genetics," *American Journal of Human Genetics* 83 (November 2008): 545.

7. **Scientific article:** Rebecca J. Garten et al., including Nancy J. Cox, "Antigenic and Genetic Characteristics of Swine-Origin 2009 A(H1N1) Influenza Viruses Circulating in Humans," *Science* 325 (July 10, 2009): 197 (published online May 22, 2009).

News release: American Association for the Advancement of Science, "*Science* Study Helps to Unravel Origins of New H1N1 Flu Virus," May 22, 2009.

Wire services: Seth Borenstein, "Experts: Mild Swine Flu Could Quickly Turn Deadly," Associated Press, May 5, 2009; Randolph E. Schmid, "Swine Flu Genes Circulated Undetected for Years," Associated Press, May 22, 2009.

Magazine: Jon Cohen, "Swine Flu Outbreak: New Details on Virus's Promiscuous Past," *Science* 325 (May 29, 2009): 1127.

8. **News release:** Wellcome Trust Sanger Institute Annual Review for 2008–2009, posted November 5, 2009, at www.sanger.ac.uk/Info/annualreview.

9. **Scientific articles:** Etienne Meylan et al., including Tyler Jacks, "Requirement for NF-B Signaling in a Mouse Model of Lung Adenocarcinoma," *Nature* 452 (November 5, 2009): 104; David A. Barbie et al., including Tyler Jacks and William C. Hahn, "Systematic RNA Interference Reveals That Oncogenic *KRAS*-Driven Cancers Require TBK1," *Nature* 462 (November 5, 2009): 108.

News releases: Ann Trafton, "Protein Is Linked to Lung Cancer Development; Drugs That Inhibit the Protein, Which Normally Helps Defend Cells from Infection, Could Target Tumors in Certain Lung Cancer Patients," MIT News Office, October 21, 2009; Nicole Davis, "Revealing Cancers' Weak Spots: Research Team Highlights Key Genetic Vulnerability in Some of the Deadliest Cancers," Broad Institute, October 21, 2009.

10. **Scientific articles:** Nathalie Cartier et al., including Patrick Aubourg, "Hematopoietic Stem Cell Gene Therapy with a Lentiviral Vector in X-Linked Adrenoleukodystrophy," *Science* 326 (November 6, 2009): 818; Luigi Naldini, "A Comeback for Gene Therapy," *Science* 326 (November 6, 2009): 805.

Wire services: Lauran Neergaard, "New Gene Therapy Halts 2 Boys' Rare Brain Disease," Associated Press, November 6, 2009; "Gene Therapy Beats Back Brain Wasting Disease: Study," Agence France Presse, November 6, 2009.

Newspapers: Paul Benkimoun, "Thérapie génique: Premiers succes," *Le Monde*, November 7, 2009, 20; Gina Kolata, "After Setbacks, Small Successes for Gene Therapy," *New York Times*, November 6, 2009, A19; Thomas H. Maugh II, "New Triumph Cited for Gene Therapy; Researchers Stabilize 'Lorenzo's Oil' Disease in Two Boys. Experts Hail the Breakthrough," *Los Angeles Times*, November 6, 2009, A13.

11. **Wire service:** Victor K. McElheny, "Personalized Medicine, a Tall Mountain," Xconomy.com, May 11, 2009.

12. **Testimony:** National Human Genome Research Institute (NHGRI), "Fiscal 2009 Budget Request," statement by Francis S. Collins, director, before the Senate Subcommittee on Labor-HHS-Education Appropriations, Genome, July 16, 2008, www.genome.gov/27527406.

News release: "R & D Spending by U.S. Biopharmaceutical Companies Reaches Record Levels in 2008 Despite Economic Challenges," PhRMA, March 10, 2009.

Wire service: "HHMI Endowment Fell 20 Percent in FY 2009," GenomeWeb .com, November 18, 2009.

13. **News release:** "Fact Sheet: Recovery to Discovery: $5 Billion Recovery Act Investment in Scientific Research and Jobs," White House, September 30, 2009, http://whitehouse.gov.

14. **News release:** "Remarks by the President on the American Recovery and Reinvestment Act, National Institutes of Health, Bethesda, Maryland," White House Office of the Press Secretary, September 30, 2009.

15. **News releases:** The Cancer Genome Atlas, National Cancer Institute (NCI), and NHGRI, "NIH Launches Comprehensive Effort to Explore Cancer Genomics. The Cancer Genome Atlas Begins with Three-Year, $100 Million Pilot," December 13, 2005; Francis S. Collins, MD, PhD, director of NHGRI, and Andrew C. von Eschenbach, MD, director of NCI, "Message from the Institute Directors," December 2005, http://cancergenome.nih.gov/about/message.asp.

Wire services: Jeff Nesmith, "Project Launched to Identify Cancer's 'Aberrant' Genes," Cox News Service, December 13, 2005; Lauran Neergaard, "Federal Researchers to Create Map of Cancer's Genetic Makeup," Associated Press, December 13, 2005.

Broadcasts: Robert Weinberg and Anna Barker, interview by Helen Palmer, *Marketplace*, December 13, 2005, transcript; Francis Collins and Anna Barker, interview by Jeffrey Brown, *Lehrer NewsHour*, PBS, December 13, 2005, transcript.

Newspapers: Rick Weiss, "NIH Launches Cancer Genome Project; Genetic Mapping Could Revolutionize Treatment and Prevention, Health Officials Say," *Washington Post*, December 14, 2005, A14; Andrew Pollack, "New Genome Project to Focus on Genetic Links in Cancers," *New York Times*, December 14, 2005, A31.

Magazines: Kevin Davies, "TCGA: The Cancer Genome Atlas Pilot Launched," *Bio-IT World*, December 13, 2005, www.bio-itworld.com; Erika Check, "Big Money for Cancer Genomics," *Nature* 438 (December 15, 2005): 894.

16. **Interviews:** Bert Vogelstein, by the author, Johns Hopkins Medical School, Baltimore, Maryland, April 9, 2007, in *Mapping Life Notebook* 5, 332–337; Bert Vogelstein, by the Academy of Achievement, Baltimore, Maryland, May 23, 1997, www.achievement.org/autodoc/page/vog0int-1.

17. **Scientific article:** Tobias Sjöblom et al., including Bert Vogelstein, Kenneth W. Kinzler, and Victor E. Velculescu, "The Consensus Coding Sequences of Human Breast and Colorectal Cancers," *Science* 314 (October 13, 2006): 268.

Newspapers: Liz Szabo, "Killer Cancer Genes ID'd; 'Achilles' Heel' of Tumors Discovered," *USA Today*, September 8, 2006, 1A; Jonathan Bor, "Genetic Code of Cancers Mapped; Hopkins Researchers Find New Leads on Breast, Colon Tumors," *Baltimore Sun*, September 8, 2006, 1A; Regina McEnery, "Case Team Helps Crack

Genetic Code of Tumors," *Plain Dealer* (Cleveland), September 8, 2006, A1; Antonio Regalado, "Research Links Scores of Genes to Two Common Cancer Types; Results Come from Efforts to Map DNA of Tumor Cells, and Find What Goes Wrong," *Wall Street Journal*, September 8, 2006, B4; Maggie Fox, "Any of 189 Mutations Found to Cause Tumours," *Globe and Mail* (Canada), September 8, 2006, A21.

Magazine: "Variations on a Theme; Cancer Genetics: There Are a Lot of Cancer Genes Around," *The Economist* (U.S. edition), September 9, 2006.

18. **Scientific articles:** The Cancer Genome Atlas Research Network, including Linda Chin and Matthew Meyerson, "Comprehensive Genomic Characterization Defines Human Glioblastoma Genes and Core Pathways," *Nature* 455 (October 23, 2008): 1061–1068 (published online September 4, 2008); Hongwu Zheng et al., including Ronald A. DePinho, "p53 and Pten Control Neural and Glioma Stem/Progenitor Cell Renewal and Differentiation," *Nature* 455 (October 23, 2008): 1129; editorial, "A Bigger Picture: Beneath Cancer's Daunting Complexity Lies a Simplicity That Gives Grounds for Hope," *Nature* 455 (September 11, 2008): 138; D. Williams Parsons et al., including Bert Vogelstein, Victor E. Velculescu, and Kenneth W. Kinzler, "An Integrated Genomic Analysis of Human Glioblastoma Multiforme," *Science* 321 (September 26, 2008): 1807; Siân Jones et al., including Bert Vogelstein, Victor E. Velculescu, and Kenneth W. Kinzler, "Core Signaling Pathways in Human Pancreatic Cancers Revealed by Global Genomic Analyses," *Science* 321 (September 26, 2008): 1801.

News releases: NCI and NHGRI, "The Cancer Genome Atlas Reports First Results of Comprehensive Study of Brain Tumors; Large-Scale Effort Identifies New Genetic Mutations, Core Pathways," September 4, 2008, www.genome.gov/27527925; "The Cancer Genome Atlas Reports First Results, Johns Hopkins Kimmel Cancer Center Collaborates," AScribe Newswire, September 8, 2008; Howard Hughes Medical Institute, "Scientists Find New Gene Alterations in Pancreatic and Brain Cancers," September 4, 2008.

Wire service: Lauran Neergaard, "Gene Domino Effect Behind Brain, Pancreatic Tumors," Associated Press, September 5, 2008.

Newspapers: Frank D. Roylance, "Genome Lights the Way to Cancer 'Milestone'; Data on Defective Genes Make Tailored Remedies Possible," *Baltimore Sun*, September 5, 2008, 1A; Jose Luis de la Serna, "Un hito clave que reorienta la lucha contra el cancer," *El Mundo*, September 5, 2008, 33.

Magazines: Jocelyn Kaiser, "Cancer Genetics: A Detailed Genetic Portrait of the Deadliest Human Cancers," *Science* 321 (September 5, 2008): 1280; Bernadine A. Healy, "Breaking Cancer's Gene Code; Understanding the Genetic Underpinnings of Cancer Is a Giant Step Toward Personalized Medicine," *U.S. News & World Report* 145 (November 3, 2008): 46.

19. **Scientific articles:** Ryan Lister et al., including Bing Ren and Joseph R. Ecker, "Human DNA Methylomes at Base Resolution Show Widespread Epigenomic Dif-

ferences," *Nature* 462 (October 14, 2009): 315; Florian Eckhardt et al., including Stephan Beck, "DNA Methylation Profiling of Human Chromosomes, 6, 20, and 22," *Nature Genetics* 38 (October 29, 2006): 1378.

News releases: Salk Institute, "What Drives Our Genes? Salk Researchers Map the First Complete Human Epigenome," October 14, 2009; University of Western Australia, "UWA Researchers Part of U.S. Consortium That Has Mapped the Human Epigenome—the Genome's 'New Clothes,'" October 15, 2009.

Wire service: Andrea Anderson, "Salk Team Generates First Map of Human Methylome," GenomeWeb.com, October 14, 2009.

Newspapers: Ian Sample, "First Came the Building Blocks of Life. Now Scientists Decode the Instruction Manual: 'Epigenome' Maps How Genes Work Together: Hope of New Drugs to Treat Cancer and Schizophrenia," *Guardian*, October 15, 2009, 3; Ian Sample, "Environment Holds the Key: Code in the Cells," *Guardian*, October 15, 2009, 3; Cathy O'Leary, "Gene Switches a Breakthrough," *West Australian*, October 15, 2009, 17; Sharon Begley, "How a Second, Secret Genetic Code Turns Genes On and Off," *Wall Street Journal*, July 23, 2004, A9; Faye Flam, "'Asterisks' in the DNA; Changeable Chemical 'tags' on Genetic Code May Be the Keys to Individuality and Emergence of Disease. Science Goes Beyond the Genome: It's Called Epigenetics," *Philadelphia Inquirer*, February 12, 2007, F1; Adam Cresswell, "Genes Begin to Yield Secrets," *Weekend Australian*, August 23, 2008, 13; Carl Zimmer, "Now: The Rest of the Genome," *New York Times*, November 11, D1; Nicholas Wade, "From One Genome, Many Types of Cells. But How?" *New York Times*, February 24, 2009, D4.

Magazines: "Methylated Spirits: The Rise of Epigenomics; the Human Genome Gets More and More Complicated," *The Economist*, October 15, 2009, 93; "Life Story: The Sequel," *The Economist*, December 24, 2005, 114; "Moving AHEAD with an International Human Epigenome Project," *Nature* 454 (August 7, 2008): 711.

20. **Newspapers:** Gautam Naik, "In Long-Awaited Maps of Cancer, the Breakthrough Is the Problem," *Wall Street Journal*, September 5, 2008, A10; Joseph Hall and Megan Ogilvie, "Rethink Cancer Care, Studies Urge; Research Shows Disease Too Complex for Cure; Money Better Spent on Prevention, Treatment," *Toronto Star*, September 5, 2008, A6.

Magazines: Erika Check Hayden, "Cancer Complexity Slows Quest for Cure; Genomic Analysis Reveals Multiple Mutations in Tumours," *Nature* 455 (September 11, 2008): 148; Sharon Begley et al., "We Fought Cancer . . . and Cancer Won; After Billions Spent on Research and Decades of Hit-or-Miss Treatments, It's Time to Rethink the War on Cancer," *Newsweek* (U.S. edition) 152 (September 15, 2008): 42.

21. See note 2 above.

Scientific articles: Elaine Mardis et al., including Timothy J. Ley, "Recurring Mutations Found by Sequencing an Acute Myeloid Leukemia Genome," *New England Journal of Medicine* 361 (September 10, 2009): 1058; James R. Downing, "Cancer

Genomes—Continuing Progress," *New England Journal of Medicine* 361 (September 10, 2009): 1111.

Author's notes: Elaine Mardis, "Comparing Acute Myeloid Leukemia Genomes with Different Outcomes Following Comprehensive Analysis of Whole Genome Re-Sequencing Data" (paper presented at Session 1, "The Biology of Genomes" meeting, Cold Spring Harbor Laboratory, May 5, 2009), in *Mapping Life Notebook* 8, 233–235, and Abstracts, Biology of Genomes 2009, 3.

News releases: Caroline Arbanas, "Decoding Leukemia Patient Genome Leads Scientists to Mutations in Other Patients," Washington University, August 5, 2009; Caroline Arbanas, "Decoding Leukemia Patient Genome Another Step Forward in Cancer Fight," Washington University, August 13, 2009.

Wire services: Andrea Anderson, "WashU Team Publishes Second AML Genome," GenomeWeb.com, August 6, 2009; Dan Koboldt, "Second Cancer Genome in New England Journal," MassGenomics.org, August 6, 2009.

Magazines: Ford Vox, "Elaine Mardis and Richard Wilson, Directors, Washington University Genome Sequencing Center," *U.S. News & World Report* 146 (August 1, 2009): 44; Allison Proffitt, "Second Cancer Genome Reveals Pros of Genome-Wide Approach," *Bio-IT World*, August 7, 2009, www.bio-itworld.com/news/2009/08/07/cancer-genome-NEJM.html; Elaine Mardis, "Cancer Genomics; DNA Sequencing Will Transform Our Understanding of Cancer," *Technology Review* (MIT) 112 (January–February 2009): 10.

22. **Interview:** Elaine Mardis, by the author, Cold Spring Harbor Laboratory, Cold Spring Harbor, New York, May 7, 2009, in *Mapping Life Notebook* 8, 293–300.

News releases: "Amersham Pharmacia Biotech's MegaBACE™ Purchased by Washington University School of Medicine in St. Louis, One of the Top Five Public Genome Sequencing Centers," PR Newswire, July 22, 1999; "Genome Technology Magazine Names the 2002 GT All-Stars™; Genomics Industry Luminaries Honored at Awards Bash," PR Newswire, October 4, 2002; Gwen Ericson, "GSC Gets Big Boost from Small Package," *Washington University in St. Louis Record*, May 20, 2005; Caroline Arbanas, "Genome Technology Wizard Mardis Stays on Top of Technology to Help Pinpoint Causes of Disease," *Washington University in St. Louis Record*, September 11, 2008; "Research-Based Undergraduate Course Expands Beyond Washington University," States News Service, November 21, 2008.

Wire service: Justin Pope, "Tiny Steps Toward the $1,000 Genome," Associated Press, September 4, 2003.

Newspapers: Matt Schofield, "Robot Saves Time, Tedium in Human Genome Efforts," *Kansas City Star*, February 25, 1998; Matt Schofield, "Teams Race to Solve Mysteries at Heart of Being Human," *Kansas City Star*, February 26, 1998; Deb Peterson, "Wash. U. Professor Takes Genetics Research to Wider Student Audience," *St. Louis Post-Dispatch*, November 18, 2008, D3.

Magazines: Deborah Janssen, "Exploiting the Potential of Genomic Sequences; Elaine Mardis Uses Her DNA Sequencing Expertise to Further Medical Science," *Genomics and Proteomics,* November 1, 2002, 18; Tariq Malik, "Taking the Work Out of Purifying Nucleic Acids; Genome Projects Spawned Robotic Purifiers, and Technological Innovation Is Propelling These Large-Scale Research Programs," *Genomics and Proteomics*, October 1, 2003, 21.

23. **Scientific articles:** Jean-Philippe Collet et al., including Gilles Montalescot, "Cytochrome P450 2C19 Polymorphism in Young Patients Treated with Clopidogrel After Myocardial Infarction: A Cohort Study," *Lancet* 373 (January 24, 2009): 309; Robert F. Story, "Clopidogrel in Acute Coronary Syndrome: To Genotype or Not?" *Lancet* 373 (January 24, 2009): 276; Jessica L. Mega et al., including Marc S. Sabatine, "Cytochrome P-450 Polymorphisms and Response to Clopidogrel," *New England Journal of Medicine* 360 (January 22, 2009): 354; Tabassome Simon et al., including Laurent Becquemont, "Genetic Determinants of Response to Clopidogrel and Cardiovascular Events," *New England Journal of Medicine* 360 (January 22, 2009): 363; Jane E. Freedman and Elaine M. Hylek, "Clopidogrel, Genetics, and Drug Responsiveness," *New England Journal of Medicine* 360 (January 22, 2009): 411; Harlan Krumholz, "Does Genotype Affect a Patient's Response to Clopidogrel? Evidence That It Does Is Mounting," *Journal Watch Cardiology*, January 21, 2009, http://cardiology.jwatch.org/cgi/content/full/2009/121/1.

Wire service: Matthew Perrone, "FDA Reviews Benefits of Plavix in Certain Patients," Associated Press, January 26, 2009.

Newspapers: Shirley S. Wang, "Gene Variant Reduces Effectiveness of Plavix," *Wall Street Journal*, December 24, 2008, D3; Jared A. Favole, "FDA Considers Updating Plavix Label," *Wall Street Journal*, December 31, 2008, D6; Jennifer Corbett Dooren, "Safety of Plavix Under Review," *Wall Street Journal*, January 26, 2009, D3; Jared A. Favole, "FDA Panel May Recommend High Fence Around Lilly Clot Drug [Prasugrel]," *Wall Street Journal*, February 2, 2009, B1; Natasha Singer, "FDA Panel Approves Lilly Drug for Clotting," *New York Times*, February 4, 2009, B3; Peter Loftus, "FDA Approves Effient, a Challenger to Plavix," *Wall Street Journal*, July 11, 2009, B5; Marie McCullough, "New Hope on Finding Better Blood Thinners," *Philadelphia Inquirer*, October 20, 2009, A1.

24. **News release:** "JCVI Cuts Staff, Consolidates Sequencing Operations," GenomeWeb.com, January 8, 2009.

Wire service: "Q4 2008 Illumina, Inc. Earnings Conference Call," FD Wire, February 3, 2009.

Newspaper: Paul C. Mathis, "Broad Institute Cuts Staff," *Harvard Crimson*, February 3, 2009.

25. **Speech:** Leroy Hood, "New Ideas Thrive on New Organizational Structures," Batten Briefings, Darden Graduate School of Business, spring 2004, www.darden.virginia.edu.

Interview: Leroy Hood, by the author, Seattle, Washington, February 14, 2004, transcript and in *Mapping Life Notebook* 2, 12–15.

26. **Magazine:** Elaine Mardis, "The Impact of Next-Generation Sequencing Technology on Genetics," *Trends in Genetics* 24 (March 2008): 133.

27. **Scientific article:** Marcel Margulies et al., including Jonathan M. Rothberg, "Genome Sequencing in Microfabricated High-Density Picolitre Reactors," *Nature* 437 (September 15, 2005): 376 (published online July 31, 2005).

Author's notes: Eric Lander, telephone interview by the author, Cambridge, Massachusetts, October 18, 2005, in *Mapping Life Notebook* 4, 6–9; Maynard Olson, telephone interview by the author, October 18, 2005, in *Mapping Life Notebook* 4, 9–14.

News releases: "454 Life Sciences Completes First Whole Genome Sequence Using First Novel Technology Since 1977 Invention of DNA Sequencing—Complete Adenovirus Sequence Submitted to Genbank® Generated by First Sequencing Method Designed to Sequence Whole Genomes Not Genes," PR Newswire, August 21, 2003; "454 Life Sciences Secures $20 Million in Equity Financing to Initiate Commercialization of Whole Genome Sequencing Technology—CuraGen Increases Majority Ownership to 66%," PR Newswire, September 18, 2003; "454 Life Sciences Publishes Breakthrough Genome Sequencing Technique; Microfabrication Allows Sequencing in Record Time," PR Newswire, August 1, 2005.

Wire services: Pope, "Tiny Steps Toward the $1,000 Genome"; "Yale Study Uses 454 Sequencing to Detect Drug-Resistant HIV Variants," GenomeWeb.com, June 18, 2007.

Newspapers: Andrew Pollack, "Company Says It Mapped Genes of Virus in One Day," *New York Times*, August 22, 2003, C5; Nicholas Wade, "DNA Machine May Advance Genetic Sequencing for Patients," *New York Times*, August 1, 2005, A10; Michael Totty, "A Better Idea," *Wall Street Journal*, October 24, 2005, R1.

Magazines: Victor K. McElheny, "The Human Genome Project + 5," *The Scientist* 20 (February 2006): 45; Jim Kling, "Ultrafast DNA Sequencing: Single-Molecule Sequencing May Be the Wave of the Future, but Don't Count Out Microfluidics Yet," *Nature Biotechnology* 21 (December 2003): 1425; Elizabeth Pennisi, "Cut-Rate Genomes on the Horizon?" *Science* 309 (August 5, 2005): 862; Alison McCook, "Seven Technologies: Where Would the Human Genome Project, Bioinformatics, and Life Sciences in General, Be Without It?" *The Scientist* 19 (August 29, 2005): 15.

28. **Interview:** Maynard Olson, telephone interview by the author, October 18, 2005, in *Mapping Life Notebook* 4, 9–14.

Magazine: McElheny, "The Human Genome Project + 5."

29. **News release:** "Illumina and Solexa Inc. Agree Stock-for-Stock Merger," M2 EquityBites, November 14, 2006.

Newspapers: Andrew Pollack, "A DNA Chip Maker Acquires Gene-Sequencing Company," *New York Times*, November 14, 2006, C3; "Hayward Firm Bought for

$600 Million," *Contra Costa Times* (California), November 14, 2006, F4; Marilyn Chase, "Gene-Scanner Price War Could Spur Research," *Wall Street Journal*, December 21, 2006, B1.

30. **Wire services:** "CuraGen Announces Definitive Agreement to Sell 454 Life Sciences to Roche; Conference Call to Be Held Today at 10:00 A.M. EDT," PR Newswire, March 29, 2007; "Roche to Acquire 454 for $155m, Plans to Use Sequencer for IVD Applications," GenomeWeb.com, March 29, 2007; Meredith Salisbury, "Next-Gen Sequencing: The Waiting Game," www.genome-technology .com, June 2007.

Newspaper: Kathleen Gallagher, "Drug Giant to Acquire NimbleGen; Roche Will Keep Firm in Madison, Where UW Scientists Started It," *Milwaukee Journal Sentinel*, June 20, 2007, D1.

31. **News releases:** "Applied Biosystems to Acquire Agencourt Personal Genomics, Privately-Held Developer of Genetic Analysis Technologies; Ultra High-Throughput Technology with Potential Application in DNA Sequencing, Gene Expression, and Genotyping," Business Wire, May 30, 2006; "Applied Biosystems Completes Acquisition of Agencourt Personal Genomics, Developer of Genetic Analysis Technologies," Business Wire, July 11, 2006.

Wire service: "Applied Biosystems to Buy APG for $120M," Associated Press, May 30, 2006.

32. **Wire service:** "Life Technologies Investor Q&A at AGBT—Final," FD Wire, February 6, 2009.

33. **Wire services:** Kevin Davies, "Complete Genomics Details Its First Human Genome Assembly," *Bio-IT World*, February 6, 2009, www.bio-itworld.com/news/ 02/06/09/Reid-complete-genomics-assembly.html; Julia Karow, "Broad Institute to Pilot Complete Genomics' Sequencing Service," *Genome Web Daily News*, February 6, 2009, www.genomeweb.com/sequencing/broad-institute-pilot-complete -genomics-sequencing-service; Julia Karow, "Illumina, ABI Say GAII, SOLiD 3 Will Yield 100Gb Runs, $10K Human Genome by End of Year," *In Sequence*, February 10, 2009, www.genomeweb.com/sequencing/illumina-abi-say-gaii-solid-3-will -yield-100-gb-runs-10k-human-genome-end-year.

Magazine: Erika Check Hayden, "Genome Sequencing: The Third Generation; Companies Unveil Data from Their Latest Technologies," *Nature* 457 (February 12, 2009): 768.

34. **Scientific articles:** Dmitry Pushkarev, Norma F. Neff, and Stephen R. Quake, "Single-Molecule Sequencing of an Individual Human Genome," *Nature Biotechnology* 27 (September 2009): 847 (published online August 10, 2009); editorial, "DNA Confidential: As the Cost of Human Genome Sequencing Plunges and Large-Scale Genome-Phenotype Studies Become Possible, Society Should Do More Than Reward Those Individuals Who Choose to Disclose Their Data, Despite the Risk," *Nature Biotechnology* 27 (September 2009): 777; Yingrui Li and Jun Wang,

"Faster Human Genome Sequencing," *Nature Biotechnology* 27 (September 2009): 820.

News releases: David Orenstein, "Professor Sequences His Entire Genome at Low Cost, with Small Team," *Stanford Report*, August 10, 2009; "Helicos Biosciences Announces the First Ever Single-Molecule View of Whole Human Genome," Business Wire, August 10, 2009.

Wire services: Stephen R. Quake, "Guest Column, Genome Mania," March 3, 2009, Opinionator, http://judson.blogs.nytimes.com/2009/03/03/guest-column-genome-mania; Julia Karow, "Helicos Co-Founder Says He Sequenced Own Genome on Company's Platform," *In Sequence*, March 10, 2009, www.genomeweb.com/sequencing/helicos-co-founder-says-he-sequenced-own-genome-companys-platform; Kevin Davies, "Quake Sequences Personal Genome Using Helicos Single-Molecule Sequencing," bio-itworld.com, August 10, 2009; Marcus Wohlsen, "Stanford Prof Sequences Own Genome in Weeks," Associated Press, August 10, 2009.

Newspapers: Nicholas Wade, "Technology Lowers Cost of Decoding a Genome to $50,000," *New York Times*, August 11, 2009, D3; Steve Connor, "Unlock the Secrets of Your Genes—for £31,000; Scientists' Breakthrough Raises Prospect of Personalised Medicine Affordable for All," *Independent*, August 11, 2009, 8.

35. **News release:** "Illumina and Oxford Nanopore Enter into Broad Commercialization Agreement; Single-Molecule DNA Sequencing System Promises Dramatic Improvement in Cost, Speed, Versatility, and Simplicity of Sequencing," Business Wire, January 12, 2009.

Wire services: "Oxford Nanopore Raises $21.1M from Illumina and Other Investors," GenomeWeb.com, January 12, 2009; Julia Karow, "With Oxford Nanopore Stake, Illumina Makes Another Claim in Third-Generation Sequencing," GenomeWeb.com, January 13, 2009; Julia Karow, "Life Tech Expects Early Access for Single-Molecule Sequencer in 2010/11, Focuses on Accuracy," GenomeWeb.com, September 22, 2009.

Magazine: Ken Rubenstein, "Sequencer Market Heats Up," *Bio-IT World*, November–December 2009, 12.

36. **Scientific articles:** Malcolm J. Gardner et al., including Bart Barrell, "Genome Sequence of the Human Malaria Parasite *Plasmodium falciparum*," *Nature* 419 (October 3, 2002): 498; Ennio De Gregorio and Bruno Lemaitre, "The Mosquito Genome: The Post-Genomic Era Opens," *Nature* 419 (October 3, 2002): 496; Carlos M. Morel et al., "The Mosquito Genome—a Breakthrough for Public Health," *Science* 298 (October 4, 2002): 79; Robert A. Holt et al., "The Genome Sequence of the Malaria Mosquito Anopheles Gambiae," *Science* 298 (October 4, 2002): 129; editorial, "Malaria's Watershed," *Nature* 455 (October 9, 2008): 707; Elizabeth Ann Winzeler, "Malaria Research in the Post-Genome Era," *Nature* 455 (October 9, 2008): 751.

Magazines: Declan Butler, "What Difference Does a Genome Make?" *Nature* 410 (October 3, 2002): 426; Elizabeth Pennisi, "Malaria Research: Parasite Genome Sequenced, Scrutinized," *Science* 298 (October 4, 2002): 33–34.

37. **News release:** "Illumina Genome Analyzer Selected by Internationally Recognized Medical Center to Sequence 50 Human Genomes," Business Wire, January 12, 2009.

38. **Newspaper:** James D. Watson, "To Fight Cancer, Know the Enemy," *New York Times*, August 6, 2009, A29.

39. **News release:** "Landmark Research Study Is Launched to Assess Impact of Personal Genetic Testing; Scripps Translational Science Institute, Navigenics, Affymetrix and Microsoft Team on Groundbreaking Health Study," PR Newswire, October 9, 2008.

Wire services: Marcus Wohlsen, "Gene-Testing Start-Up's Study Responds to Critics," Associated Press, October 9, 2008; "Scripps, Navigenics, Affy, and Microsoft Launch Long-Term Study on Behavioral Effects of Personal Genetic Testing," GenomeWeb.com, October 9, 2008; Thomas Goetz, "What Good Is Your Genome, Really?" wired.com, October 9, 2008, http://blog/wiredscience/2008/10/what-good-is-yo.html.

Newspapers: Shirley S. Wang, "Study to Track How People React to Disease-Risk Data," *Wall Street Journal*, October 9, 2008, D6; Terri Somers, "Will Disease Data Change Patients' Habits? Scripps to Conduct Study on Genetic Predispositions," *San Diego Union-Tribune*, October 9, 2008, A1.

Magazine: "Will Knowing Your Genes Change Your Behaviour? Whether People Act on News of Their Genetic Predisposition to Disease, or Simply Ignore It, Is the Subject of a 20-Year Study of 10,000 Americans," *New Scientist*, October 18, 2008, 6.

40. See Chapter 6.

41. **Scientific articles:** Patrick S. Schnable et al., including Richard K. Wilson, "The B73 Maize Genome: Complexity, Diversity, and Dynamics," *Science* 326 (November 20, 2009): 1112; Richard A. Gore et al., including Edward S. Buckler, "A First-Generation Haplotype Map of Maize," *Science* 326 (November 20, 2009): 1115; Ruth A. Swanson-Wagner, "Paternal Dominance of Trans-eQTL Influences Gene Expression in Maize Hybrids," *Science* 326 (November 20, 2009): 1118; Jean-Philippe Vielle-Calzada et al., "Brevia: The Palomero Genome Suggests Metal Effects on Domestication," *Science* 326 (November 20, 2009): 1078; Catherine Feuillet and Kellye Eversole, "Perspectives: Plant Science: Solving the Maze," *Science* 326 (November 20, 2009): 1071; and others in *PLoS Genetics*.

News releases: by National Science Foundation; U.S. Department of Agriculture; Cold Spring Harbor Laboratory; Washington University; University of Arizona; University of California, Davis; University of Minnesota; and University of Wisconsin.

Newspaper: David Brown, "Scientists Have High Hopes for Corn Genome," *Washington Post*, November 20, 2009, A8.

Magazines: Tina Hesman Saey, "Corn Genome a Maze of Unusual Diversity; Multiple Teams Announce Complete Draft of the Maize Genome, with a Full Plate of Surprises," *Science News* 176 (December 19, 2009): 9; Elie Dolgin, "Maize Genome Mapped; Sequence Should Help Corn Breeders Meet Global Demands for Food and Fuel," nature.com, November 19, 2009, www.nature.com/news/2009/091119/full/news.2009.1098.html?s=news_rss; Katherine Harmon, "Cracked Corn: Scientists Solve Maize's Genetic Maze; Boasting More Genes Than Humans, the Corn Genome Proved Difficult to Decode," scientificamerican.com, November 19, 2009, www.scientificamerican.com/article.cfm?id=corn-genome-cracked.

42. **News release:** Royal Society, "£2 Billion Needed for Science 'Grand Challenge' to Help Feed the World," October 21, 2009.

Newspapers: Louise Gray, "GM 'Will Be Only Way to Feed the World,'" *Daily Telegraph*, October 20, 2009, 13; Sean Poulter, "Anger As Leading Scientists Call for GM Push," *Daily Mail*, October 21, 2009.

Websites: Richard Black, "U.K. Urged to Lead on Future Food," *BBC News*, October 21, 2009, http://news.bbc.co.uk/2/hi/8317511.stm; John Travis, "Royal Society Report Backs GM Crops, Other Measures to Boost Food Production," Science Insider, October 21, 2009, http://blogs.sciencemag.org/scienceinsider/2009/10/royal-society-r.html#at.

43. **Scientific articles:** Daniel G. Gibson et al., including Clyde A. Hutchison III and Hamilton O. Smith, "Complete Chemical Synthesis, Assembly, and Cloning of a *Mycoplasma genitalium* Genome," *Science* 319 (February 29, 2008): 1215; Carole Lartigue et al., "Creating Bacterial Strains from Genomes That Have Been Cloned and Engineered in Yeast," *Science* 325 (September 25, 2009): 1693.

News releases: "Venter Institute Scientists Create First Synthetic Bacterial Genome; Publication Represents Largest Chemically Defined Structure Synthesized in the Lab; Team Completes Second Step in Three Step Process to Create Synthetic Organism," PR Newswire, January 24, 2008; "J. Craig Venter Institute Researchers Clone and Engineer Bacterial Genomes in Yeast and Transplant Genes Back into Bacterial Cells," J. Craig Venter Institute, August 20, 2009.

Wire services: "Scientists Synthesize a Bacterium's Complete DNA," Associated Press, January 24, 2008; Tom Randall and Bob Drummond, Bloomberg.com, January 25, 2008; Marlowe Hood, "Life, Version 2.0: Is the World Ready for Synthetic Biology?" Agence France Presse—English, January 24, 2008.

Newspapers: Andrew Pollack, "Researchers Take Step Toward Synthetic Life," *New York Times*, January 25, 2008, A15; Rick Weiss, "Md. Scientists Build Bacterial Chromosome," *Washington Post*, January 25, 2008, A4; Gautam Naik, "Scientists Advance in Effort to Create Synthetic Organism," *Wall Street Journal*, January 25, 2008, B1; Faye Flam, "Building a Better Genome? In a first, a Bacterium's Entire

Genetic Code Was Re-Created. Experts See Benefits Down the Road," *Philadelphia Inquirer*, January 25, 2008, A1; Roger Highfield, "Scientists Create the World's First Artificial Genome," *Daily Telegraph*, January 25, 2008, 18; Alok Jha, "Biologist Claims Significant Step Towards Artificial Life: Creation of Synthetic Chromosome Announced; Final Step Will Be to Put Manufactured DNA in Cell," *Guardian*, January 25, 2008, 14; Jesus Avila, "Una obra de titanes con la firma de un gran artista," *El Mundo*, January 25, 2008, 38; Hanno Charisius, "Schritt zum künstlichen Leben; Biotechnikern gelingt es, ein komplettes Genom nachzubauen," *Süddeutsche Zeitung*, January 25, 2008, 1; Mark Henderson, "Life Made in the Lab Just Months Away, Say Biologists; Man-Made Bacteria Could Produce Fuels and Vaccines," *The Times* (of London), August 21, 2009, 13.

Magazines: "Nearly There; Artificial Life," *The Economist*, January 26, 2008, 76–77; Elizabeth Pennisi, "Two Steps Forward for Synthetic Biology," *Science* 325 (August 21, 2009): 928; Erika Check Hayden, "Yeast Cell Surrogate May Help Scientists to Engineer Synthetic Life," Nature.com, August 20, 2009, www.freerepublic.com/focus/news/2322277/posts; Emily Singer, "A Step Forward for Microbial Machines; a Novel Approach to Genetic Engineering Could Aid in the Creation of Fuel-Producing Bacteria—and Edge Closer to Artificial Life," Technologyreview.com, August 21, 2009, www.technologyreview.com/biomedicine/23297.

44. See Chapter 5.

ACKNOWLEDGMENTS

This history of genomics was spurred by two encounters in 2003. During the celebration at Cold Spring Harbor of the fiftieth anniversary of the DNA double helix discovery, the noted biologist Sydney Brenner said I should take up the history of the genome project. To back up his point, he gave me, on the spot, an exciting micro version of the main elements of the story. Three months later, Eric Lander, the mathematician turned biologist, spoke at the twentieth anniversary conference of the Knight Science Journalism Fellowships at MIT. Patterns of tiny red and green squares summarizing data from "DNA chips," he said, tell whether genes are switched on or off. Such information, he said, already revealed that cancer cells appearing identical in the microscope actually were different genetically. Hence, doctors would eventually use this information to identify patients who would be helped or harmed by a particular anticancer drug. So a history of genomics, aimed at the general reader, might help explain where revolutionary new insights into human health and disease were springing from.

A broader obligation derives from many decades of a continuing education, centered on interviews with dozens of biologists, into the revolution in genetics that began over a century ago and increasingly influences both medicine and agriculture. I am grateful for the willingness of researchers such as Francis Ryan, Arthur Kornberg, Vincent Alfrey, George Klein, Sydney Brenner, Arne Tiselius, John Kendrew, Francois Jacob, Andre Lwoff, Elie Wollman, James Watson, Walter Gilbert, Mark Ptashne, Matthew Meselson, Salvador Luria, Gobind Khorana, David Baltimore, Joshua Lederberg, Norton Zinder, Bruce Ames, Paul Berg, Joseph Sambrook, John Cairns, David Botstein, Richard Roberts, Phillip Sharp, Louis Kunkel, and Eric Lander to explain basic concepts and developments over many years.

Very important in this book has been the work of journalists around the world who have kept covering developments in biology whose effects on daily life may be distant. Among them are John Pfeiffer; Leonard Engel of *Life*; Francis Bello of *Fortune*; Richard Saltus of the *Boston Globe*; Margaret Munro of the *Ottawa Citizen*; Magie Fox of the *Globe and Mail*; Natalie Angier, Amy Harmon, Gina Kolata, Andrew Pollack, Harold Schmeck Jr., Nicholas Wade, and Carl Zimmer of the *New York Times*; Sharon Begley, Jerry Bishop, Marilyn Chase, Gautam Naik, David Stipp, Michael Waldholz, and Ron Winslow of the *Wall Street Journal*; Robert Cooke of *Newsday*; Faye Flam of the *Philadelphia Inquirer*; Douglas Birch and Frank Roylance of the *Baltimore Sun*; David Brown, Victor Cohn, Justin Gillis, Philip Hilts, Boyce Rensberger, Rob Stein, and Rick Weiss of the *Washington Post*; Steve Sternberg and Dan Vergano of *USA Today*; Byron Spice of the *Pittsburgh Post Gazette*; Tina Hesman Saey of the *St. Louis Post-Dispatch* and *Science News*; Todd Ackerman of the *Houston Chronicle*; Thomas Maugh II of the *Los Angeles Times*; Scott LaFee of the *San Diego Union-Tribune*; Tom Abate, Carl T. Hall, David Perlman, Sabin Russell, and Bernadette Tansey of the *San Francisco Chronicle*; Lisa Krieger of the *Contra Costa Times*; Tom Paulson of the *Seattle Post-Intelligencer*; Paul Rayburn, Paul Recer, and Malcolm Ritter of the Associated Press; Dick Ahlstrom of the *Irish Times*; Roger Highfield of the *Daily Telegraph*; Steve Connor of *The Independent*; Mark Henderson of *The Times*; Ian Sample of *The Guardian*; Clive Cookson of the *Financial Times*; Paul Benkimoun of *Le Monde*; Andy Coghlan of *New Scientist*; John Carey of *Business Week*; Robert Lamgreth and Matthew Herper of *Forbes*; Emily Singer of *Technology Review*; Fred Golden, Leon Jaroff, Michael Lemonick, and Dick Thompson of *Time* magazine; Richard Preston of *The New Yorker*; Ingrid Wickelgren of *Scientific American*; Kevin Davies of *Bio-IT World*; Jennifer Couzin, Jocelyn Kaiser, Jean Marx, Elizabeth Pennisi, Leslie Roberts, and Robert Service of *Science* magazine; Declan Butler, Meredith Wadman, and Erika Check Hayden of *Nature*; Jefrey L. Fox of *Nature Biotechnology*; Joseph Palca and Ira Flatow of National Public Radio; Charlie Rose of the Public Broadcasting System; and Andrea Anderson, Julia Karow, Matt Jones, and Edward Winnick of GenomeWeb.com; Georgina Ferry (*The Common Thread*), Horace Freeland Judson (*The Eighth Day of Creation*), Matt Ridley (*Genome*), and James Shreeve (*The Genome War*).

Interviews are crucial to the quest for a balanced account. I thank the many researchers who have patiently given time to me, my tape recorder, and my notebook, including Phil Green, Mary-Claire King, Maynard Olson, and Robert Waterston of the University of Washington, Seattle; Nitin Baliga and Leroy Hood of the Institute for Systems Biology; David Botstein and Leonid Krugliak of the Lewis-Sigler Institute for Integrative Genomics at Princeton University; David Bentley and Glenn Powell of Illumina, Inc.; Pieter de Jong of the Children's Hospital of Oakland Research Institute; Michael Eisen of the University of California, Berkeley; David Gilbert and Eddy Rubin of the U.S. Department of Energy Joint Genome Institute; Bruce Alberts, Joe De Risi, and Neil Risch of the University of California, San Francisco; Kelly Frazer of the University of California, San Diego; Michael Hunkapiller of Alloy Ventures; Patrick Brown, Ronald Davis, and Steven Quake of Stanford University; David Haussler, James Kent, and Sofie Salama of the University of California, San Diego; Richard Myers of the HudsonAlpha Institute for Biotechnology; Francis Collins, Eric Green, Mark Guyer, Elke Jordan, and Jane Peterson of the National Human Genome Research Institute; David Lipman of the National Center for Biotechnology Information; Hamilton Smith and J. Craig Venter of the J. Craig Venter Institute; Claire Fraser-Liggett of the University of Maryland; Aravinda Chakravarti, Victor McKusick, and Bert Vogelstein of the Johns Hopkins University School of Medicine; Sean Eddy, Gerald Rubin, and Eugene Myers of the Howard Hughes Medical Institute; Thomas Gingeras, Greg Hannon, Rob Martienssen, Richard McCombie, Jonathan Sebat, Bruce Stillman, and Michael Wigler of Cold Spring Harbor Laboratory; Thomas Hudson and Lincoln Stein of the Ontario Institute for Cancer Research; Elaine Mardis and George Weinstock of Washington University, St. Louis; Richard Gibbs, Donna Muzny, Michelle Rives, and Steve Scherer of Baylor College of Medicine; David Altshuler, Bruce Birren, Eric Lander, and Chad Nusbaum of the Broad Institute; and John Sulston of the Wellcome Trust Sanger Institute.

I am grateful for the many who have helped me gain access to archives and images. These include Geoff Spencer of the National Human Genome Research Institute, Anita Weber and Abigail Dixon of History Associates; Janice Goldblum of the National Academies archives; Claire Clark and Ludmila Pollock of Cold Spring Harbor Laboratory; Walter Gilbert; David Botstein of Princeton University; Richard Myers of the HudsonAlpha Institute; Don Powell of the Wellcome Trust Sanger Institute; Louise Crane of the Wellcome Trust; David Gilbert of the Joint Genome Institute; Alec Jeffries of the University of Leicester; Charles Gershman and William Hathaway of Yale University; Heather Kowalski of the J. Craig Venter Institute; Lynn

Brazil, Jorge Garcia, and David Haussler of the University of California, Santa Cruz; Patty Zamora of Life Technologies; Joann Rogers and Vanessa Wasta of Johns Hopkins Medicine; Peter Tarr of Cold Spring Harbor Laboratory; Clare Hagerty and Paula Kassos of the University of Washington; Christine Emswiler of the Institute for Systems Biology; Eugene Myers and Dianne Palmer of the Howard Hughes Medical Institute; Nicole Davis of the Broad Institute of Harvard and MIT; Kim Worley of Baylor College of Medicine; Caroline Arbanas of Washington University School of Medicine; and Herbert Ragan of the William J. Clinton Presidential Library.

It is a great pleasure, once again, to express my deep gratitude to my agent, Jill Kneerim, and my editor, Merloyd Lawrence, who have believed in and fostered my work over twenty years. Of vital help in the preparation of this book have been Melissa Veronesi, project editor, and Jennifer Kelland, copy editor. Above all, I must thank my wife, Ruth, cherished editor and companion in all my work.

—Cambridge, Massachusetts, March 8, 2010

INDEX

ABOUT THE AUTHOR

In the sixties, as a fledgling science reporter, Victor McElheny began hearing that the double-helix structure of DNA, discovered barely a decade earlier, was the greatest breakthrough in biology in the twentieth century. In the years since, working at the *Charlotte Observer* in North Carolina, *Science* magazine (while based in London), the *Boston Globe*, and the *New York Times*, covering such subjects as the Apollo moon missions, the Green Revolution, and early days in Silicon Valley, he has never gotten very far away from the DNA revolution, which continues into the present. He covered advances in knowledge of how genes are switched on and off, the transfer of genes from animals into microbes, and the deciphering of the exact order of subunits in DNA. During four years at the Cold Spring Harbor Laboratory on Long Island, he organized some twenty conferences on the interaction between human genetics and toxins in the environment. As the founding director of the Knight Science Journalism Fellowships at MIT from 1982 to 1998, he organized numerous seminars for visiting journalists on advances in biology and the early stages of planning what became the Human Genome Project. Victor McElheny is also the author of *Insisting on the Impossible* (1998), a biography of Edwin Land, inventor of instant photography, and *Watson and DNA* (2003), about the life of James Watson, discoverer with Francis Crick of the double-helix structure of DNA.